Brain-Machine Interface

Xilin Liu • Jan Van der Spiegel

Brain-Machine Interface

Closed-loop Bidirectional System Design

Springer

Xilin Liu
Department of Electrical
and Systems Engineering
University of Pennsylvania
Philadelphia, PA, USA

Jan Van der Spiegel
Department of Electrical
and Systems Engineering
University of Pennsylvania
Philadelphia, PA, USA

ISBN 978-3-319-67939-6 ISBN 978-3-319-67940-2 (eBook)
https://doi.org/10.1007/978-3-319-67940-2

Library of Congress Control Number: 2017953003

Printed on acid-free paper

This Springer imprint is published by Springer Nature
The registered company is Springer International Publishing AG
The registered company address is: Gewerbestrasse 11, 6330 Cham, Switzerland

Preface

Brain–machine interfaces (BMIs) create an artificial pathway between the brain and the external world. BMIs have broad applications in fundamental neuroscience research, neuroprosthetics development, neural disease treatment, and may eventually change the way that humans interact with the world. In the past decade, the research and application of BMIs have received enormous attention from the scientific community as well as the public. However, conventional medical instrumentation used in existing BMI research is not capable of studying the complex and dynamically changing behavior of the brain. Moreover, many neuroscience experiments need to be conducted on freely behaving animals during locomotion and social interaction. The key goal of this book is to address these challenges by the design of next-generation BMIs with innovative solutions from the neuron-electronics interface level up to the system architecture level.

This book provides an introduction to the emerging area of BMIs, with an emphasis on the electrical circuit and system design. This book can help electrical engineers, bioengineers, as well as neuroscience investigators to understand the next-generation bidirectional closed-loop BMIs. Background information, comprehensive surveys and reviews, and design specifications are presented, which will be beneficial for researchers who are new to this area or readers with general interests in this research. In addition, the in-depth discussion of circuit and system design methods, trade-offs, practical issues, and animal experiments will also be valuable for experienced researchers.

Design innovations have been proposed in neural recording front-end (Chap. 2), neural feature extraction module (Chap. 3), and neural stimulator (Chap. 4). Practical design issues of bidirectional closed-loop neural interface (Chap. 5) and overall system integration (Chap. 6) have been carefully studied and discussed. To the best of our knowledge, this work presents the first reported portable system to provide all required hardware for a closed-loop sensorimotor neural interface, the first wireless sensory encoding experiment conducted in freely swimming animals, and the first bidirectional study of the hippocampal field potentials in freely behaving animals. The circuit and system design details are presented with bench testing and animal experimental results. The methods, circuit techniques, system topology, and experimental paradigms proposed in this work can be used in a wide range of relevant neurophysiology research and neuroprosthetics development, especially in experiments with freely behaving animals.

We would like to express our appreciation and gratitude to many individuals who have contributed to this book. During this research, we collaborated closely with Prof. Timothy H. Lucas, Dr. Andrew G. Richardson, and their team at the Department of Neurosurgery, University of Pennsylvania. Dr. Lucas directed the overall brain–machine interface research project with his expertise in neuroengineering and therapeutics. Dr. Richardson organized the research, especially the design, conduction, and analysis of the animal experiments presented in Chap. 6. Prof. Milin Zhang, former postdoc in our group, now at Tsinghua University, jointly organized the research. Dr. Zhang supervised the overall electronics design, and implemented the digital compressed sensing module presented in Chap. 2 and the stimulator's digital control module presented in Chap. 4.

We would like to thank our colleagues and friends for their help and insightful discussion during this research; they are Prof. Nader Engheta, Prof. Firooz Aflatouni, Prof. Naveen Verma, Dr. Xiaotie Wu, and Dr. Matt Hongjie Zhu. We would also like to thank Dr. Tao Xiong, Prof. Peter S. Chin, Prof. Trac D. Tran, and Prof. Ralph Etienne-Cummings at Johns Hopkins University for their collaboration in the work of the compressed sensing recording front-end presented in Chap. 2. Students from the Center for Sensor Technologies, University of Pennsylvania, helped in the embedded system programming, RTL implementation, and physical design; they are Rohit Dureja, Hanfei Sun, Tian Qiu, Wanqing Xin, Basheer Subei, and Jacob Sacks. The students and fellows in the Translational Neuromodulation Laboratory (TNL), University of Pennsylvania, carried most of the animal experiments presented in this book; they are Srihari Y. Sritharan, Yohannes Ghenbot, Sam DeLuccia, Pauline K. Weigand, Ivette Planell-Mendez, Solymar Torres Maldonado, and Gregory Boyek. We also appreciate Prof. Dengteng Ge and Prof. Shu Yang for their help in the device development presented in Chap. 6.

We would like to thank the National Science Foundation (NSF), National Institutes of Health (NIH), Defense Advanced Research Projects Agency (DARPA), and Catalyst Foundation for the grants that supported this research. We would like to thank the MOSIS Educational Program for fabricating the chips. We would also like to thank Shurui Zhou, Han Hao, Xinyi Chang, and Hao Gu for their reviews and suggestions during the writing of this book. Finally, we would like to thank the staff at Springer International Publishing AG for their efforts in producing this book.

Philadelphia, PA, USA

Xilin Liu
Jan Van der Spiegel

Contents

About the Authors

Xilin Liu received a B.S. degree in electrical engineering from the Harbin Institute of Technology, China, in 2011. He received M.S. and Ph.D. degrees in electrical engineering from the University of Pennsylvania, USA, in 2013 and 2017, respectively. He joined Qualcomm Inc., USA, in 2017. His research interests include analog and mixed-signal integrated circuits and system design, implantable medical electronics, brain–machine interface, high-performance data converters, and CMOS sensors.

Dr. Liu is a member of the IEEE, the IEEE Solid-State Circuits Society, and the IEEE Circuits and Systems Society. He has served as a reviewer for the IEEE Transactions on Circuits and Systems (TCAS), the IEEE Transactions on Biomedical Circuits and Systems (TBioCAS), the IEEE Sensors Journal, and the Journal of Engineering (JoE). He has also served on several program and technical committees (ISCAS, BioCAS, MWSCAS, SOCC, and ISVLSI). Dr. Liu received the IEEE Solid-State Circuits Society (SSCS) 2015-16 Predoctoral Achievement Award, the Best Student Paper Award and the Best Track Award of the 2017 International Symposium on Circuits and Systems (ISCAS), the Best Paper Award (first place) of the 2015 Biomedical Circuits and Systems Conference (BioCAS), and the Best Paper Award of the BioCAS Track of the 2014 International Symposium on Circuits and Systems (ISCAS). He is also the recipient of the Student-Research Preview Award (Honorable Mention) of the 2014 IEEE International Solid-State Circuits Conference (ISSCC).

Jan Van der Spiegel is a professor of electrical and systems engineering and the director of the Center for Sensor Technologies at the School of Engineering and Applied Science at the University of Pennsylvania. He is the former chair of the Department of Electrical Engineering and interim chair of the Department of Electrical and Systems Engineering. Dr. Van der Spiegel received his master's degree in electromechanical engineering and his Ph.D. degree in electrical engineering from the University of Leuven, Belgium, in 1974 and 1979, respectively. His primary research interests are in mixed-mode VLSI design, bio-inspired CMOS vision sensors, biologically based image sensors and sensory information processing systems, and brain–machine interface electronics. He has published over 250 journal and conference papers and is the coauthor of eight US patents.

He is a life fellow of the IEEE, received the IEEE Major Educational Innovation Award, and is the recipient of the IEEE Third Millennium Medal, the UPS Foundation Distinguished Education Chair, and the Bicentennial Class of 1940 Term Chair. He received the Christian and Mary Lindback Foundation Distinguished Teaching Award, the S. Reid Warren Award for Distinguished Teaching, the IBM Young Faculty Development Award, and the Presidential Young Investigator Award. He has served on several IEEE program committees (IEDM, ICCD, ISCAS, and ISSCC) and was the technical program chair of the 2007 International Solid-State Circuits Conference (ISSCC 2007).

Dr. Van der Spiegel is an associate editor of the *IEEE Transactions on Biomedical Circuits and Systems*, member of the Editorial Board of the Proceedings of the IEEE, and section editor of electrical and electronic engineering of the *Journal of Engineering* of the IET, and former editor of *Sensors and Actuators A* for North and South America. He has been the chair of the IEEE SSCS Chapters Committee from 1998 to 2015. He is currently the president of the IEEE Solid-State Circuits Society. He is a member of Phi Beta Delta and Tau Beta Pi. He spent a 6-month sabbatical at Tsinghua University, Beijing, as a distinguished visiting Professor in 2017–2018.

Acronyms

AP	Action Potential
ASIC	Application-Specific Integrated Circuit
BMI	Brain–Machine Interface
CMRR	Common-Mode Rejection Ratio
CN	Cuneate Nucleus
CR	Compression Ratio
CS	Compressed Sensing
DAC	Digital to Analog Converter
DBS	Deep brain stimulation
DNL	Differential Non-linearity
ECoG	Electrocorticography
EEG	Electroencephalogram
ENOB	Effective Number of Bits
FES	Functional Electrical Stimulation
FoM	Figure of Merit
INL	Integral Non-Linearity
LFP	Local Field Potential
LNA	Low Noise Amplifier
NEF	Noise Efficiency Factor
NI	Neural Interface
OTA	Operational Transconductance Amplifier
PEF	Power Efficiency Factor
PGA	Programmable Gain Amplifier
PID	Proportional-Integral-Derivative
PSD	Power Spectrum Density
SAR	Successive Approximation Register
SFDR	Spurious-Free Dynamic Range
SNR	Signal to Noise Ratio
SoC	System on Chip
TIA	Transimpedance Amplifier

List of Figures

List of Tables

Chapter 1
Introduction

1.1 Background and Motivation

Since the dawn of human civilization, people have started the attempts to study the brain, with the hope that it will give us answers to fundamental questions like who we are, where is the consciousness from. However, even with the science and technology advancements nowadays, many mechanisms of brain functions remain unclear. There are about a hundred billion neurons in a human brain [1], approximating the number of stars estimated in our galaxy [2]. Each neuron establishes connections with seven thousand other neurons on average, forming a massive neural network. Interestingly, neurons represent the information in terms of electrical signals by distributing ions with different charges [3]. This gives electrical engineers a unique opportunity to design artificial devices for collecting the neural signal, and more importantly, generating electrical signals imitating the neural signal. The direct communication pathway between the brain and the external world is named brain–machine interface (BMI), brain–computer interface (BCI) or neural interface [4].

The first BMI experiment was conducted by Jacques Vidal from University of California, Los Angeles in 1973 [5], for an observation and detection of brain events in electroencephalogram (EEG). The first intracortical BMI was built by Phillip Kennedy from Georgia Institute of Technology in 1987 [6]. The first demonstration of controlling a physical object using EEG signal was reported by S. Bozinovski in 1988 [7]. In 1999, Yang Dan and researchers at University of California, Berkeley decoded neuronal firings to reproduce images seen by cats [8]. The same year, John K. Chapin and researchers from MCP Hahnemann School of Medicine and Duke University demonstrated the first direct control of a robotic manipulator by decoding an assembly of cortical neurons [9]. In 2000, Miguel A. Nicolelis and his colleagues from Duke University developed BCIs that decoded brain activity in monkeys and used the devices to reproduce monkey movements in robotic arms [10]. The same year, Gerwin Schalk from the Wadsworth Center of New

X. Liu, J. Van der Spiegel, *Brain-Machine Interface*,
https://doi.org/10.1007/978-3-319-67940-2_1

York State Department of Health developed a general–purpose system for BCI research named BCI2000 [11]. In 2002, Richard A. Andersen and researchers from California Institute of Technology demonstrated decoding high-level cognitive plans for movement from posterior parietal cortex. In 2005, the first human trial of BMI implant for controlling artificial hand was demonstrated by Leigh R. Hochberg from Harvard Medical School and a few collaborating institutes. In 2012, Leigh R. Hochberg, John Bonoghue and fellow investigators at Brown University helped two people with tetraplegia to reach for and grasp objects in three-dimensional space using robotic arms from decoding the motor cortex [12]. In 2016, Sharlene N. Flesher, Andrew B. Schwartz, and researchers from University of Pittsburgh helped a paralyzed man experience the sense of touch in his mind-controlled robotic arm by intracortical microstimulation of the somatosensory cortex [13].

These achievements are promising and encouraging for the development of future generation BMI systems. However, several bottlenecks still need to be overcome before this technology can be extensively used in neuroscience experimentation and clinical therapies, such as the development of a robust bidirectional neural interface for closed-loop operation [14]. The importance and motivation of a bidirectional closed-loop neural interface can be understood from three perspectives: (1) the development of neuroprosthetic devices with sensory feedback, (2) the treatment of neural disorders, and (3) the study of neuroscience and neurology. Each perspective is analyzed as follows.

1. Firstly, a bidirectional closed-loop neural interface is important for the development of the neuroprosthetic and BMI device with sensory feedback. Sensations and actions are inextricably linked. Behavioral goals are achieved by sampling the environment with the available sensory modalities and modifying actions accordingly. Recent developments in hand prosthetics with motor pathway replacement alone do not lead to the adequate use of a paralyzed hand [15]. Artificial sensation restoration is needed for this technology to meet the performance required for clinical adoption. The sensation may be restored by direct electrical microstimulation of the brain [16]. Figure 1.1 illustrates an envisioned bidirectional clinical hand neuroprosthesis with motor function restored through brain-controlled stimulation of hand muscles, and somatosensation restored through sensor-controlled electrical stimulation of the brainstem.
2. Secondly, a bidirectional closed-loop neural interface is important for the treatment of neural disorders. Deep brain stimulation (DBS) is an FDA approved treatment for essential tremor, Parkinson's disease, dystonia, and obsessive-compulsive disorder (OCD) [17]. Despite the long history and success in the clinical use, the underlying mechanism of DBS remains not clear [18]. However, recent research has shown that a closed-loop stimulation can achieve a better performance than conventional open-loop treatments. In 2011, the research conducted by B. Rosin and his colleagues from Hebrew University-Hadassah Medical Association School of Medicine shows that the closed-loop

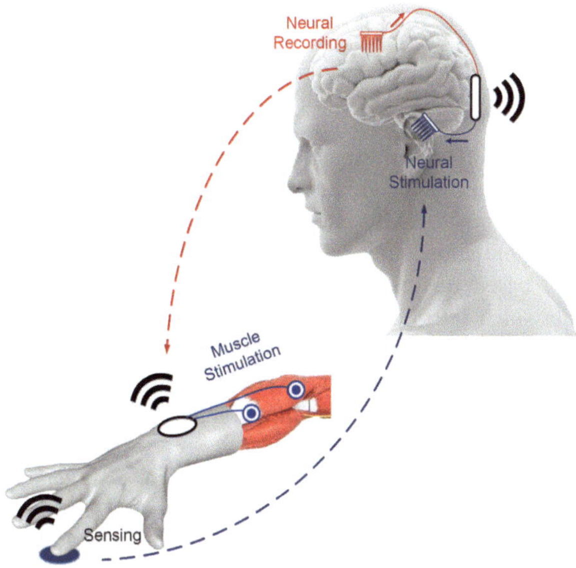

Fig. 1.1 Envisioned bidirectional clinical hand neuroprosthesis. Motor function is restored through brain-controlled electrical stimulation of hand muscles, and somatosensation is restored through sensor-controlled electrical stimulation of the brain

stimulation has a greater effect than the conventional open-loop stimulation paradigms, and has the potential to be effective in other brain disorders [19]. A. Berenyi et al. from Rutgers University presented the closed-loop control of epilepsy by transcranial electrical stimulation in 2012 [20]. The seizure-triggered transcranial electrical stimulation can effectively reduce the pathological brain pattern while leaving the other aspects of brain functions unaffected. In 2013, J. Paz and researchers from Stanford University showed that a closed-loop optogenetic control of thalamus can immediately interrupt electrographic and behavioral seizures [21]. All of these successful implementations encourage the implementation of a fully integrated closed-loop bidirectional neural interface for clinical treatments of neural disorders.

3. Last but not least, a bidirectional closed-loop neural interface is an essential approach for the study of neuroscience and neurology. The fundamental goal of neuroscience research is to better understand the operational principles of the brain. Brain activities consist of complex interactions of both internal state and external stimuli [22]. This is reflected from a single neuron level to a recurrent neuronal network level, and is important for both in vitro and in vivo studies [23]. Examples of studies using bidirectional BMI include bridging lost biological connection [24], generating synaptic plasticity and strengthening weak synaptic connections [25], reinforcing the activity that generates the stimulation [26], and so on.

1.2 Review of Prior Work

This section presents a comprehensive survey and review of bidirectional brain–machine interface designs published to date. BMI systems can be categorized from several different perspectives: (1) based on the electrodes' location, BMI systems can be classified into non-invasive and invasive systems; (2) based on the signal and control flows, BMI systems can be classified into one-directional BMIs (recording or stimulation alone) and bidirectional BMIs (both recording and stimulation); (3) based on the study and characterization approaches, BMI systems can be classified into open-loop and closed-loop BMIs. Figure 1.2 shows the historical trend for the publications of BMI systems over the past 15 years. Specific categories of BMIs including invasive studies, closed-loop studies, and bidirectional neural interface studies are plotted for comparison. The data was retrieved from the Scopus database [27].

Non-invasive approach doesn't require a surgery for implanting electrodes and thus has a significantly less safety concern. However, invasive approach gives a more direct interaction with neurons, thus has advantages in recording resolution and bandwidth as well as stimulation effectiveness and accuracy. This book mainly focuses on the study of invasive BMI systems. In addition, most existing neuroscience research and BMI circuits and system development are based on an open-loop one-directional signal flow: either neural recording or neural stimulation.

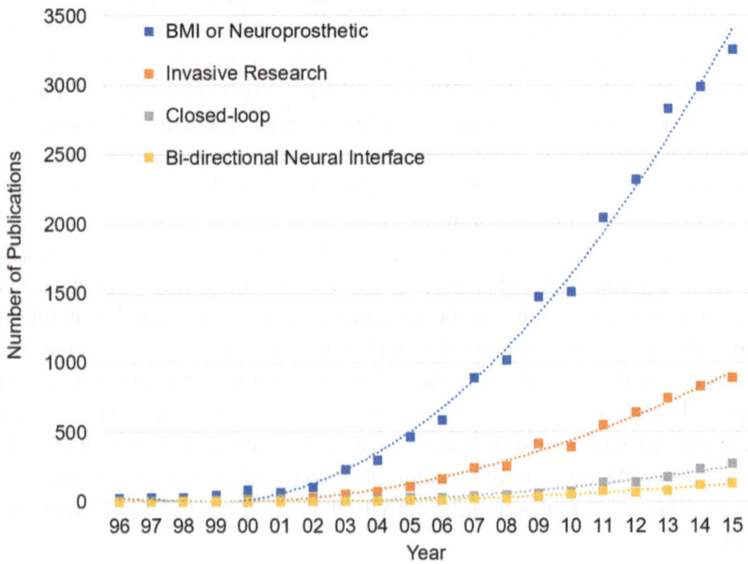

Fig. 1.2 The historical trend for publications of BMIs over the past 15 years. Specific categories of BMIs including invasive studies, closed-loop studies, and bidirectional studies are plotted for comparison

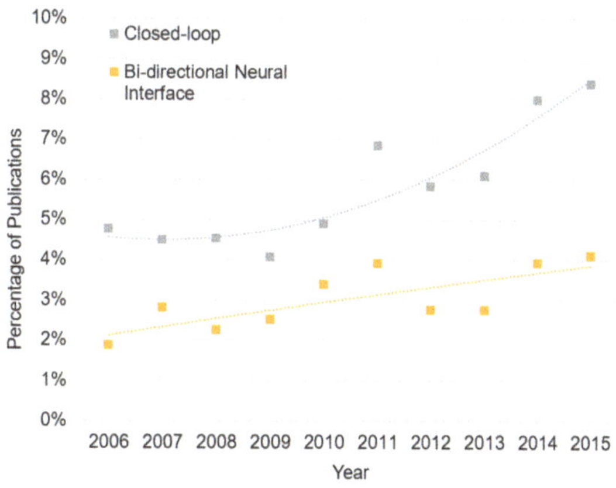

Fig. 1.3 Percentages of closed-loop and bidirectional designs among all BMI publications. Second order polynomial fitting curves are plotted for showing the trends

Figure 1.3 illustrates the percentages of BMI research papers and designs using bidirectional signal flow and closed-loop approach. Second order polynomial fitting curves are plotted for showing the trend. There is a clearly increasing trend for using closed-loop approach and bidirectional neural interface. It should be noticed that this is based on the exponential increase of the overall publications of BMIs. However, the overall percentage is still very low. This book mainly focuses on the study of bidirectional closed-loop BMI systems.

In order to have a comprehensive understanding of the progress of bidirectional BMI research, especially from the electrical engineering perspective, a survey of bidirectional BMI designs is given below. Tables 1.1 and 1.2 list bidirectional BMI and neural interface electronic designs with key design features. The tables include five of the publications from the authors of this book [43, 51, 52, 57, 63]. The selected features include recording and stimulation channel-counts, neural features, closed-loop operation, wireless communication, target applications, and animal experimental methods.

Among all bidirectional BMI systems, designs with high channel-counts have been reported [34, 47, 53]. It should be noticed that most of the high channel-count designs were for in vitro studies, which have less restraint on power consumption. Neural feature extraction has been performed in computers [31, 32], general–purpose microcontrollers [35–37], and application-specific integrated circuits (ASICs) [42, 48, 54, 60]. Commonly used neural features include: neural energy in specific frequency bands [31, 33, 37, 48, 52], action potentials [28, 32, 36, 46], wavelet domain features [54, 60], entropy [35, 44], and phase synchrony [42]. Commonly used closed-loop methods include: simple trigger [28, 31, 50], linear mapping [46, 51, 63], classifiers [35, 40, 54], PI or PID control [48, 52, 57]. Wireless communication modules integrated in BMI systems include commercial solutions [33, 35, 43] and ASIC designs [44, 48, 63]. Target applications of these papers include generalized neuroscience research [39, 45, 45, 49, 55], in vitro neuronal studies [38, 47, 53], deep brain stimulation (DBS) treatments [33, 48] especially neural disorder control [35, 42, 54, 60], and sensory encoding studies [63]. The

Table 1.1 Survey of bidirectional neural interface designs (Part I)

Ref.	Publication	Affiliation	Lead author	Rec.	Stim.
2005 [28]	J. Neurosci. Methods	Univ. of Washington	J. Mavoori	1ch	1ch
2006 [29]	JSSC	ETH Hoenggerberg	F. Heer	128ch	128ch
2007 [30]	TCAS-I	Georgia Tech.	R. Blum	16ch	16ch
2009 [31]	TBE	UC Berkeley	S. Venkatraman	16ch	16ch
2009 [32]	EMBC	Emory Univ.	J. Rolston	60ch	60ch
2010 [33]	JSSC	Univ. Michigan, Ann Arbor	J. Lee	8ch	64ch
2010 [34]	TBioCAS	Univ. of Toronto	F. Shahrokhi	128ch	128ch
2010 [35]	EMBC	National Cheng Kung Univ.	S. Liang	1ch	2ch
2011 [36]	TNSRE	Univ. of Washington	S. Zanos	3ch	3ch
2011 [37]	JNE	Washington Univ.	A. Rouse	4ch	8ch
2011 [38]	JSSC/TBE	Case Western Reserve Univ.	M. Azin	4ch × 2	1ch × 2
2011 [39]	TBioCAS	Univ. of Cagliari	D. Loi	8ch	8ch
2012 [40]	TNSRE	Medtronic/MIT	S. Stanslaski	12ch	8ch
2013 [41]	ISCAS	Univ. of Ulm	U. Bihr	1ch	1ch
2013 [42]	JSSC	Univ. of Toronto	K. Abdelhalim	64ch	64ch
2014 [43]	ISCAS	Univ. of Penn	This work	4ch	2ch
2014 [44]	JSSC	National Chiao Tung Univ.	W. Chen	8ch	1ch x2
2014 [45]	CICC	Case Western Reserve Univ.	K. Limnuson	1ch	1ch
2014 [46]	Scientific Reports	Italian Institute of Tech.	G. Angotzi	8ch	8ch
2014 [47]	JSSC	ETH Zurich	M. Ballini	1024ch	1024ch
2014 [48]	JSSC	Univ. Michigan, Ann Arbor	H. Rhew	4ch	8ch
2014 [49]	ESSCIRC	Medtronic/Washington Univ.	P. Cong	8ch	32ch
2014 [50]	JNE	Imec	T.K.T. Nguyen	32ch	1ch
2015 [51]	TBioCAS	Univ. of Penn	This work	4ch	2ch
2015 [52]	BioCAS	Univ. of Penn	This work	12ch	12ch
2015 [45]	AICSP	Case Western Reserve Univ.	K. Limnuson	1ch	1ch
2015 [53]	TBioCAS	Univ. of Toronto	R. Shulyzki	256ch	64ch
2015 [54]	JSSC	Masdar Inst. of Sci. Tech.	M. Altaf	16ch	1ch
2015 [55]	VLSI	Univ. Michigan, Ann Arbor	A. Mendrela	8ch	4ch
2015 [56]	JSSC	UC Berkeley	W. Biederman	8ch	2ch
2016 [57]	TBioCAS	Univ. of Penn	This work	16ch	16ch
2016 [58]	Microele-ctronics J	Seoul National Univ. Sci. Tech.	A. Abdi	1ch	1ch
2016 [59]	JSSC	Univ. Michigan, Ann Arbor	A. Mendrela	8ch	4ch
2016 [60]	VLSI	Cal Tech./ UCLA	M. Shoaran	16ch	1ch
2016 [61]	Sensors	Wuhan Univ.	Y. Su	32 ch	4 ch
2016 [62]	BioCAS	Univ. of Ulm	M. Haas	1ch	1ch
2017 [63]	ISCAS	Univ. of Penn	This work	16ch	16ch

Table 1.2 Survey of bidirectional neural interface designs (Part II)

Ref.	Neural feature ex.	Closed-loop	Wireless	Application	Animal exp.
2005 [28]	Spike	Trigger	No streaming	Generalized	Free behaving
2006 [29]	–	–	–	Generalized	–
2007 [30]	–	–	–	Generalized	–
2009 [31]	Energy[a]	Trigger	–	Generalized	Awake
2009 [32]	Spike[a]	Trigger	–	Generalized	Free behaving
2010 [33]	Energy	Trigger	Off-chip	DBS treatment	Anesthetized
2010 [34]	–	–	–	Generalized	Anesthetized
2010 [35]	Entropy/spectrum[b]	Classifier	Zigbee	Seizure ctrl.	Free behaving
2011 [36]	Spike[b]	Classifier	–	Generalized	Free behaving
2011 [37]	Energy[b]	Classifier		Generalized	–
2011 [38]	Spike	Trigger	–	Neuronal study	Anesthetized
2011 [39]	–	–	–	Generalized	Anesthetized
2012 [40]	Spectrum	Classifier	–	Generalized	Awake
2013 [41]	–	–	–	Generalized	–
2013 [42]	Phase synchrony	Trigger	UWB	Seizure ctrl.	Anesthetized
2014 [43]	Energy/spike	Trigger	2.4 GHz[c]	Generalized	Awake
2014 [44]	Entropy/spectrum	Trigger	OOK	Seizure ctrl.	Awake
2014 [45]	–	–	–	Generalized	–
2014 [46]	Spike[b]	Mapping	2.4 GHz[c]	Generalized	Free behaving
2014 [47]	–	–	–	Neuronal study	In Vitro
2014 [48]	Energy	PI ctrl.	Back-scattering	DBS treatment	–
2014 [49]	Spectrum	Unknown	–	Generalized	Awake
2014 [50]	Spike	Trigger	–	Generalized	Awake
2015 [51]	Spike/Energy	Mapping[b]	2.4 GHz[c]	Generalized	Free behaving
2015 [52]	Spike/Energy	PID ctrl.	2.4 GHz[c]	Generalized	Free behaving
2015 [45]	–	–	–	Generalized	–
2015 [53]	–	–	–	Neuronal study	Anesthetized
2015 [54]	Freq-time	Classifier	–	Seizure ctrl.	–
2015 [55]	–	Unknown	–	Generalized	–
2015 [56]	Spike	Trigger	–	Generalized	Anesthetized
2016 [57]	Spike/Energy	PID ctrl.	2.4 GHz[c]	Generalized	Free behaving
2016 [58]	–	–	–	Generalized	–
2016 [59]	–	–	–	Generalized	Anesthetized
2016 [60]	Freq-time	Trigger	–	Seizure ctrl.	Free behaving
2016 [61]	–	–	2.4 GHz[c]	Generalized	Free behaving
2016 [62]	Spectrum	–	–	Generalized	–
2017 [63]	–	Mapping	UWB	Sensory encoding	Awake

[a]Off-chip, in computer or workstation
[b]Off-chip, in commercial microcontroller
[c]Off-the-shelf electronic solution

animal experiments and validation methods can be categorized as: anesthetized animals [33, 34, 42], awake but restrained animals [44, 49, 50], and freely behaving animals [57, 60, 61]. Anesthetized and restrained awake animal experiments can be conducted using wire connected instrumentation, which has less concerns than those experiments conducted in completely freely behaving animals [57].

In addition to the general survey, a few key designs that can be considered as milestones in the development of BMIs are reviewed. The selected works are from (in an alphabetical order) Brown University, Case Western Reserve University, Duke University, National Chiao Tung University, Medtronic Inc., Stanford University, University of California, Berkeley, University of Michigan, Ann Arbor, University of Toronto, University of Washington, and Washington University, St Louis. The selected papers are focused on the design from an electrical engineering perspective. The system architecture and circuit implementation of these papers are very helpful in understanding bidirectional BMI designs. Major innovations and contributions are highlighted. The first and corresponding authors with associated laboratories mentioned in this section have years of experience in BMI development, thus are very valuable resources for tracking the trends of BMI designs.

Stavros Zanos and Eberhard E. Fetz et al. from University of Washington designed an autonomous head-fixed computer (the Neurochip-2) for recording and stimulating in freely behaving monkeys in 2011 [36]. The first generation of the device developed in this group was published in 2008 [28]. The device has three recording and three stimulating channels. Digital filtering and action potential discrimination can be performed in the hardware, and action potential triggered stimulation was demonstrated. An accelerometer was integrated in the system. The device had a wireless interface for uploading data and setting device configuration, but real–time data streaming was not supported. An 8MB on-board memory was used to store the recorded data. This work was among the early demonstrations of long-term bidirectional recording and stimulation in freely behaving monkeys.

A.G. Rouse and T.J. Denison et al. from Washington University, St. Louis and Medtronic Inc. designed a chronic generalized bidirectional BMI in 2011 [37, 40]. The system incorporated neural recording and processing subsystems into a commercial neural stimulator. The system can perform spectral analysis, algorithm processing, and event-based data logging. A three-axis accelerometer was also included in the system. The prototype underwent verification testing to ensure reliability. The system included a wireless link for data upload and configuration, but real–time data streaming was not supported. The device integrates an 8 MB SRAM for storing data.

Subramaniam Venkatraman and Jose M. Carmena et al. from University of California, Berkeley designed a system for neural recording and closed-loop intracortical microstimulation in awake rodents [31]. This work also demonstrated the first real–time whisker tracking system. The system employed commercial recording and stimulation instrumentation and a custom PCB interface board. An on-board circuit was designed to reduce stimulus artifacts. This work performed signal processing on a computer and didn't support wireless communication. William Biederman et al. from the same group proposed a fully integrated neuromodulation SoC in 2015 [56]. This work consisted of 64 acquisition channels and dual stimulation channels. The work also featured on-chip digital compression and presented the lowest area and power for the highest integration complexity achieved to the date of publication.

Meysam Azin and Pedram Mohseni et al. from Case Western Reserve University designed a battery-powered activity-dependent intracortical microstimulation IC in 2011 [38]. The chip consisted of two modules, each module integrated four recording channels and one stimulating channel. The chip was designed and fabricated in 0.35 μm CMOS technology, powered by a 1.5 V battery and provided a stimulation voltage up to 5.05 V. This design was among the early demonstrations of on-chip action potential discrimination and spike-triggered stimulation. Follow-up works from the same group added an on-chip stimulation artifact rejection feature in 2014 [45].

Farzaneh Shahrokhi and Roman Genov et al. from University of Toronto designed a 128 channel fully differential digital integrated neural recording and stimulation interface in 2010 [34]. The chip was designed and fabricated in 0.35 μm CMOS technology. The same group developed a 320-channel bidirectional neural interface chip in 2015 [53]. A seizure onset detector was implemented off-chip.

Wei-Ming Chen et al. from National Chiao Tung University, Taiwan, designed a fully integrated closed-loop neural prosthetic CMOS SoC for real–time epileptic seizure control in 2014 [44]. The SoC consisted of eight recording channels, one stimulating channel, a digital seizure detection processor, and a wireless transceiver. The SoC was fabricated in 0.18 μm CMOS technology. The developed system and the seizure detection algorithm were verified in Long-Evans rats.

Hyo-Gyuen Rhew and Michael P. Flynn et al. from University of Michigan, Ann Arbor designed a fully self-contained logarithmic closed-loop deep brain stimulation SoC in 2014 [48]. This work was the first reported implantable SoC with an on-chip closed-loop DBS algorithm. Logarithmic ADC and logarithmic filters were used in this work. A digital PI controller was implemented as the closed-loop controller. This work also integrated an ultra-low-power backscattering wireless transceiver. Adam E. Mendrela et al. from the same group developed a bidirectional neural interface circuit with active stimulation artifact cancellation in 2016 [59]. This work also featured cross-channel common-mode noise suppression.

Peng Cong and Tim Denison et al. from Medtronic Inc. and Washington University designed a 32-channel modular bidirectional BMI with embedded DSP for closed-loop operation in 2014 [49]. The system performed on-chip digital FFT, and a Cortex M3-based microcontroller was used for implementing closed-loop algorithms. The sensing performance of the developed system was demonstrated with 2-D cursor control experiments in non-human primates.

There are also a few publications that describe custom designs for freely behaving animal experiments. These papers address a lot of practical design challenges. Krishna V. Shenoy et al. from Stanford University developed a wireless recording system for freely behaving animals, namely Hermes system, reported in 2007 [64], 2009 [65], 2010 [66], and 2012 [67]. The most recently reported HermesE system featured a 96-channel full data rate direct neural signal recording.

Ming Yin and Arto V. Nurmikko et al. from Brown University developed a wireless neurosensor for full-spectrum electrophysiology recording during free behavior in 2014 [68]. This work supported 96 channel full-spectrum data wireless streaming in a short distance. The wireless data rate was up to 200 Mbps. David A. Schwarz and Miguel A.L. Nicolelis et al. from Duke University developed a chronic wireless recording system for freely behaving monkeys in 2014 [69]. This work featured 3-D multielectrode implants and was capable of isolating up to 1800 neurons from an animal. The design was validated in several monkeys, and the work reported the highest number of neurons wireless recorded from freely behaving animals to the date of publication.

1.3 Overview of the Bidirectional Closed-Loop Brain–Machine Interface System

This section gives an overview of the developed bidirectional BMI system for closed-loop neuroscience experiments. In general, BMI systems should be optimized for safety, reliability, functionality, miniaturization, and long-term operation. To achieve this goal, design optimizations need to be performed from the neuron–electronics interface level up to the system architecture level. The key design requirements are summarized as follows:

1. **Safety**: Tissue damage from implanted electrodes and electronics must be minimized. This requires the design of the neural interface electronics to have proper input and output impedances, a proper stimulation power density based on the electrode material and surface area, a sufficient stimulation charge balancing, and so on. In addition, the packaging and housing of the electronics and batteries also pose important safety requirements;
2. **Performance and Reliability**: Both performance and reliability are critical for BMI systems. The requirements usually include a good signal quality, a reliable wireless data link or a local data storage, reliable signal processing and on-chip closed-loop operation, reliable electrode connection and electronic assembling, and so on. A dependable and robust performance in recording and stimulation is of great importance for both neuroprosthetics and neuroscience investigation;
3. **Interfaces**: BMI systems should provide multiple functional interfaces for neural signal recording, neural stimulation, and various sensing including wearable sensors and supervision. The interfaces should also include the user interface for researchers and investigators to use the BMI system for experiments and data analysis;
4. **Flexibility**: BMI systems should have programmable configurations for recording and stimulation, for example, amplifier gain, bandwidth, sampling rate, and simulation parameters. In addition, the system may also offer programmable neural feature extraction, machine learning, modulation algorithms to support real–time closed-loop operation;

5. **Portable**: One of the key requirements for animal behaving experiments is the ability to record and stimulate wirelessly while the animal is freely moving, such as locomotion and social interaction. Conventional rack-mounted medical instrumentation doesn't support these experiments. So custom BMI systems should be lightweight and can be carried by the animals without disturbing their normal behavior;
6. **Low Power**: A sufficient battery life is important for studying animals' long-term neural activities, such as during sleep. Another typical experiment is to build plasticity, which requires a consistent closed-loop operation. For implantable devices, low power is important to minimize the tissue damage due to the generated heat. Usually, wireless data streaming or programming flash memory consumes most of the power consumption in BMI devices. A high stimulation current also demands a high peak power.

The key features and specifications of the proposed BMI system are listed in Table 1.3. The neural recording front-end of the proposed system is designed for invasive recording, including local field potential (LFP) and action potential (AP) signals. Major building blocks of the recording front-end include low-noise neural amplifiers, programmable neural filters, programmable gain amplifiers, and analog-to-digital converters. On-line data compression, including compressed sensing and action potential detection can be used to reduce the wireless data rate. The stimulator back-end is designed for functional electrical stimulation (FES). A high compliance voltage is required when using high impedance electrodes. On-chip neural feature extraction for both local field potentials and action potentials is implemented. Closed-loop operation is supported using the on-chip PID controller or by the off-chip general–purpose microcontroller. Various body-area sensors are designed to monitor animals' behavior and sensory inputs. The battery-powered devices should

Table 1.3 Key features of the described bidirectional BMI system

Analog front-end		*Stimulator back-end*	
Channel count	16	Channel count	16
Input referred noise	<5 μV	Full-scale current	0–255 μA/4 mA
Bandwidth (LFP)	1–200 Hz	Current resolution	6 bit
Bandwidth (AP)	300–10 kHz	Pulse width	1–255 μs
Gain	1000–8000	Time interval	8 ms–2 s
ADC resolution	10–12 bit	Compliance voltage	<5 V
Neural feature ex.		*Closed-loop operation*	
Local field potential	Energy envelope	Directions	rec/sensor-stim
Action potential	Detection and discrim.	Closed-loop controller	On-chip
Amplitude-phase	Matched filter	Machine Learning	Off-chip
Wireless		*Power*	
Wireless protocol	Bluetooth/FSK	Chip power	<1 mW
Wireless datarate	2 Mbps	System power	<30 mW
Micro-SD Card	FAT32	Total battery life	>12 h

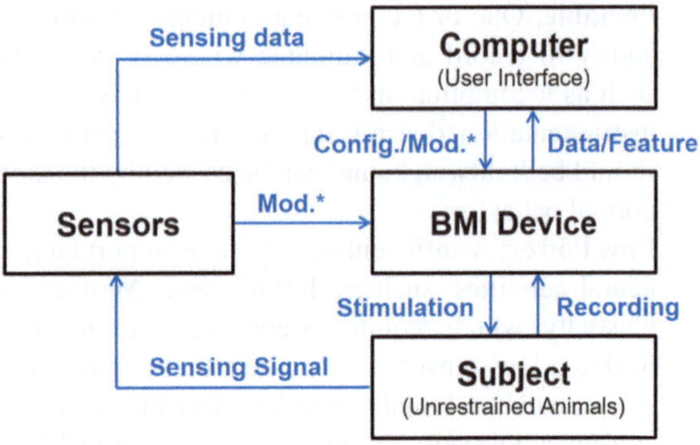

Fig. 1.4 The block diagram of the proposed generalized closed-loop, bidirectional BMI system

be able to support a continuous operation as the target experiment requires. The data loss over the wireless link should be minimized, and the on-board flash memory would need to support the data storage during the long-term recording. Other features including custom packaging of the device are also important aspects of the system design.

The block diagram of the proposed general–purpose bidirectional closed-loop BMI system is illustrated in Fig. 1.4. The main building blocks of the system include: a bidirectional BMI device, various sensors, and a computer with a user interface. The data and control signal flows are marked in the figure. As the core of the BMI system, the BMI device features a bidirectional neural interface and a duplex wireless communication with the computer. The bidirectional neural interface enables both neural signal recording and neural stimulation. The duplex wireless communication allows the BMI device to stream data back to the computer, and to read commands from the computer or the sensor nodes. In addition, the BMI device can process certain neural feature extraction and some closed-loop algorithms. Sensors are also important elements in the system. There are two types of sensors used in this system: wearable sensors and surveillance sensors. Wearable sensors may include a pressure sensor, flex sensor, accelerometers, goniometer, etc. A sensor node is built using commercial sensors and a wireless transceiver. Surveillance sensors include a video recorder and a motion tracking sensor. A computer station provides a user interface for data display, device configuration, and can also perform closed-loop algorithms in certain applications. A standard Bluetooth module is a custom designed wireless dongle is used as the computer interface. The graphic user interface is designed based on MATLAB. The wireless communication between all blocks uses a customized command protocol.

It should be noticed that not all components are necessary for a certain experiment. The system can be configured to work in various closed-loop operating modes. Figure 1.5 shows four commonly used configurations. For example, if the sensor is a camera, Fig. 1.5a shows the operation of the sensory augmentation experiment presented in Sect. 6.3. In this experiment, the camera tracks the animal, sends the location information to the computer. The computer performs

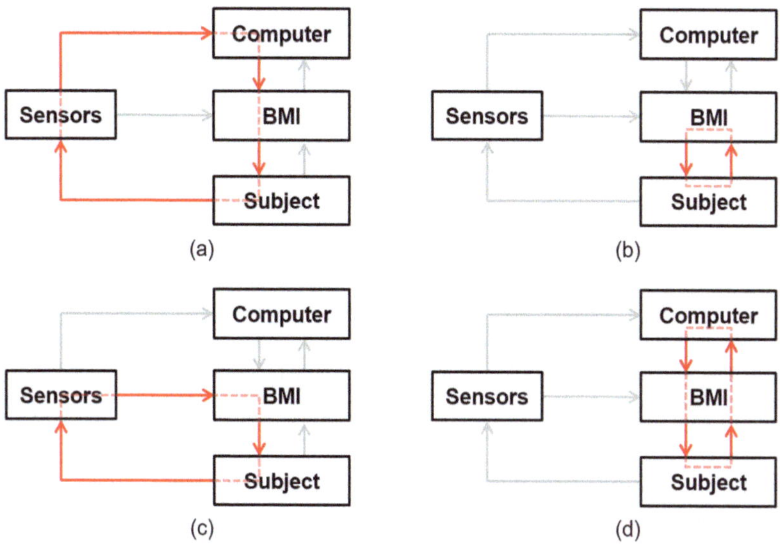

Fig. 1.5 Various closed-loop configurations of the proposed BMI system

the mapping, then sends the stimulation commands to the wireless stimulator. Figure 1.5b shows the operation of the bidirectional recording and stimulation experiment in freely behaving monkeys described in Sect. 6.4. Similarly, the BMI system can also be configured to perform the operations as illustrated in Fig. 1.5c and d.

1.4 Outline of This Book

The book presents the design and analysis of BMI and neural interface systems, with emphasis on bidirectional closed-loop system design. Figure 1.6 highlights the major blocks in a typical bidirectional closed-loop BMI system. The corresponding chapters in this book are marked in the figure.

The book is organized as follows. Chapter 2 presents the analysis and design of neural recording front-ends. Background of neural recording and signal characteristics are first introduced, followed by a review of prior work and an analysis of the key design trade-offs. The design and testing of a general–purpose low-noise amplifier and a low-power SAR ADC are presented. Then, a novel pre-whitening neural front-end design is proposed. The pre-whitening amplifier design takes advantage of neural signal's frequency characteristic. It significantly reduces the dynamic range requirement of the neural amplifier and the ADC resolution requirement. Detailed analysis, simulation, and measurement results are presented. In the end, a fully integrated neural signal acquisition system that features compressive sensing is developed for chronic recording and BMI.

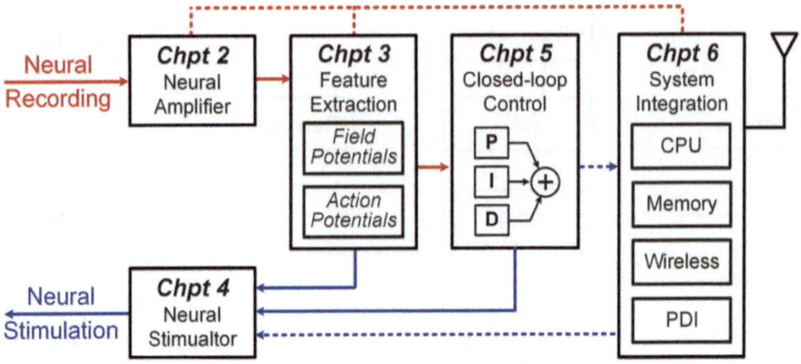

Fig. 1.6 The building blocks of a typical bidirectional closed-loop BMI system, and the organization of this book

Chapter 3 discusses neural feature extraction from three perspectives: local field potential energy extraction, action potential detection, and amplitude-phase coupled feature extraction. Several novel circuits and algorithms have been proposed to improve the performance and power efficiency of conventional designs. A novel natural logarithmic domain tuning scheme is proposed for energy extraction in favor of the brain oscillation bands. An ultra low power current-mode action potential discrimination module is designed. A matched filter is proposed to work with the pre-whitening neural amplifier, which significantly improves the detection accuracy in amplitude-phase coupled neural features.

Chapter 4 presents the analysis and design of high efficiency electrical neural stimulators. Background of neural stimulation and physicochemical properties of the electrode-electrolyte interface are reviewed, and the key design requirements are summarized. A general–purpose neural stimulator design is presented. Next, a novel stimulation strategy is proposed to achieve charge balancing even in existence of irreversible electrochemical processes and unrecoverable charge injection. A high-efficiency net-zero charge neural stimulator has been designed using the proposed strategy. Bench testing, in vitro and in vivo experimental results are given to verify the operation of the designed stimulator.

Chapter 5 addresses the issues in bidirectional closed-loop neural interfaces from two perspectives: the stimulation artifacts and the closed-loop operation. A study of stimulation artifacts in different recorder and stimulator configuration is presented with experimental results. Design suggestions are given in different cases. Next, the mechanisms of various closed-loop neural interface and BMI systems are summarized. A commonly used PID controller is designed and tested. The PID controller has been integrated in a bidirectional neural interface system-on-chip (SoC). Experimental results are presented.

Chapter 6 presents the BMI system integration with a focus on experiments in freely behaving animals. A general–purpose experimental platform, namely the PennBMBI, is presented, featuring wireless recording, stimulation, and sensing ability. A user-friendly computer interface has been developed with a custom wireless communication protocol. Next, a watermaze experiment is designed and conducted for the study of augmenting perception through modulated electrical stimulation of somatosensory cortex. A waterproofed wireless neural stimulator and a complete animal tracking and neuromodulation experimental system have been designed. In the end, custom BMI devices are developed and used in long-term bidirectional experiments in freely behaving monkeys. A study of hippocampal gamma-slow oscillation coupling using the developed system is presented.

To the best of our knowledge, this book presents the first comprehensive study of bidirectional BMI design for freely behaving animal experiments featuring closed-loop operation. The book gives the most complete survey of bidirectional BMI designs published to date. The survey gives insights on the BMI systems' development progress with emphasis on real–time feature extraction, closed-loop operation, and validation in animal experiments. The key contributions of the book are summarized as follows:

1. **System Level**: A complete wireless bidirectional BMI system capable of on-chip neural feature extraction and closed-loop operation has been developed. This work is the first reported portable system to provide all necessary hardware for a closed-loop sensorimotor neural interface. The book also gives a comprehensive study of stimulation artifacts in various BMI configurations, which is a critical issue in bidirectional BMI design. In addition, the book presents the first review and summary of the mechanisms for real–time closed-loop BMI operation.
2. **Circuit Level**: Novel circuits have been proposed to improve the cutting-edge designs. Several innovative designs are highlighted here: a pre-whitening recording front-end is proposed to improve the dynamic range of neural recording front-end; a natural logarithmic domain neural energy extraction unit is designed to improve the efficiency; a matched filter is proposed to improve the detection accuracy of amplitude-phase coupled neural features; a novel net-zero charge neural stimulator is designed for achieving a high safety and power efficiency. Moreover, custom circuits have developed and optimized in support of the system integration and closed-loop operation.
3. **Application and Experiment Level**: Research and investigation have been conducted using the developed bidirectional BMI system. Novel animal experiment paradigms and methods have been proposed and implemented. The presented watermaze experiment is the first wireless sensory encoding experiment conducted in freely swimming animal. Bidirectional neuroscience experiments have been conducted in macaques using the developed device, including the first study that directly compares the hippocampal field potentials in sleep and sedation.

In summary, the system architecture, design methods, circuit techniques, and experimental paradigms presented in this book can be used in a wide range of neurophysiology research and neuroprosthetics development, especially bidirectional closed-loop experiments in freely behaving animals.

Chapter 2
Neural Recording Front-End Design

2.1 Introduction

Neural signal recording revolutionizes our understanding of the human brain. Since the first extracellular recording pioneered by the investigators Ward and Thomas in the 1950s [70], neural recording has revealed the fundamental structure and organization of the brain. The number of simultaneously recorded neurons doubled approximately every 7 years [71]. The exponential growth in the recording ability is to a large extent driven by innovations in CMOS technology, circuits and systems design, microelectrode fabrication, and bio-compatible packaging techniques.

The large-scale neural recording also provides a unique opportunity for the research in brain–machine interface (BMI), which builds the interface between brain and artificial devices [70, 72]. However, recent studies estimate that a simultaneous invasive recording of 100,000 neurons is needed for decoding full-body movements [69], which is beyond the recording ability of the cutting-edge BMI devices. At the same time, multi-channel recording from freely behaving animals in a natural environment is important for both neuroscience research and neuroprosthetic development. However, most of the research to-date still relies on rack-mount instrumentation with restrained cables. The requirements of recording high bandwidth neural signal from multi-channel, in multi-brain areas, via wireless miniature devices place a significant challenge on existing electronic technology and design techniques. The design optimization of a fully integrated neural recording front-end is thus highly desirable.

In the last two decades, a large number of neural recording front-end designs have been reported with improvements from many different aspects. The major innovations have come from novel circuit and system topologies [73–75], low-noise design techniques [76–79], large number of channel-count [76, 80–82], energy efficient designs [74, 76, 82], and low-power wireless interfaces including ISM band FSK [76, 83, 84], FM [85, 86], UWB [66, 81], and backscattering [82, 87, 88].

X. Liu, J. Van der Spiegel, *Brain-Machine Interface*,
https://doi.org/10.1007/978-3-319-67940-2_2

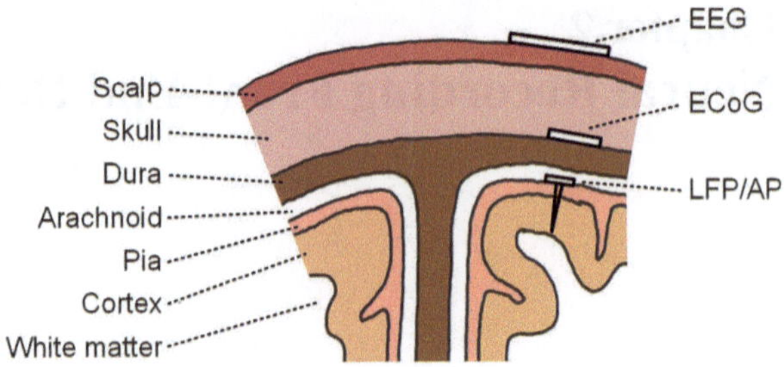

Fig. 2.1 The sources of the neural signals and their locations relative to the brain (not drawn to scale)

In addition, several systems have been used in freely behaving animal experiments [68, 69], and some of the prototype devices are fully integrated and are potentially implantable [80, 82, 89].

This chapter presents the analysis and design of neural recording front-end. Several novel circuit and system designs are proposed to improve the state-of-the-art. The chapter is organized as follows. Section 2.1 introduces the characteristics of the neural signal. The design specifications of neural recording circuit and systems are summarized. Section 2.2 reviews prior work, and analyzes the key trade-offs in the neural amplifier design, followed by a design of a general–purpose low-noise neural amplifier. Section 2.3 proposes a novel pre-whitening neural amplifier design, which exploits the frequency characteristics of the neural signal to relax the dynamic range and linearity requirement of the recording front-end. Section 2.4 presents the design of a 10-bit low-power SAR ADC for neural signal digitization. Section 2.5 presents the design of a complete neural signal acquisition front-end with compressive sensing for long-term neural signal recording in freely behaving animals.

2.1.1 *Signal Characteristics*

Neural signal can be recorded via invasive or non-invasive electrodes. Figure 2.1 shows the most common used neural signal sources and the corresponding electrode placements [90]. The electroencephalography (EEG) is the electrical brain activity recorded from the scalp, the electrocorticography (ECoG) is the electrical brain activity recorded beneath the skull, and local field potential (LFP) and action potential (AP) are electrical signals recorded within the parenchyma. The AP is the individual neuron activity, and the LFP is the activities from multiple nearby neurons.

Figure 2.2 illustrates the amplitude and frequency characteristics of different types of neural signals [91]. Main noise sources are also marked in this figure, including the thermal and flicker noise from the electrodes and electronic recording device, and the interferences from the environment.

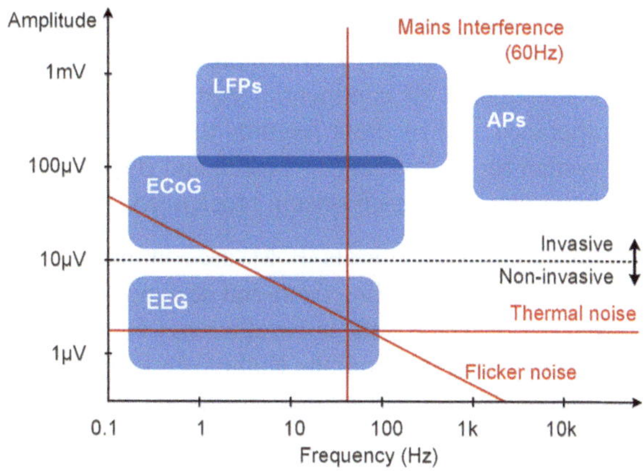

Fig. 2.2 The amplitude and frequency characteristics of different types of neural signals, in comparison with noise sources

Table 2.1 Summary of specifications for neural recording

Requirement	Range	Unit
Input dynamic range	10	mVpp
Electrode offset	±300	mV
Input impedance	>10	Mohm
Common mode rejection	60, at 50–60 Hz 30, at 100–120 Hz	dB
Gain accuracy	Error <10% and <±10	μV
Gain stability over 24 h	<3	%
Noise	50	μV
Crosstalk	<0.2 <5	mV %
Timing accuracy	<30 (over 24 h)	s
Temporal alignment	Error <20	ms

2.1.2 Design Specifications

The key specifications for a neural recording front-end include: input-referred noise, dynamic range, input impedance, linearity, common-mode rejection ratio (CMRR) and power-supply rejection ratio (PSRR), and so on. The minimum requirements are listed in Table 2.1, cited from International Electro-Technical Commission (IEC) medical electrical equipment standard 60601-2-47 [92]. It should be noticed that the requirement for a specific application will usually be higher than the general standard.

In addition to the requirements of the neural amplifier circuit, there are several key requirements of a successful chronic invasive neural recording system:

1. Longevity requirement: safe electrode interface, minimum tissue damage or infection;
2. Noise, bandwidth, and channel-count requirement for the target signal source;
3. Reliable data storage or wireless transmission;

4. Low power for minimum heat damage and long-term recording. In addition, the BMI research usually requires the front-end to be highly programmable, wireless compatible with commercial equipment and sensors, and also can be easily upgraded. All of these features together are required for a practical recording system for neuroscience research and BMI development. A balance between the requirements of each system block needs to be carefully considered.

Several figure-of-merits (FoM) are commonly used for evaluating and comparing different neural recording front-end designs. The most important FoM for noise and power performance is the noise efficiency factor (NEF). The NEF was first proposed by M. Steyaert et al. from the Katholieke Universiteit Leuven in 1987 [93], and was resurrected by R. Harrison et al. from the University of Utah in 2003 [73]. The NEF is defined as:

$$\mathrm{NEF} = \overline{V_{\mathrm{ni,rms}}}\sqrt{\frac{2I_{\mathrm{tot}}}{\pi\Phi_t \cdot 4kT \cdot \mathrm{BW}}} \tag{2.1}$$

where $\overline{V_{\mathrm{ni,rms}}}$ is the input-referred rms noise voltage of the amplifier, I_{tot} is the amplifier's total supply current. For a single bipolar transistor, the input-referred rms noise is:

$$\overline{V_{\mathrm{ni,rms}}} = \sqrt{\frac{4kT\Phi_t}{I_{\mathrm{tot}}} \cdot \frac{\pi\mathrm{BW}}{2}} \tag{2.2}$$

So, the NEF of a single bipolar transistor is 1 [73]. Paper [73, 93] predicted that all practical circuits must have a NEF greater than 1, however, later developed techniques overcame this limitation [44, 94].

It should be noticed that the NEF leaves the supply voltage out of the trade-off. As a result, two amplifier designs with different supply voltages but a same supply current and noise performance may have the same NEF. To mitigate this tissue, R. Muller et al. from the University of California, Berkeley proposed a FoM named power efficiency factor (PEF) in 2012 [95]. The PEF is defined as:

$$\begin{aligned}\mathrm{PEF} &= \mathrm{NEF}^2 V_{DD} \\ &= \frac{V_{\mathrm{ni,rms}}^2 \cdot P_{\mathrm{tot}}}{\pi \cdot kT/q \cdot 4kT \cdot \mathrm{BW}}\end{aligned} \tag{2.3}$$

The PEF gives a direct trade-off between power and noise, and two amplifiers with the same input rms noise and power consumption should have the same PEF. Reducing supply voltage significantly reduces the power consumption resulting a better PEF. However, reducing supply voltage usually comes at a cost of lowering the dynamic range. In order to compare the overall system's efficiency, D. Han et al.

from the Nanyang Technological University further proposed a FoM named system efficiency factor (SEF) in 2013 [96]. The SEF is defined as:

$$\text{SEF} = \frac{\text{PEF}}{\text{DR}_{\text{out}}} \tag{2.4}$$

where

$$\text{DR}_{\text{out}} = 10\log\frac{V_{\text{amp,max}}^2}{2\cdot G_{\text{AFE}}^2 V_{\text{ni,rms}}^2} \tag{2.5}$$

where $V_{\text{amp,max}}$ is the maximum voltage swing of the amplifier, and G_{AFE} is the voltage gain of the amplifier. SEF takes noise, power, and dynamic range performance into account, thus is more suitable for a system level performance comparison.

2.2 Design of a Low-Noise Neural Amplifier

2.2.1 Review of Prior Work

Numerous designs of neural amplifier have been reported in literature. The motivation of this section is not to give a comprehensive survey of prior work, but to analyze the key design trade-offs with featured examples. Review, tutorial, and comprehensive surveys for neural amplifier designs can be found in [70, 97–99].

2.2.1.1 System Topology

Many electrical engineers are familiar with the classical instrumentation amplifier using a 3-opamp topology. The 3-opamp instrumentation amplifier has a high input impedance, a good CMRR, but at a low power efficiency. Commonly used low-power CMOS neural amplifiers use capacitor and resistor elements to set the closed-loop gain. Typical block diagrams are shown in Fig. 2.3. Using capacitive gain elements, the design is inherently AC coupled [73]. Thus, the input common-mode range is from ground to VDD, limited by the ESD protection circuits. The only active component is the operational transconductance amplifier (OTA). The CMRR is mainly limited by the mismatch of the capacitors. The input impedance is limited by the size of the input capacitor and is usually frequency dependent. Using resistive gain elements, the design is inherently DC coupled [100]. The input common-mode range is less than VDD. The DC headroom is VDD/gain, which is much smaller than the capacitive counterpart. The input impedance is limited by the

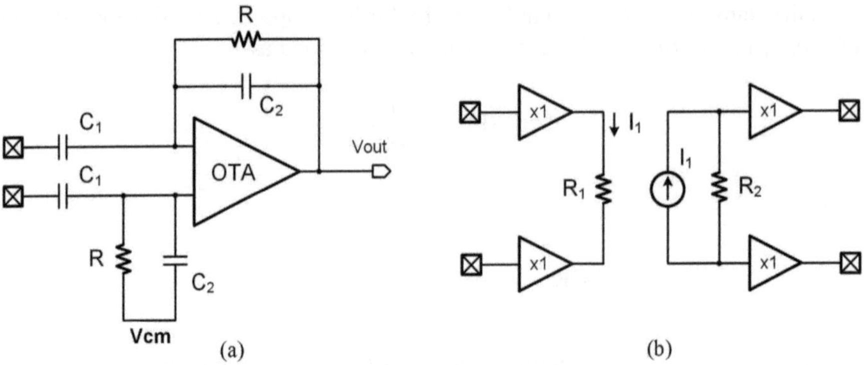

Fig. 2.3 The block diagram of the typical neural amplifiers with (**a**) capacitive gain element and (**b**) resistive gain element. The x1 symbol is for a unity gain buffer

Table 2.2 Comparison of neural amplifiers with capacitive and resistive gain elements

Gain element	Capacitive	Resistive
Gain	C_1/C_2	R_2/R_1
Noise PSD	$v^2_{\text{Thermal}}, v^2_{\text{Flicker}}$	$v^2_{\text{Thermal}}, v^2_{\text{Flicker}}, v^2_R$
Input impedance	$1/j\omega C_1$	$1/j\omega C_p$
DC headroom	VDD	VDD/gain
Input CM range	VDD	<VDD

parasitic capacitance, which is usually much higher than the capacitive counterpart. The CMRR is mainly limited by the mismatch of the analog buffers. Unlike the capacitive gain elements, the resistors also contribute to the overall noise. Several key features of the capacitive and resistive amplifiers are listed in Table 2.2 [101]. In summary, the topology using capacitive gain elements enjoys several inherent advantages over the resistive counterpart. Although a lot of techniques have been reported to address these problems [74, 100, 101], the capacitive gain element topologies are the mainstream designs for neural amplifiers.

In addition to passive gain elements, active feedback topologies have also been used to shape the frequency response [95, 102]. With an active feedback, the large input capacitor can be replaced by a small integrating capacitor, and a high input impedance can be achieved. However, the active feedback adds to the power consumption, and also contributes to the overall noise of the system.

Moreover, open-loop amplifiers have also been reported in literature [103, 104]. Compared with the closed-loop topologies, open-loop amplifiers can achieve a higher power efficiency, but usually suffer from a poor linearity. But since the neural signal has a small amplitude, the linear input range may be good enough in certain applications.

2.2.1.2 Low-Noise OTA

Operational transconductance amplifier (OTA) usually serves as the core of a low-noise neural amplifier. Commonly used OTA structures include: (a) current mirror OTA, (b) two-stage OTA with Miller compensation, (c) folded cascode OTA, and (d) telescopic OTA. Current mirror OTA usually has two stages with the dominant pole located at the second stage. No compensation capacitor is required to maintain stability. A detailed noise analysis is presented in [73]. However, the current mirror OTA has a limited gain, and there exists a trade-off between the noise and the phase margin. A gain boosting circuit can be used to enhance the gain of this amplifier, which is important in low voltage design in advanced CMOS technology. Two-stage OTA with a Miller compensation capacitor is also widely used in neural amplifiers. A detailed design analysis can be found in [105, 106]. The folded cascode OTA can achieve a high gain in a single stage, at the price of a higher power consumption. Paper [77] describes the strategy in choosing the parameters in a folded cascode OTA design for the optimal power-noise efficiency. Telescopic OTA can achieve the highest gain in a single-stage. But telescopic OTAs have very limited input range and voltage swing, which make the design very challenging especially in a low supply voltage.

In summary, all circuit topologies have pros and cons. Modified and improved versions have been widely reported. The voltage gain and input-referred noise of different OTA topologies are derived and summarized in [98, 99]. For thermal noise, increasing the transconductance of the input devices is usually critical. Thus, maximizing the transconductance for a given supply current is important for achieving an optimal power-noise efficiency. Besides, the supply current can be programmed to optimize the power efficiency in different noise conditions [44].

2.2.1.3 Other Noise Reducing Techniques

Many circuit techniques have been proposed in the literature to reduce the noise in the amplifier circuits. Commonly used low-noise techniques include chopping [78, 79, 107], auto-zeroing [108], digital assisted trimming [44, 109], analog and digital filtering, and so on.

For example, chopping is a very popular technique among neural amplifier designs, especially for EEG recordings. Figure 2.4 illustrates the concept of chopping. Before amplification, the input signal is modulated by a chopping frequency f_{chop}, which is much higher than the signal frequency. The modulated signal is then located to a frequency higher than the filter noise. After the amplification, the signal is converted back to the baseband frequency, at the same time, the flicker noise will be up-converted to the chopping frequency, which can be removed by a lowpass filter. Chopping technique reduces both flicker noise and DC offset, and the circuits after the chopping switches can achieve an excellent CMRR. However, it should be noticed that chopping also causes extra non-idealities, including offset, ripple, charge injection, clock feed-through, switch noise, and so on. Many techniques have

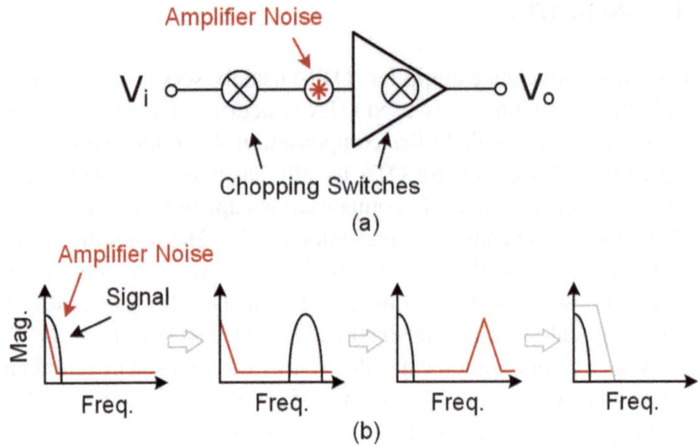

Fig. 2.4 (**a**) The block diagram of a chopping amplifier. (**b**) Illustration of the signal and noise spectrum before and after chopping

been proposed to suppress these problems, including: chopping within the feedback loop [78], chopping at the virtual ground [107], ripple reduction techniques [79], input impedance compensation [110], offset cancellation [82, 111], and so on.

2.2.2 *Circuit Implementation*

This section presents the analysis and design of a general–purpose neural recording front-end. Figure 2.5 shows the high-level block diagram of the proposed design. The building blocks of neural recording front-end consist of: a low-noise neural amplifier, a programmable gain amplifier (PGA), a multiplexer, an ADC, and a control module. This section mainly discusses the design of the neural amplifier.

The neural amplifier uses a fully differential, capacitor feedback topology. The input capacitors block the electrode offset and the half-cell potential from the electrode–tissue interface. The closed-loop differential gain is set to be 40 dB by C_{IN}/C_{FB}. A relative high gain is used to relieve the noise requirement of the following stages. A large MOS pseudo-resistor is used in the feedback loop. The highpass time constant is determined by $R_{pseudo} \cdot C_{FB}$. The circuit schematic of the pseudo-resistor is shown in Fig. 2.6. Compared with the MOS-bipolar resistor implemented in [73], this resistor has a larger linear range. Besides, setting the gate voltage to ground can short the feedback loop and force the input gate to mid-supply. This is a useful feature to implement a fast recovery from motion or stimulation artifact, which will be discussed in Chap. 5. A simulation of the MOS resistor in IBM 180 nm CMOS technology is shown in Fig. 2.7. The W/L of the MOS used in this simulation is 2 μm/2 μm. The simulated impedance is in the order of 100 GΩ.

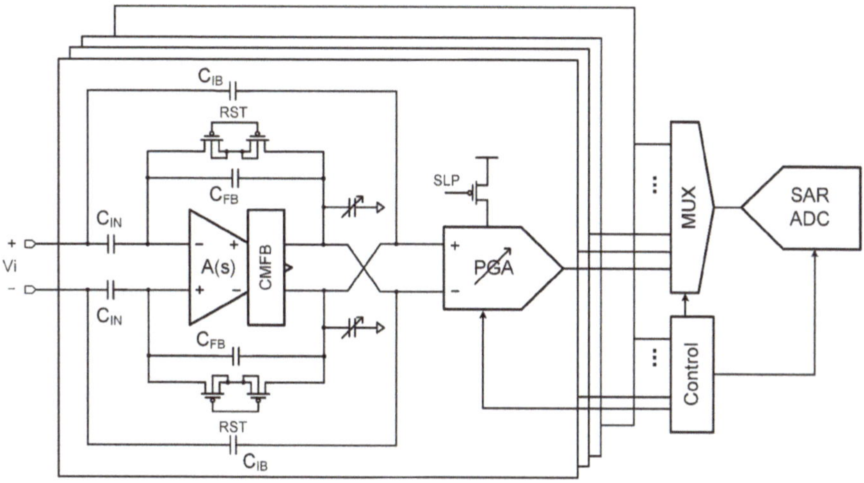

Fig. 2.5 The block diagram of the proposed low-noise neural recording front-end

Fig. 2.6 The circuit schematic of the MOS pseudo-resistor

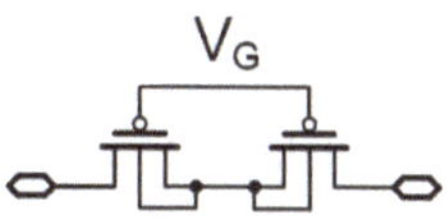

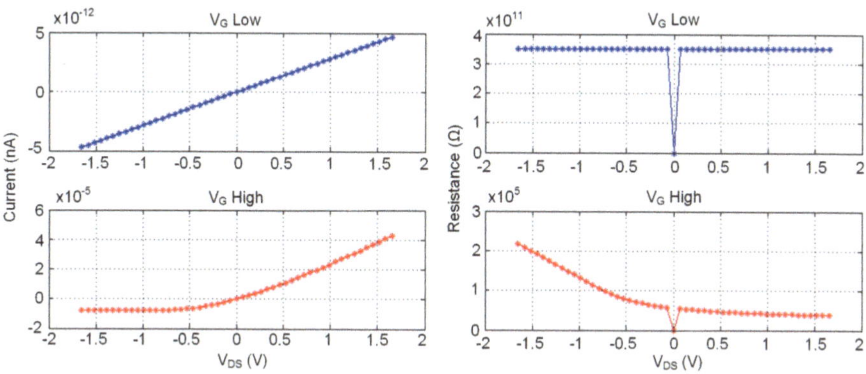

Fig. 2.7 The simulated resistance of the MOS pseudo-resistor. The *left column* shows the current versus the voltage applied, and *right column* shows the derived resistance

A very accurate resistance usually cannot be derived from the simulation. In practice, the highpass frequency is usually set to be much lower than the required signal frequency band to prevent the resistor's noise from rolling into the signal. Additional highpass or bandpass filter can be implemented in the following stage, if a better frequency shaping is necessary.

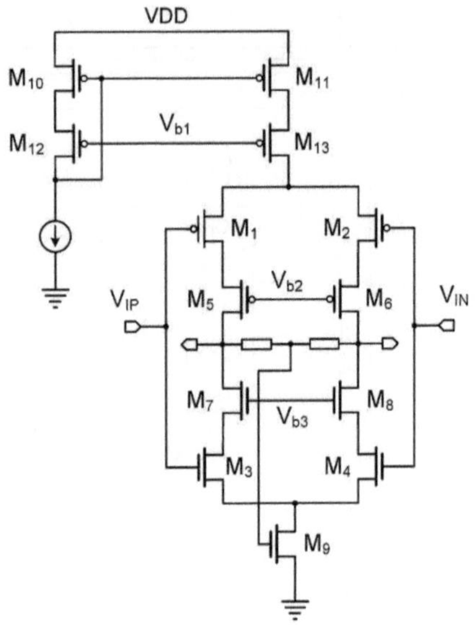

Fig. 2.8 The circuit schematic of the fully differential low-noise OTA with a complementary input stage

The circuit schematic of the designed OTA is shown in Fig. 2.8. The OTA has been designed to maximize the noise and power efficiency. A single-stage amplifier with a high gain is used to avoid the stability compensation in two-stage structures. The overall gain of the amplifier is given by:

$$A_v = (g_{m1} + g_{m3})(g_{m5}r_{o5}r_{o1}||g_{m7}r_{o7}r_{o3}) \tag{2.6}$$

where g_{mX} is the transconductance of the transistor M_X, and r_{oX} is the output resistance of the transistor M_X. The output thermal noise is:

$$\overline{i_{no}^2} = 4kT\gamma(g_{m1} + g_{m2} + g_{m3} + g_{m4})\Delta f \tag{2.7}$$

where $k = 1.38 \times 10^{-23}$ J/K is the Boltzmann constant. The input-referred thermal noise is:

$$\overline{v_{\rm ni}^2} = \frac{4kT\gamma(2g_{m1} + 2g_{m4})}{(g_{m1} + g_{m3})^2}\Delta f \tag{2.8}$$

Taking the flicker noise into account, the total input-referred noise power of the OTA can be expressed as:

$$\overline{v_{\text{ni,tot}}^2} = \frac{1}{(g_{m1}+g_{m3})^2}\left[8kT\gamma(g_{m1}+g_{m3}) + 2\left(\frac{K_N g_{m3}}{C_{\text{ox},N} f W_N L_N} + \frac{K_P g_{m1}}{C_{\text{ox},P} f W_P L_P}\right)\right]\Delta f \tag{2.9}$$

The flicker noise can be reduced by increasing the size of the input transistors or using techniques like chopping. If only thermal noise is considered in the following design optimization, the input-referred noise voltage equals to

$$\overline{V_{\text{ni,rms}}} = \sqrt{\frac{8kT\gamma}{g_{m1}+g_{m3}}\frac{\pi}{2}\text{BW}} \tag{2.10}$$

The noise efficiency factor (NEF) [93] for this amplifier can be derived as:

$$\begin{aligned} \text{NEF} &= \overline{V_{\text{ni,rms}}}\sqrt{\frac{2I_{\text{tot}}}{\pi\Phi_t \cdot 4kT \cdot \text{BW}}} \\ &= \sqrt{\frac{8kT\gamma}{g_{m1}+g_{m3}}\frac{\pi}{2}\text{BW}\frac{2I_{\text{tot}}}{\pi\Phi_t \cdot 4kT \cdot \text{BW}}} \\ &= \sqrt{\frac{2\gamma I_{\text{tot}}}{(g_{m1}+g_{m3})\Phi_t}} \end{aligned} \tag{2.11}$$

Thus, a lower NEF (the lower the better) can be expected if a higher power efficiency (g_m/I_{tot}) is achieved.

In this work, complementary input devices are used. The overall transconductance can be approximately doubled without increasing the quiescent current. Besides, all input transistors are biased in the sub-threshold region to achieve a high power efficiency [112]. In the conventional operation (above threshold):

$$g_m = \sqrt{2\mu C_{\text{ox}}\frac{W}{L}I_D} \quad \text{and} \quad g_m \propto \sqrt{I_D} \tag{2.12}$$

In the sub-threshold operation:

$$g_m = \frac{\kappa I_D}{\Phi_t} \quad \text{and} \quad g_m \propto I_D \tag{2.13}$$

where Φ_t is the thermal voltage. Thus, sub-threshold operation gives a higher transconductance than conventional above threshold operation in a same drain current I_D. It should be noticed that sub-threshold operation has a limited bandwidth due to the low biasing current and the relative large device parasitic capacitance. But neural signal has a low bandwidth in nature, so it is usually not a limiting factor in neural amplifier design. A simulation result shows that 98% of the noise is from the four input transistors, and flicker noise contributes more than thermal noise in the frequency range from 1 to 10 kHz.

Cascode transistors (M5–M8) are used to increase the voltage gain. However, this is usually at the cost of limiting the voltage headroom, and can be a challenge in a low-supply voltage using advanced CMOS technology. The simulation shows an open-loop gain of 90 dB is achieved in this OTA in a biasing current of 1 μA. The common mode feedback (CMFB) loop is merged in the main current path to avoid additional biasing current. Pseudo-resistors are used to get the common mode voltage without loading the amplifier. A drawback of this design is the threshold dependence of the common mode voltage.

A high input impedance reduces the signal attenuation. In a practical neural recording using a multi-channel electrode array, the recording electrode and the reference electrodes are usually not the same type of electrode, and may have a large impedance difference. Thus, even if the neural amplifier achieves a perfect common mode rejection, it cannot reject the conversion of common-mode signal to differential-mode due to the electrode mismatch. This problem can be relieved by increasing the input impedance of the neural amplifier. Positive current feedback [79] can be used to boost the input impedance by providing the driving current required at the input stage. A post-layout simulation of the input impedance boosting circuit is shown in Fig. 2.9.

A programmable capacitor array (C_L) is put at the output of the low-noise OTA. The bandwidth of the closed-loop amplifier is given by:

$$\mathrm{BW} = \frac{g_m}{C_L}\frac{C_2}{C_1} \tag{2.14}$$

The C_L can be programmed. The bandwidth of the OTA can also be tuned by changing the biasing current.

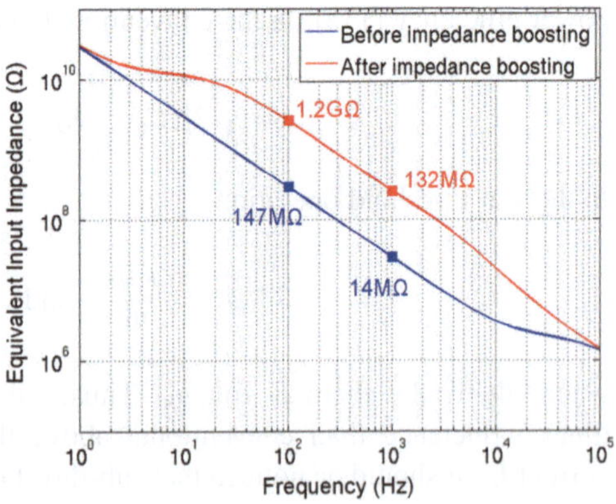

Fig. 2.9 A post-layout simulation of the effects of the input impedance boosting

The programmable gain amplifier (PGA) implemented in this work (2.5) is a classical 3-opamp amplifier. The gain is set by the resistors' ratio and can be chosen from 7, 10, and 19. Thus, the maximum gain of a recording channel is 1900. Additional analog buffers are added to the debugging points to drive the IO pads directly.

2.2.3 Measurement Results

The design has been fabricated in IBM 180 nm CMOS technology. The micrograph of the chip is shown in Fig. 2.10. The occupied silicon area of the full chip is $4.5 \times 1.5\,\text{mm}^2$, including IO pads. One recording channel has a dimension of $400\,\mu\text{m} \times 320\,\mu\text{m}$.

Bench testing was conducted to verify the function and performance of the fabricated chip. Figure 2.11 shows a measurement of the neural amplifier's output with a 1 kHz sinusoidal input signal. A resistor divider consists of a $2\,\text{k}\Omega$ and $1\,\Omega$ was applied at the output of the function generator to scale the signal amplitude, resulting a gain of 1/2001. The neural amplifier was configured to have the maximum gain of 1900. The measured gain was 1892.94, which corresponds to an absolute gain error of 0.37%.

The measured differential-mode and common-mode frequency responses of the low-noise amplifier are shown in Fig. 2.12. The closed-loop gain is set to be 60 dB. The highpass frequency corner is approximately 0.5 Hz. The measurement shows a CMRR above 110 dB.

The input-referred noise spectrum is shown in Fig. 2.13. An integration under this curve from 1 Hz to 7 kHz yields an rms noise voltage of $2.55\,\mu\text{V}$. This noise

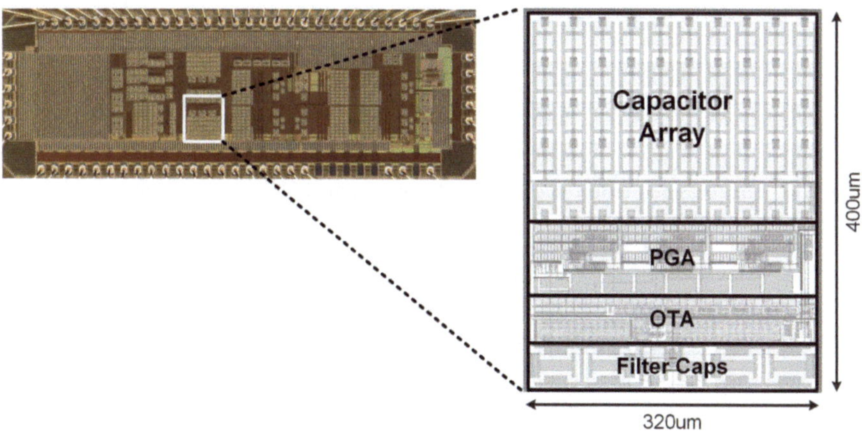

Fig. 2.10 The microphotograph and layout of one channel of the neural recording front-end. Major building blocks are highlighted in the layout

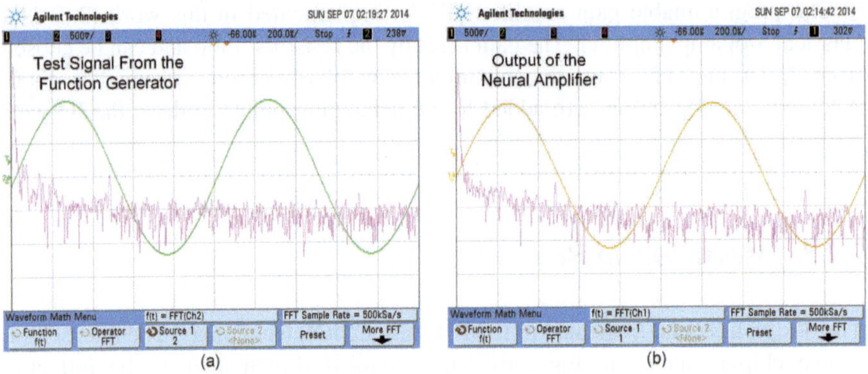

Fig. 2.11 The measured response of the neural amplifier with a 1 kHz sinusoidal input signal. The amplifier is configured with the maximum gain of 1900. The midband gain error is 0.37%

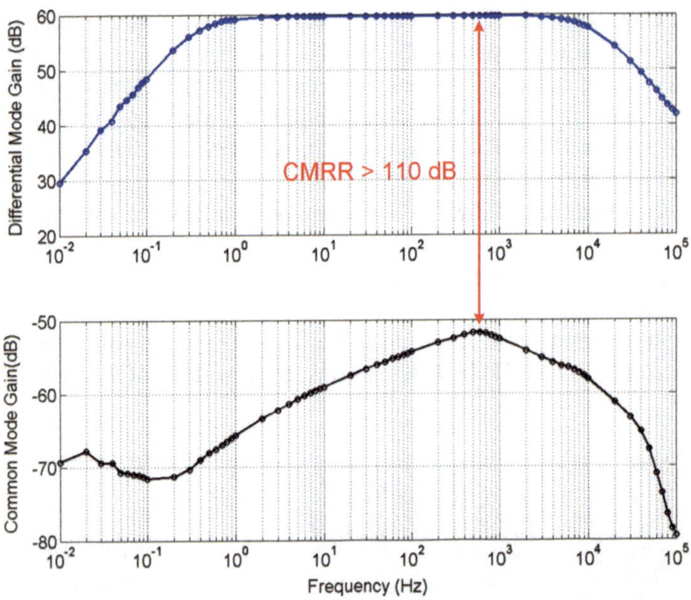

Fig. 2.12 The measured differential-mode and common-mode frequency responses of the low-noise neural amplifier

level was measured with a closed-loop gain of 60 dB, and the inputs were shorted using an internal switch. The noise density was calculated as:

$$\text{Noise Density} = \frac{V_{\text{rms}}}{\sqrt{\text{BW}\pi/2}} \tag{2.15}$$

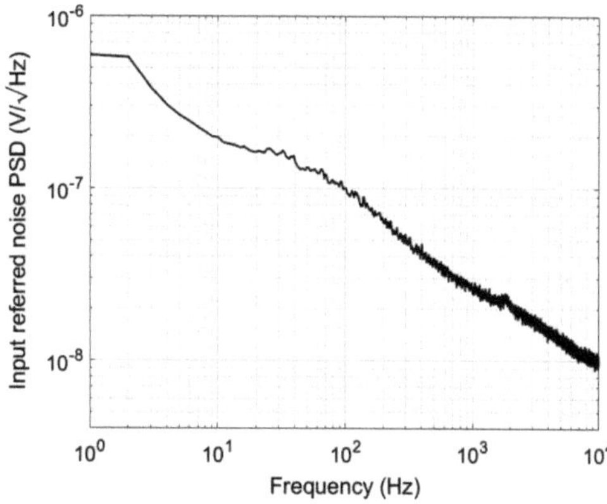

Fig. 2.13 The measured input-referred voltage noise spectrum. An integration under this curve from 1 Hz to 7 kHz yields an rms noise of 2.55 μV

Table 2.3 The neural front-end specifications summary

Parameter	Value
Process	180 nm CMOS
Supply voltage	3.3 V
LNA current	2 μA (biasing current not included)
Closed-loop gain	40 dB
Gain error	0.37%
Bandwidth	1–7 kHz
Integrated noise	2.55 μV
Noise density	24.3 nV/rtHz
NEF (Eq. (2.1))	1.68
PEF (Eq. (2.2))	9.38
Input range	4 mV
CMRR	>110 dB

The noise density in the 7 kHz bandwidth is 24.3 nV/rtHz. The calculated NEF is 1.68, and the PEF is 9.38.

The summarized measured specifications of the design are listed in Table 2.3.

In summary, this section presents the design of a general–purpose low-noise neural amplifier. The design achieves a low noise floor, an accurate gain, a good CMRR in a good power efficiency. The design was later used in in vivo neural signal acquisition.

Table 2.4 compares the measured performance of this work with prior published neural recording front-end design. This work achieves a comparable performance among the state-of-the-art designs.

Table 2.4 Comparison with prior works

Work	'03 [73]	'07 [78]	'07 [74]	'10 [107]	'13 [111]	'14 [44]	'17
Publication	JSSC	JSSC	JSSC	JSSC	JSSC	JSSC	This work
Technology	1.5 μm	0.8 μm	0.5 μm	180 nm	180 nm	180 nm	180 nm
Noise (μV)	2.2	0.95	2.26	1.3	0.91	5.23	2.55
BW (Hz)	0.025–7.2k	0.05–100	0.5–1k	100	100	7k	1–7k
Current (μA)	16	1	11.1	3.5	NA	0.97	2
Supply (V)	5	1.8–3.3	3	1	1	1.8	3.3
NEF	4.03	4.6	9.2	9.4	5.1	1.77	1.68
PEF[a]	81.2	38.1	253.9	88.4	26.2	5.6	9.3

[a]Not provided by the author, but calculated using Eq. (2.2)

2.3 A Pre-whitening Neural Amplifier

2.3.1 *Introduction*

The power spectrum of electrocorticography (ECoG) and local field potential (LFP) have a characteristic $(1/f)^n$ drop with frequency [31]. This phenomenon has been observed in multiple species including humans [113]. At frequencies around 1Hz, the signal amplitude can be as large as a few millivolts, and attenuates at $1/f^2$ until 80 Hz, then attenuates at $1/f^4$ [114]. At the same time, the noise power density of the CMOS front-end is usually inversely proportional to the frequency [115].

$$\overline{V_n^2} = \frac{K}{C_{\mathrm{ox}}WL} \cdot \frac{1}{f} \tag{2.16}$$

where K is a process-dependent parameter on the order of 10^{-25} V^2F. This suggests that the SNR of the recording front-end improves as the frequency decreases, as illustrated in Fig. 2.14a. Intuitively, if a wideband recording front-end is designed to achieve the voltage swing requirement for the low-frequency signal, at the same time preserves the SNR for the high-frequency signal, it needs to be designed with an ultra-high dynamic range. An ultra-high dynamic range wideband low-noise amplifier and a high-resolution ADC design are challenging and will cost a high power consumption.

In this work, a pre-whitening amplifier is proposed to address this problem. The basic idea of the pre-whitening processing is illustrated in Fig. 2.14b. If we reduce the gain for the low-frequency content, a sufficient SNR may still be preserved for the recording purpose, and the dynamic range requirement of the system can be significantly relaxed. Since the frequency shaping processing is similar to a whitening filter, which turns the signal into a near white signal, the amplifier is named pre-whitening amplifier in this work. The simplest way to implement this pre-whitening amplifier is via a highpass filter. If the cut-off frequency of the highpass filter is known, the amplitude and phase of the original signal can be recovered in the post-recording processing.

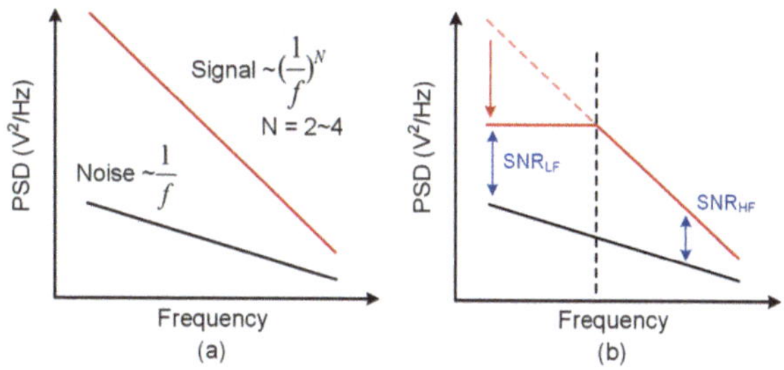

Fig. 2.14 Illustration of the pre-whitening filter. (**a**) The neural signal displays a $1/f^n$ power characteristic, while the recording front-end has a $1/f$ noise power characteristic. (**b**) The pre-whitening filter shapes the frequency response of the recording front-end to reduce the overall dynamic range requirement, while still preserves a sufficient SNR

Fig. 2.15 A first order RC highpass filter with noise source

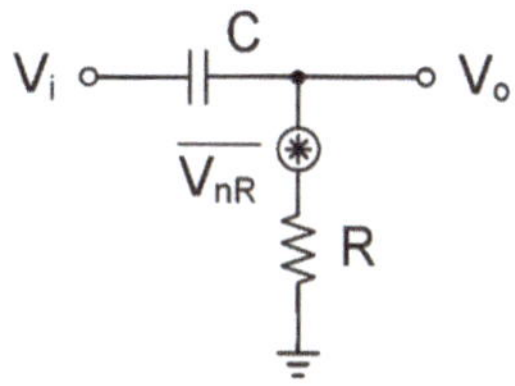

In summary, a pre-whitening neural recording front-end is proposed. In the pre-whitening amplifier, the frequency response of the neural amplifier is shaped according to the characteristic of the neural signal. The design significantly reduces the dynamic range requirement of the neural amplifier and the ADC resolution without sacrificing the signal quality. In the following sections, possible circuit implementations of the pre-whitening neural amplifier are analyzed. The key design trade-offs are described, and the simulation and experimental results of the proposed design are presented.

2.3.2 *Analysis of Pre-whitening Neural Amplifier Design*

The thermal noise power spectral density of a resistor is given here for convenience:

$$\overline{V_{nR}} = \sqrt{4kTR} \tag{2.17}$$

where $k = 1.38 \times 10^{-23}$ is the Boltzmann's constant, T is the absolute temperature in Kelvin. If a recording electrode has an impedance of 100 kΩ, it should have a noise density of 40.7 nV/$\sqrt{\text{Hz}}$. Assume the neural signal's frequency band of interest is from 1 Hz to 10 kHz, the electrode gives an integral thermal noise of 4.07 μV in this frequency range.

Consider the simplest case of a first order RC highpass filter. Figure 2.15 shows the circuit and the noise source.

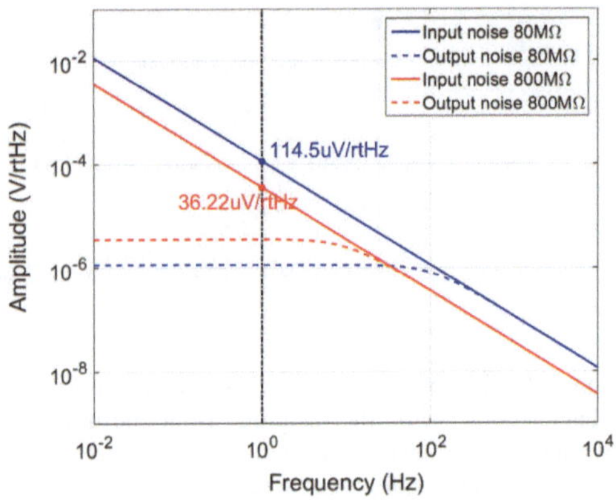

Fig. 2.16 Noise simulation of RC highpass filters with frequency corners at 10 and 100 Hz. The capacitor value is set to be 20 pF

The equivalent output noise of the RC highpass filter is given by:

$$\begin{aligned} \overline{V_{n,o}} &= \frac{1}{1 + s\mathrm{RC}} \cdot \overline{V_{nR}} \\ &= \frac{\sqrt{4kTR}}{1 + s\mathrm{RC}} \end{aligned} \tag{2.18}$$

While the input-referred noise of the RC highpass filter is given by:

$$\begin{aligned} \overline{V_{n,i}} &= \frac{1}{s\mathrm{RC}} \cdot \overline{V_{nR}} \\ &= \frac{\sqrt{4kTR}}{s\mathrm{RC}} \end{aligned} \tag{2.19}$$

The input-referred noise increases with a decreasing frequency. This is an important observation and provides some intuition for the following analysis. Figure 2.16 shows the simulation of a first order RC highpass filter. Both the output and the input-referred noise are plotted. The capacitor value is set to be 20 pF, and the resistor values are set to be 800 MΩ and 80 MΩ, and the cut-off frequency is 10 Hz and 100 Hz, respectively. The input-referred noise densities at the 1 Hz are marked in the figure. It should be noticed that using a larger capacitor value with the same cut-off frequency can achieve a lower noise density. However, large capacitors take a lot of silicon area, thus is not suitable for multiple channel recording front-end integration.

In summary, a simple RC filter is not suitable for implementing the proposed pre-whitening amplifier. In the following section, two methods of implementation are discussed: (1) pre-whitening after a wideband LNA, and (2) pre-whitening at the direct neural interface.

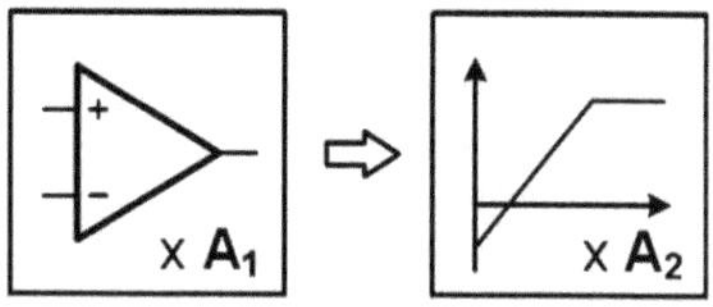

Fig. 2.17 The block diagram of a pre-whitening filter after a wideband low-noise amplifier

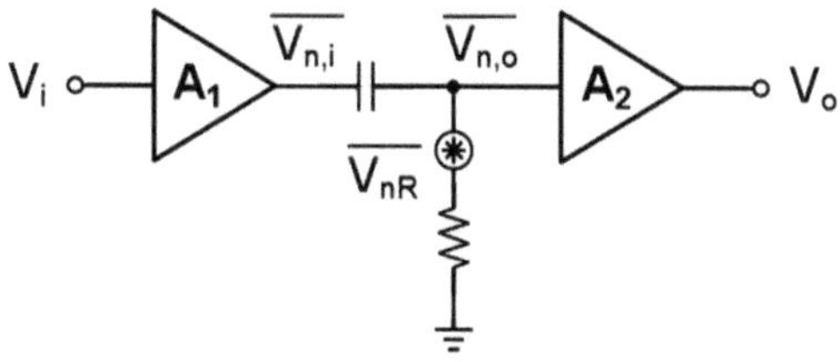

Fig. 2.18 The circuit schematic of a pre-whitening filter after a wideband low-noise amplifier. Noise source from the resistor is shown

2.3.2.1 Pre-whitening After a Wideband Low-Noise Amplifier

The block diagram of a pre-whitening filter after a wideband low-noise amplifier is shown in Fig. 2.17. Since the filtering is implemented after the wideband amplifier, the input-referred noise from the filter will be attenuated by the gain of the wideband amplifier. Again, assume using a simple RC filter, Fig. 2.18 shows the circuit diagram and the noise source.

The input-referred noise of the recording front-end from the filter is then given by:

$$\begin{aligned}\overline{V_{i,\mathrm{rms}}} &= \frac{1}{A_1}\sqrt{\int_{f_L}^{f_H}\overline{V_{n,i}^2}\cdot df} \\ &= \frac{1}{A_1\pi C}\sqrt{\frac{kT}{R}\left(\frac{1}{f_L}-\frac{1}{f_H}\right)}\end{aligned} \tag{2.20}$$

Assume the frequency band of interest is from 1 Hz to 10 kHz, and the first stage wideband neural amplifier has a gain A_1 of 100. If we set the highpass frequency of the second stage to be 10 Hz, the integral noise is 0.11 μV, and if we set the highpass frequency to be 100 Hz, the integral noise is 1.14 μV. In both cases, the integral noise is lower than the thermal noise of an electrode with an impedance of 100 kΩ (Sect. 2.3.1).

Several active highpass filters can achieve a lower input-referred noise than the simple RC filter. Consider the circuit with a capacitive feedback as shown in Fig. 2.19. The signal's transfer function can be expressed as:

$$\begin{aligned}H_{\mathrm{sig}}(s) &= \frac{s\mathrm{RC}_1}{s\mathrm{RC}_2+1} \\ &= \frac{C_1}{C_2}\cdot\frac{s}{s+\frac{1}{\mathrm{RC}_2}}\end{aligned} \tag{2.21}$$

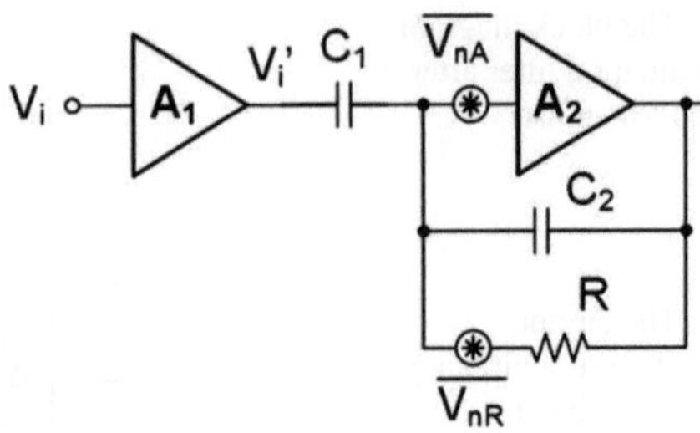

Fig. 2.19 An implementation of an active highpass filter. Noise sources from the resistor and the second stage amplifier are shown

The midband gain of the amplifier A_{CL} is $\frac{C_1}{C_2}$, and the highpass frequency is set by $\frac{1}{\mathrm{RC}_2}$. The amplifier A_2's noise transfer function can be expressed by:

$$\begin{aligned} H_{nA}(s) &= 1 + \frac{s\mathrm{RC}_1}{s\mathrm{RC}_2 + 1} \\ &= 1 + \frac{C_1}{C_2} \cdot \frac{s}{s + \frac{1}{\mathrm{RC}_2}} \end{aligned} \tag{2.22}$$

Since the amplifier's noise transfer function and the signal's transfer function have the same highpass frequency $\frac{1}{\mathrm{RC}_2}$, the amplifier's noise is shaped in the same way as the signal. Thus, the filter won't cause frequency dependent SNR degradation.

Let's look at the resistor's noise transfer function:

$$\frac{V_o - V_{nR}}{R} = sC_2 V_o = 0 \tag{2.23}$$

$$H_{nR}(s) = \frac{1}{s\mathrm{RC}_1} = \frac{1}{A_{\mathrm{CL}}} \frac{1}{s\mathrm{RC}_2} \tag{2.24}$$

where $\frac{1}{\mathrm{RC}_2}$ is the signal's highpass frequency, and A_{CL} is the closed-loop gain of the second stage. So compared with the implementation in Fig. 2.18, the noise is further suppressed by the gain of the second stage. The overall input-referred noise density from the resistor is given by:

$$\overline{V_{nR,i}} = \frac{1}{A_1 A_{\mathrm{CL}}} \frac{\sqrt{4kTR}}{s\mathrm{RC}_2} \tag{2.25}$$

In a practical design, the sum of $\overline{V_{nR,i}}$ and the input-referred noise of the first stage should be lower than the noise and the SNR requirement of the recording system. Again, assume the cut-off frequency of the pre-whitening amplifier is 10 Hz, the capacitor C_2 is 2 pF, the closed-loop gain of the first and second stage is 100 and 40, respectively. The $\overline{V_{nR,i}}$ at 1 Hz is 28.6 nV$\sqrt{\mathrm{Hz}}$, which is lower than the thermal noise density of a 100 $k\Omega$ electrode (Sect. 2.3.2). If the cut-off frequency of the pre-whitening amplifier is 100 Hz, the noise density $\overline{V_{nR,i}}$ at 1 Hz is 90.5 nV/$\sqrt{\text{Hz}}$, which is still lower than most low-noise neural amplifier designs at 1 Hz, and is sufficient for the SNR requirements in most intracortical neural recordings.

2.3.2.2 Low-Noise Neural Amplifier with Integrated Pre-whitening Filter

This section discusses the possible methods of integrating the pre-whitening filter into the first stage low-noise amplifier. It is more challenging to design the pre-whitening filter at the first stage because of the noise increase with decreasing frequency due to the filter's response. But there are also some advantages. The electrode interface usually has an offset caused by half-cell potential up to several hundred millivolts, as reviewed in Sect. 2.1. The recording amplifier will need to reject this large offset, typically accomplished by using a highpass filter with a cut-off frequency below 1 Hz. However, it is difficult to implement such a large time-constant on-chip. One solution is to use MOS pseudo-resistor, as described in Sect. 2.2.2. But the pseudo-resistors have reliability problem for the use in implanted medical devices, and they are susceptible to electromagnetic interface and degradation over time [31]. If the pre-whitening filter can be integrated into the first stage amplifier, the sub-Hertz filter can be avoided.

Consider the capacitor-coupled neural amplifier in Fig. 2.20. The signal's transfer function is:

$$H_{\text{sig}}(s) = \frac{C_1}{C_2} \cdot \frac{s}{s + 1/RC_2} \tag{2.26}$$

Thus, the highpass corner frequency is determined by $1/\text{RC}_2$. The transfer function of the amplifier's noise is:

$$H_{nA}(s) = 1 + \frac{C_1}{C_2} \cdot \frac{s}{s + 1/\text{RC}_2} \tag{2.27}$$

And the resistor noise's transfer function is:

$$H_{nR}(s) = \frac{1}{s\text{RC}_1} \tag{2.28}$$

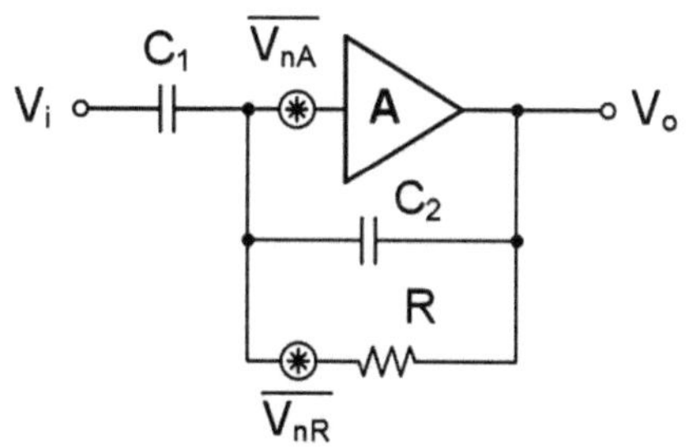

Fig. 2.20 A typical capacitor-coupled neural amplifier. Noise sources are marked in the figure

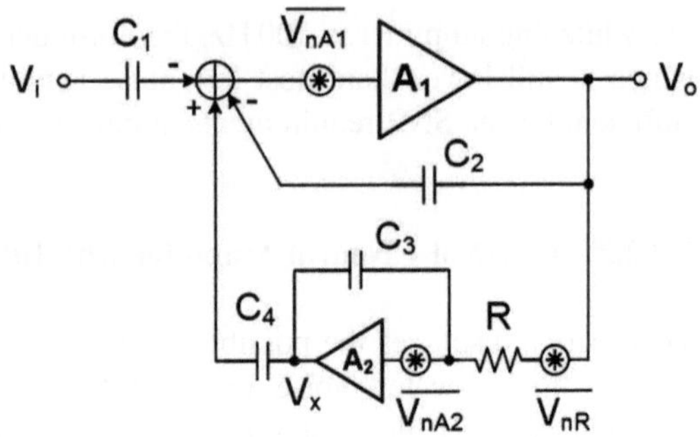

Fig. 2.21 A capacitor-coupled instrumentation amplifier with a DC servo loop

The input-referred noise density from the resistor is:

$$\overline{V_{nR,i}} = \frac{1}{A_{\mathrm{CL}}}\frac{\sqrt{4kTR}}{s\mathrm{R}C_2} \tag{2.29}$$

Compared with Eq. (2.25), the only difference is that this input-referred noise is no longer attenuated by preamplifier. If A_{CL} is designed to be the product of the gain of the two stages in previous section, it can achieve the same noise performance. But it is difficult in practical designs.

There are other circuit topologies to implement a highpass frequency response. One method is to use a DC servo loop. A typical example is shown in Fig. 2.21.

$$sC_1V_i(s) + sC_2V_o(s) - sC_4V_x(s) = 0 \tag{2.30}$$

$$Vx(s) = -\frac{1}{s\mathrm{R}C_3}V_o(s) \tag{2.31}$$

The signal's transfer function can be expressed as:

$$\begin{aligned} H_{\mathrm{sig}}(s) &= \frac{s\mathrm{R}C_1C_3}{s\mathrm{R}C_2C_3 + C_4} \\ &= \frac{C_1}{C_2}\cdot\frac{s}{s + C_4/\mathrm{R}C_2C_3} \end{aligned} \tag{2.32}$$

The mid-band gain of this circuit is $\frac{C_1}{C_2}$. Compared with Eq. (2.26), the high-pass frequency corner is $\frac{C_4}{C_2}\frac{1}{\mathrm{R}C_3}$, where $\frac{1}{\mathrm{R}C_3}$ is the frequency corner of the integrator in the feedback loop.

The noise transfer function of the amplifier A_1 is:

$$C_1V_{nA1} = C_2(V_o - V_{nA1}) + C_4(V_x - V_{nA1}) \tag{2.33}$$

$$H_{nA1}(s) = \frac{C_1 + C_2 + C_4}{C_2}\cdot\frac{1}{1 + C_4/sRC_2C_3} \tag{2.34}$$

The noise transfer function of the amplifier A_2 is:

$$\frac{V_o - V_{na2}}{R} = sC_3(V_{na2}) = sC_4Vx = sC_2V_o \tag{2.35}$$

$$H_{nA2}(s) = \frac{s\mathrm{RC}_3 + 1}{s\mathrm{RC}_3(C_2/C_4) + 1} \tag{2.36}$$

The noise transfer function of the resistor is:

$$\frac{V_o - V_{nR}}{R} = sC_3V_x \tag{2.37}$$

$$sC_4V_x = sC_2V_o \tag{2.38}$$

$$H_{nR}(s) = \frac{1}{1 + s\mathrm{RC}_3\frac{C_2}{C_4}} \tag{2.39}$$

The input-referred noise density from the amplifier A_1 is:

$$\overline{V_{nA1,i}} = \frac{C_1 + C_2 + C_4}{C_1}V_{nA1} \tag{2.40}$$

The input-referred noise density from the amplifier A_2 is:

$$\overline{V_{nA2,i}} = \frac{C_4}{C_1}\left(1 + \frac{1}{s\mathrm{RC}_3}\right)V_{nA2} \tag{2.41}$$

The input-referred noise density from the resistor is:

$$\overline{V_{nR,i}} = \frac{C_4}{s\mathrm{RC}_1C_3}\sqrt{4kTR} \tag{2.42}$$

Compared with Eq. (2.29), the noise contribution from the resistor also depends on the ratio of C_4/C_3. However, reducing the ratio of C_4/C_3 decreases the input voltage headroom.

The resistor can be further replaced by a switched capacitor circuit. A simplified circuit schematic is shown in Fig. 2.22. In the switched capacitor circuit, the time-constant can be better controlled by the ratio of the capacitors and the switching

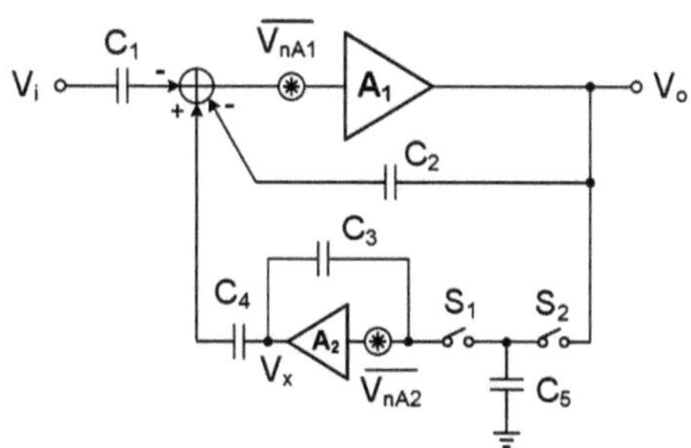

Fig. 2.22 A capacitor-coupled instrumentation amplifier with a DC servo loop implemented by switched capacitor circuits

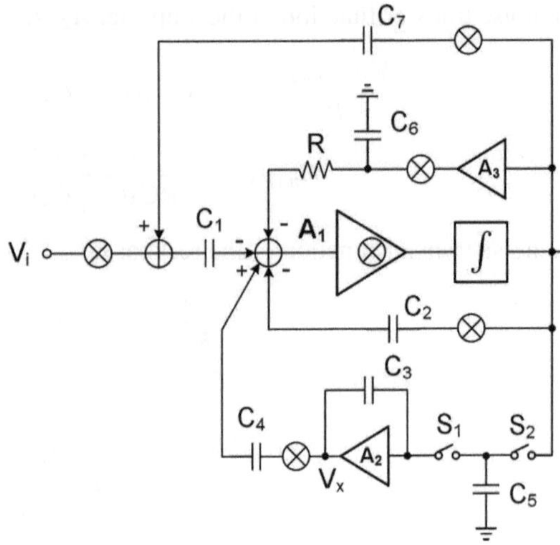

Fig. 2.23 A capacitor-coupled chopping amplifier with a DC servo loop and an input impedance boosting loop

frequency. However, in order to achieve the required large time-constant, a large capacitor tank is to be implemented. Several techniques have been proposed to reduce the required capacitor size [78, 116].

If an ultra low-noise is required for the low-frequency signal component, chopping technique can be combined in the pre-whitening amplifier design. An example of a chopping pre-whitening amplifier is shown in Fig. 2.23. With chopping, the flicker noise can be removed, and the amplifier can guarantee a good SNR for the signal even with the lower gain at the low frequency. However, there are also many trade-offs involved with the chopping amplifier design [79, 107, 110]. The effects it takes may counteract the benefits from the pre-whitening.

2.3.3 Circuit Implementation

A pre-whitening amplifier using the architecture presented in Sect. 2.3.2 is designed to demonstrate the idea. The circuit schematic of the designed pre-whitening amplifier is shown in Fig. 2.24. A single-ended architecture is used in this work. A conventional low-noise current mirror OTA is used in both stages [73]. A T-connected pseudo-resistor (TPR) proposed in [44] is used as the feedback resistor in the first stage. If the equivalent resistance of the transistor X is R_X, the total equivalent resistance of the TPR is $R_1+R_2+R_1 \cdot R_2/R_3$. The pseudo-resistor in the second stage is the same as the one used in Sect. 2.2. The gate voltage can be used to tune the resistance R_B over a large range, which is used to tune the cut-off frequency of the pre-whitening filter. The first stage has a closed-loop gain of 100, and the second stage has a closed-loop gain of 40, which are the same as the assumptions in the previous analysis.

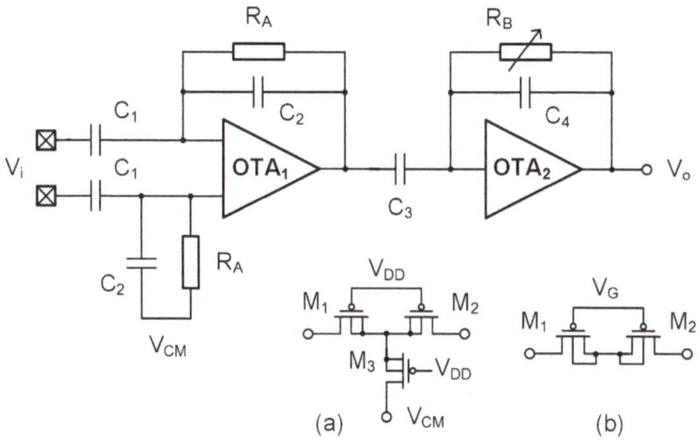

Fig. 2.24 The circuit schematic of the designed pre-whitening amplifier. The pseudo-resistors R_A and R_B used in the first and second stage are shown in subplot (**a**) and (**b**), respectively

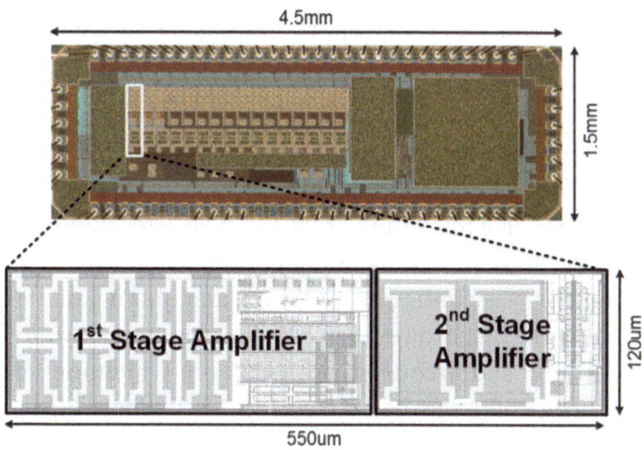

Fig. 2.25 The microphotography and layout of one channel of the pre-whitening amplifier

2.3.4 Measurement Results

The design has been fabricated in IBM 180 nm CMOS technology. The micrograph of the chip and the layout of one recording channel are shown in Fig. 2.25. The full chip occupies a silicon area of $4.5 \times 1.5\ \text{mm}^2$, including IO pads. One recording channel has a dimension of $550\ \mu\text{m} \times 120\ \mu\text{m}$.

Bench testing was conducted to evaluate the performance of the fabricated chip. Figure 2.26 shows both the amplitude and phase frequency response of the pre-whitening amplifier. The simulated frequency response is also plotted in dashed lines for comparison. With the information of the frequency response, the original signal can be recovered from the pre-whitened recording.

Synthetic neural signal was generated using an arbitrary function generator 33521A from Agilent to the test the pre-whitening amplifier. A 1-min neural signal

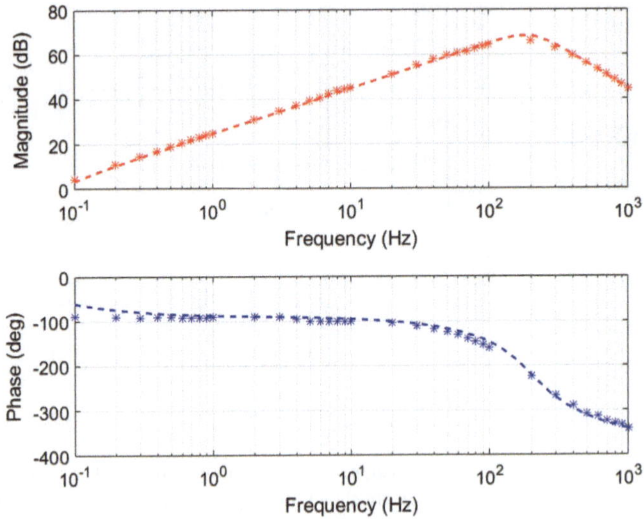

Fig. 2.26 The measured frequency response of the pre-whitening amplifier in comparison with the simulation result

containing local field potentials was used for testing. The signal was recorded using RZ2 workstation from Tucker-Davis Technologies. The signal was sampled at 24.4 kSps in a resolution of 24-bit. A resistor divider consisting of 2 kΩ and 1 Ω was applied at the output of the function generator, gives a gain of 1/2001. The neural amplifier was configured to have a maximum gain of 4000.

The designed amplifier can be configured to do both conventional wideband recording and frequency shaping pre-whitening recording. The power spectral density (PSD) was calculated for 24 channels of the LFP recordings. Figure 2.27 shows a comparison of the PSD of the conventional wideband recording and the pre-whitening recording. The result clearly shows that the spectrum of the pre-whitening recording was flattened in the low-frequency range, which saves the voltage headroom by more than an order of magnitude. The reduction in the dynamic range relaxes the requirement of the linear range of the low-noise amplifier and the ADC design.

The reconstruction of the signal was performed in Matlab. Figure 2.28 shows a comparison of a 10-s segment of the conventional recording, the pre-whitening recording, and the reconstruction from the pre-whitening recording. Pearson correlation coefficient is used here to evaluate the accuracy of the reconstruction [117]. The correlation coefficient is defined as:

$$\rho(x, y) = \frac{1}{N-1}\sum_{i=1}^{N}\left(\frac{\overline{x_i - \mu_x}}{\sigma_x}\right)\left(\frac{\overline{y_i - \mu_y}}{\sigma_y}\right) \tag{2.43}$$

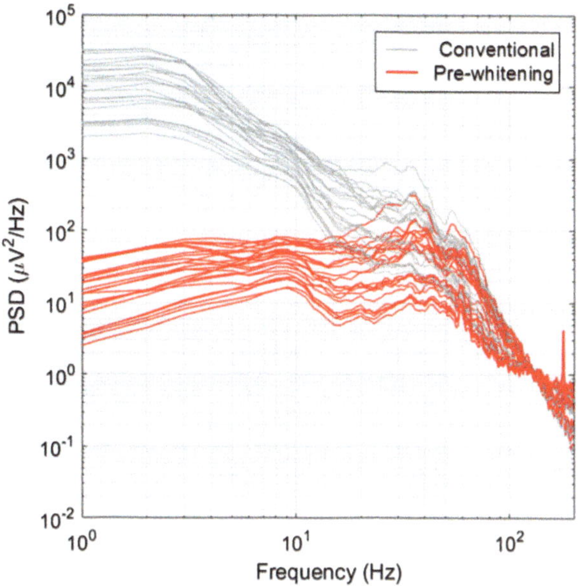

Fig. 2.27 Comparison of the PSD of the pre-whitening amplifier and the original signal

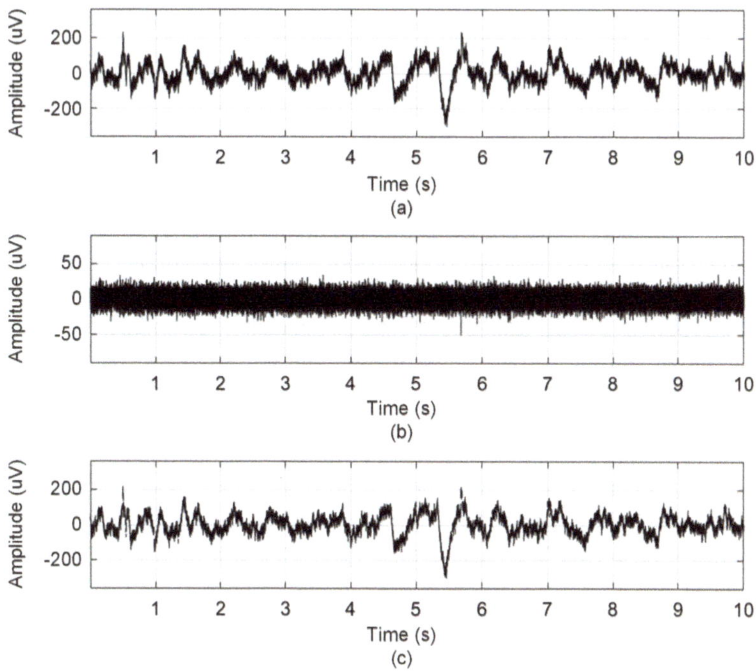

Fig. 2.28 Comparison of (**a**) the wideband signal, (**b**) the measured output of the pre-whitening amplifier, and (**c**) the reconstructed signal from the pre-whitening amplifier's recording

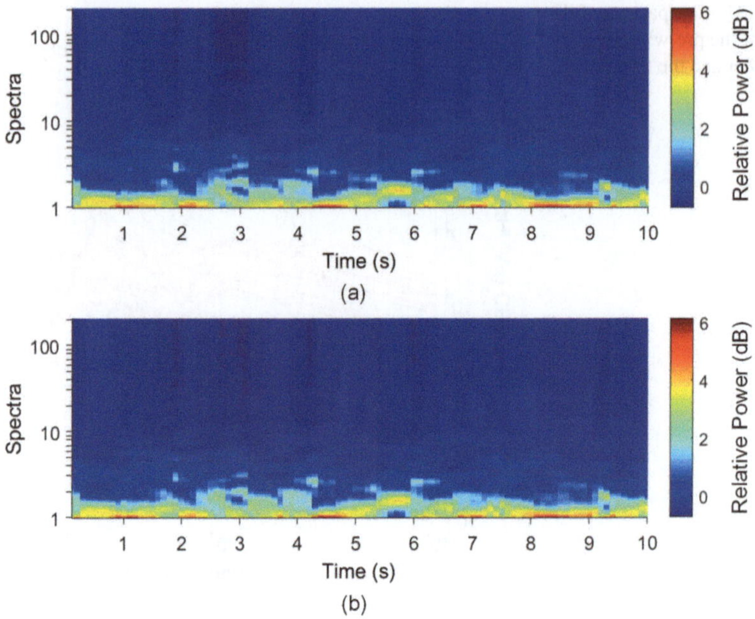

Fig. 2.29 Comparison of the spectrum of (**a**) the original signal and (**b**) the reconstructed signal from the pre-whitening amplifier's recording

where μ_x and σ_x are the mean and standard deviation of the signal x, and μ_y and σ_y are the mean and standard deviation of the signal y. The result shows a correlation coefficient of 97.6%, which indicates a faithful recovering of both phase and amplitude. A high-order zero-phase digital filter of 1–200 Hz was applied before the comparison. Even in this case, both phase and amplitude will need to be recovered at the same time to completely reconstruct the original signal.

Power spectral density estimation was calculated by periodgram for both the original signal and the reconstructed signal from the pre-whitening recording, as shown in Fig. 2.29. The result clearly shows that the pre-whitening processing can provide a faithful reconstruction of the spectrum content.

In summary, the pre-whitening amplifier design takes advantage of the characteristics of the neural signal. Since the power density of the neural signal including ECoG and LFP drops faster with frequency than the filter noise of the CMOS recording front-end, there is an opportunity to design a recording front-end with less gain at low frequency while preserving a sufficient SNR for the wideband signal. The design significantly reduces the dynamic range and linearity requirement of the low-noise amplifier and the ADC. The circuit implementation of the pre-whitening front-end is analyzed in this section with a detailed noise analysis. A prototype was designed and fabricated in CMOS technology. Experimental results are presented in comparison with simulation and theoretical computation. The proposed pre-whitening amplifier provides an opportunity to improve the performance of neural recording-end without a power penalty, and it can be advantageous to integrate it into a high channel-count neural recording front-end system.

2.4 Design of a Low-Power Analog-to-Digital Converter

2.4.1 Introduction

A low-power analog-to-digital converter (ADC) is an essential component in a neural interface system. In a typical bidirectional neural interface system, ADCs can be used to digitize neural signal, sensory signal, extracted neural features, and stimulation compliance voltage. Among all ADC topologies, successive approximation register (SAR) ADCs have advantages in accuracy and power efficiency at a moderate sampling rate. Firstly, a SAR ADC does not require a high gain and high bandwidth opamp for high accuracy and linearity. Secondly, SAR logic mainly consists of digital circuits, so the speed and power scales down with deep submicron CMOS technologies. Thirdly, if a capacitive DAC is used, no static power is consumed, thus the power scales with the sampling rate. Comprehensive reviews and tutorials of SAR ADC design can be found in [118–120].

Recently, a lot of techniques for power-efficient SAR ADC designs have been reported. Among these techniques are split capacitor array [121, 122], monotonic capacitor switching [123], partial floating capacitor switching [124], step-charging design [125], reference free design [126], asynchronous timing [127], and so on. In addition to techniques for general–purpose SAR ADC designs, several techniques have been reported to optimize the design particularly for neural or sensory signal. Among them are:

- Adaptive resolution or dynamic range: including programming the number of bits [128], or adding additional programmable gain amplifier before the ADC [129]
- Data dependent or data-driven sampling: for example, combine action potential detection and digitization together [130]. Besides, the sampling rate can also be adapted to the activity using a continuous time level-crossing sampling [131].
- Delta difference sampling: since the neural signal has both slow and fast oscillations over time, normal sampling during a slow activity period is not energy efficient. So digitize only the difference [44, 132], using a bypass window [133], or using LSB-first approach [134] can achieve a better power efficiency.

In this section, the design of a voltage-mode 10-bit SAR ADC is presented. Specifications are analyzed, circuit design details are described, and measurement results are presented.

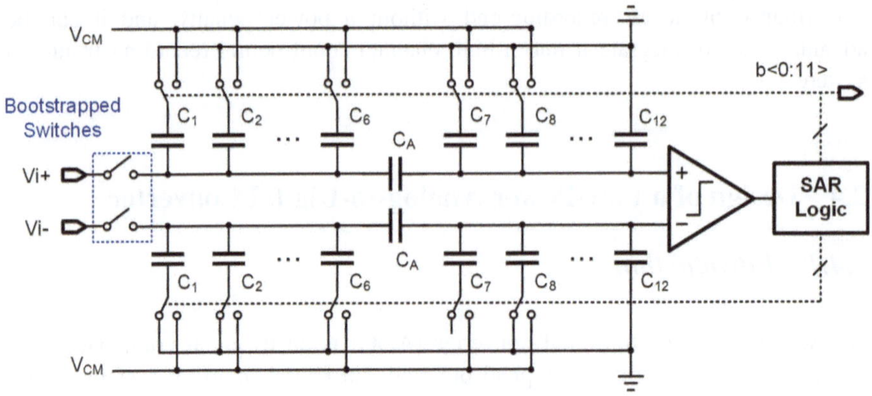

Fig. 2.30 The circuit diagram of the 10-bit voltage-mode SAR ADC

2.4.2 Circuit Implementation

The circuit diagram of the 10-bit voltage-mode SAR ADC is shown in Fig. 2.30. The major building blocks are: (1) comparator, (2) SAR logic, (3) DAC, and (4) sample and hold switch. Fourteen clock cycles are used to complete one conversion, allowing four clock cycles for sampling.

A commonly used capacitive DAC is employed in this SAR ADC. Since the required capacitor size in a conventional binary capacitor array can be very small without compromising the ENOB, custom designed capacitors are often used to achieve a minimum total input capacitance with an ultra low-power consumption [123, 135]. However, these designs usually require a custom characterization for a specific fabrication process. In this work, a split capacitor array is adopted to reduce the total capacitance, lowering the power consumption and area. The capacitors are realized as a standard metal-insulator-metal (MIM) structure available in the standard PDK.

A monotonic switching procedure is applied to minimize the power consumption from unnecessarily charging and discharging of the capacitor array [123]. In the monotonic switching procedure, the first comparison is performed without switching, and the total capacitance is half of the conventional capacitive SAR ADC's DAC array [123].

Figure 2.31 shows the circuit schematic of the SAR timing generation module. A global reset signal is used to synchronize the start of the conversion, and the control logic generation is cyclic. clk is the input clock, clks is the signal for the sampling switch, clkc is the clock for the comparator, and clk[x] is for the bit[x] of the DAC.

The sample and Hold (S/H) circuit is critical in achieving good SFDR for an ADC design. The bootstrapped switch is commonly used since it provides a constant small on-resistance [136]. The circuit schematic of the bootstrapped switch implemented

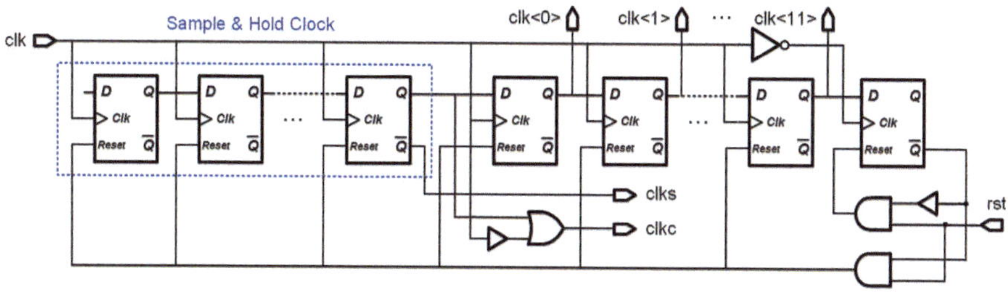

Fig. 2.31 The circuit schematic of the SAR timing generation module

Fig. 2.32 The circuit schematic of the bootstrapped switch

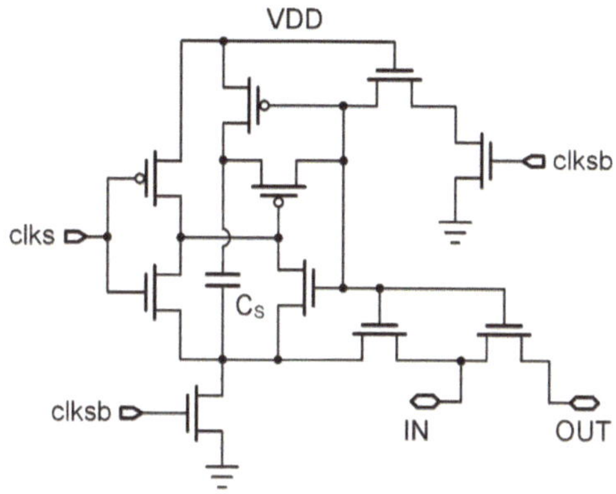

in this work is shown in Fig. 2.32. The gate to source voltage of the switch transistor is fixed at the supply voltage by the capacitor C_S.

Figure 2.33 shows the circuit schematic of the comparator. The comparator consists of a pre-amplifier and a dynamic latch. Since the input voltage has a range from the ground to V_{cm}, the comparator uses a PMOS input stage. The current source NMOS are used in parallel with the diode-connected NMOS for increasing the gain [115]. The pre-amplifier provides moderate gain to reduce the input referred mismatch from the latch. The latch consumes no static current. When clkc (as shown in Fig. 2.31) is high, the outputs are reset to high. When clkc goes to low, the regeneration latch forces one output to high and the other to low. The SAR logic only takes the V_{OP} and generates an inverted signal V'_{ON} to avoid the metastability problem.

A Class-AB output stage has been designed to drive the sample-hold circuits of the following ADC stage. To digitize a single-ended signal, a single-to-differential converter (S2D) can be integrated. An example of the S2D circuit is shown in Fig. 2.34. The resistor values are designed to be R_1=R_3=R_4, and the voltage gain is 2(1+R_2/R_1). R_2, which can be programmed by a shift register.

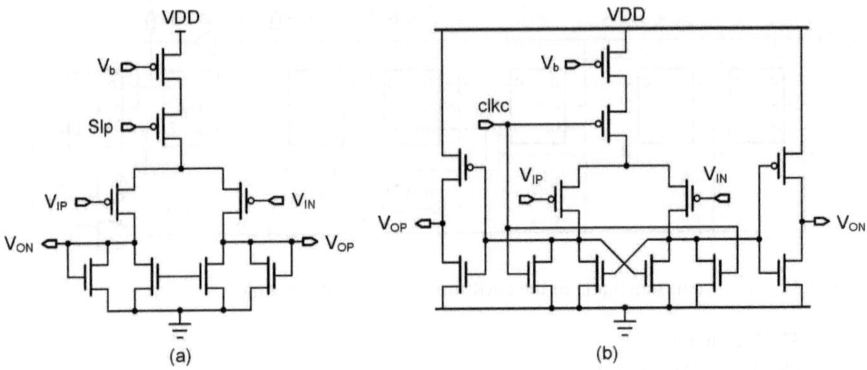

Fig. 2.33 The circuit schematic of the comparator. (**a**) A pre-amplifier, and (**b**) a dynamic latch

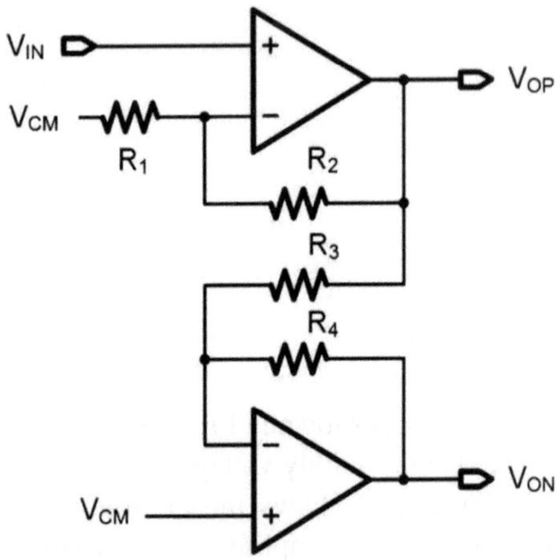

Fig. 2.34 The circuit schematic of a single-to-differential converter

2.4.3 *Measurement Results*

The ADC has been fabricated in IBM 180nm CMOS technology. The layout of the 10-bit SAR ADC is shown in Fig. 2.35 with major building blocks highlighted. The total occupied silicon area is 220 µm × 190 µm. The measurement results of the design are presented below.

The differential nonlinearity (DNL) and integral nonlinearity (INL) of the ADC were measured using slow ramps. The result is shown in Fig. 2.36. The peak DNL and INL are −0.49/+0.56 LSB and −0.82/+0.77 LSB, respectively.

The SAR ADC's dynamic performance was measured with a low-frequency input tone and a near Nyquist frequency input tone. The output spectrums are shown in Figs. 2.37 and 2.38, respectively.

The spurious-free dynamic range (SFDR) achieved in these tests was 76.54 dB and 71.6 dB, respectively. The signal-to-noise and distortion ratio (SNDR) was measured to be 56 dB and 54.6 dB, respectively. The effective number of bit (ENOB) is defined as:

Fig. 2.35 The layout of the 10-bit SAR ADC with major building blocks highlighted

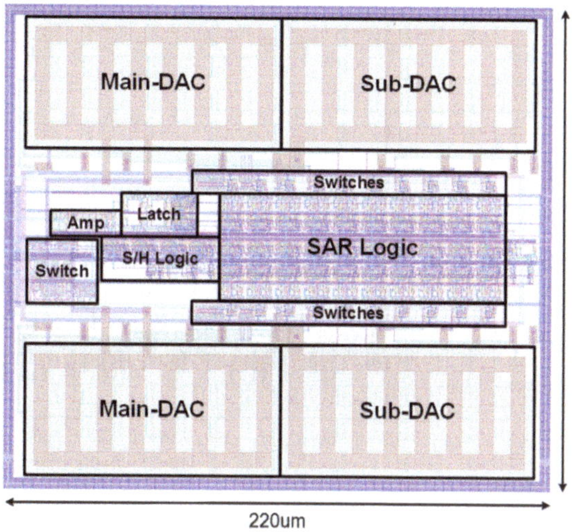

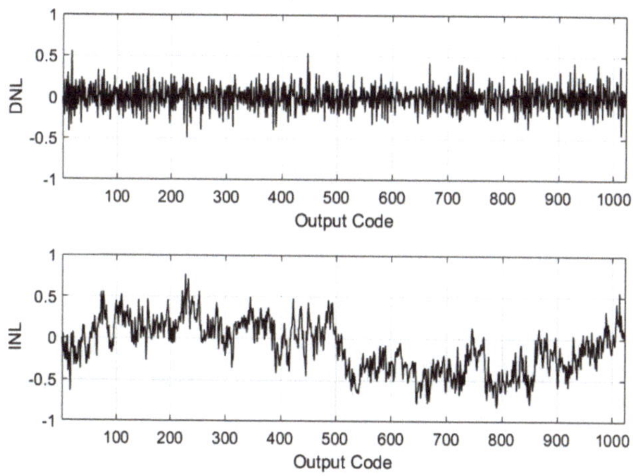

Fig. 2.36 The measured DNL and INL of the 10-bit SAR ADC. The worst DNL is −0.49/+0.56 LSB, and the worst INL is −0.82/+0.77 LSB

$$\text{ENOB} = \frac{\text{SNDR} - 1.76}{6.02} \tag{2.44}$$

The ENOB of the designed ADC was measured to be 9.01 and 8.77, respectively. The figure-of-merit (FoM) is calculated using:

$$\text{FoM} = \frac{\text{Power}}{2^{\text{ENOB}} \times f_s} \tag{2.45}$$

The FoM of the ADC is 98 fJ/conv-step at 1 MSps with a supply of 1.8 V. The measured specifications of the ADC were summarized in Table 2.5.

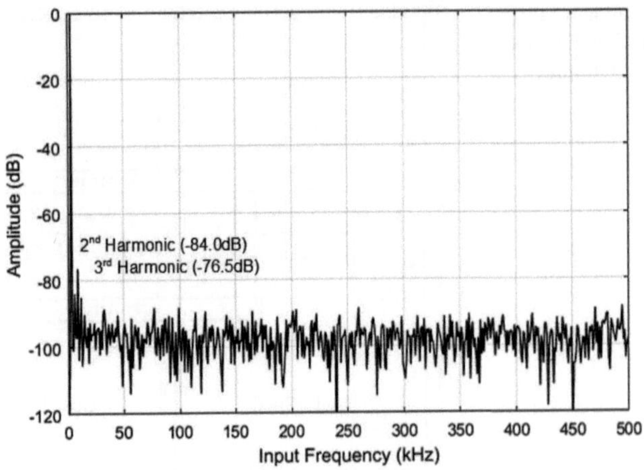

Fig. 2.37 The measured FFT spectrum at 1 MS/s with an input tone of 3 kHz. The SFDR is 76.54 dB and the SNDR is 56 dB. The ENOB at 3 kHz is 9.01

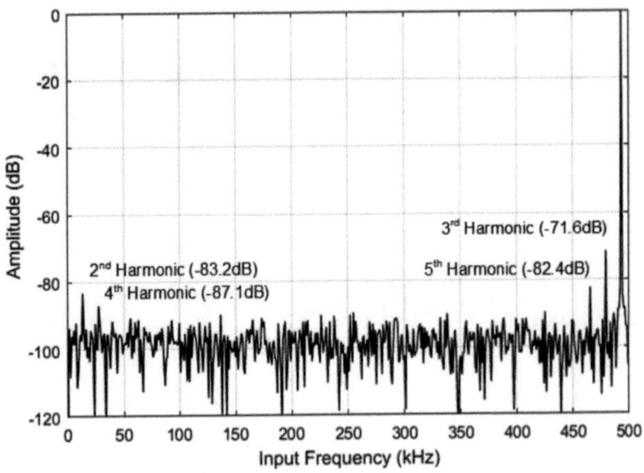

Fig. 2.38 The measured FFT spectrum at 1 MS/s with an input tone of 493 kHz. The SFDR is 71.6 dB and the SNDR is 54.6 dB. The ENOB at 493 kHz is 8.77

In summary, a 10-bit SAR ADC was presented in this section. A prototype was fabricated in 180 nm CMOS technology. The design uses an energy efficient switching procedure and a split-capacitor array. The measurement results successfully meet the design specifications, with a comparable performance among the state-of-the-art ADC designs for the neural recording purposes. As a part of the neural interface system, the power consumption of the ADC was usually not the bottleneck. So this work didn't seek to aggressively minimize the ADC's power using techniques like charge recycling [124], asynchronous timing [127], or step charging [125]. The supply voltage was kept at 1.8V to be compatible with the

Table 2.5 The measured specifications of the 10-bit SAR ADC

Specification	Measurement result
Technology	180 nm
Supply voltage	1.8 V
Input range	3 Vp-p
Sampling rate	1 MSps
Active area	0.042 mm^2
INL	−0.82/+0.77 LSB
DNL	−0.49/+0.56 LSB
SNDR	54.6 dB
SFDR	71.6 dB
ENOB	8.77
FoM	98 fJ/conv-step

neural recording front-end. The designed ADC was later integrated in a bidirectional neural interface system-on-chip, and used in long-term recording experiments in freely behaving animals.

2.5 A Compressed Sensing Neural Signal Acquisition System

2.5.1 Introduction

The wireless telemetry is the power bottleneck of most wireless neural recording systems [137]. On-chip data compression is an effective solution to reduce the power consumption by reducing the data rate. Various on-chip data compression techniques for neural signal acquisition have been proposed. For single or multi-units recording, action potential detection [138] or classification [139] is the most effective way to reduce the wireless data rate, and can also be used to drive prosthetics directly. The hardware implementation of an action potential detection unit can be as simple as a comparator with a pre-defined threshold. A compression ratio higher than 100× can be achieved with a minimum power consumption [51]. However, the action potential detection based compression drops most of the raw waveform, and is vulnerable in long-time recording since the spike waveform may change due to the electrode impedance drifting or electrode displacement. For EEG, ECoG, or LFP, wavelet transformation is an effective solution, given its high compression ratio and good reconstruction quality [140, 141]. However, the hardware implementation of a wavelet transformation is non-trivial and usually takes a considerable amount of area and power consumption. Moreover, the custom design for a specific signal type and sampling frequency significantly limit the applications of these recording systems.

Compressed sensing is an emerging signal processing technique that enables sub-Nyquist sampling and near lossless reconstruction of a signal [142, 143].

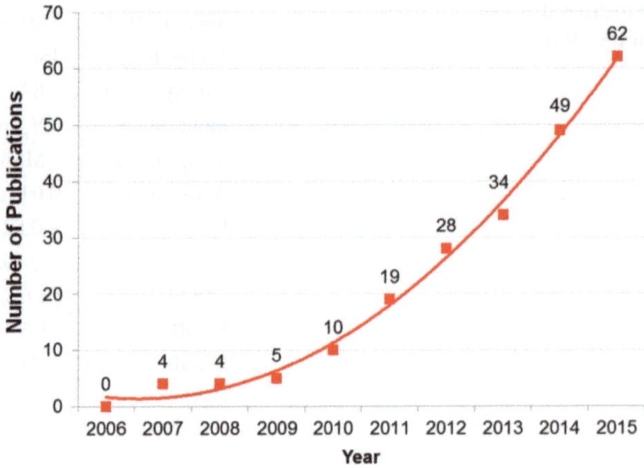

Fig. 2.39 Historical trend for publications using compressed sensing technique in biomedical signal acquisition in the past decade. Data retrieved from Web of Science

Since its introduction in 2006 [144], the compressed sensing technique has been successfully applied to rapid MRI [144], computational image sensors [145], biomedical sensors [137, 146], high frequency receivers [147], and many other applications. Compressed sensing is especially attractive to neural signal recording given its minimum hardware cost in the front-end favoring the power constraint of implanted devices.

Prior research shows the sparsity of neural signal in different frequency bands [146, 148–150]. Since an on-chip transformation using a random matrix usually achieves a sufficient incoherence with a restricted isometry property (RIP) [151], a general-purpose recording device can be designed without the knowledge of the target signal. In addition, the compressed sensing measurements can be used in signal processing (e.g., machine learning classifiers) [152]. Without a full reconstruction of the raw signal, the processing in the compressed domain can be easily implemented in a low-power embedded system.

Figure 2.39 shows a survey of publications related to compressed sensing's applications in neural recording. It clearly shows that compressed sensing has a fast growth trend, and plays an increasingly important role in the neural signal acquisition system design. This section presents the design and analysis of a fully integrated wireless compressed sensing neural signal acquisition system for chronic recording and brain–machine interface. All experimental procedures used in this study were approved by the institutional animal care and use committee (IACUC) of the University of Pennsylvania. Some of the figures and tables presented in this section were originally published in [153] ©IEEE. Reused, with permission.

2.5.2 A Brief Background of Compressed Sensing

Compressive sensing (CS) is a signal processing technique that enables sub-Nyquist sampling and near lossless reconstruction of a signal with sparsity in a certain domain. The technique is particularly appealing for low-power high channel-count neural signal recording. This section gives a brief introduction to the compressive sensing theory. Detailed explanation and rigid mathematical proof can be found in [143, 144, 151].

2.5.2.1 Compression Process

Assume the digitized signal $\mathbf{x}$ has a dimension of N, denoted by $\mathbf{x} \in \mathbb{R}^{N\times 1}$. Consider a general linear measurement process that computes $\mathbf{y}$ with a full row-rank matrix denoted by $\Phi \in \mathbb{R}^{M\times N}$, and $M \ll N$

$$\mathbf{y} = \Phi\mathbf{x} \tag{2.46}$$

where $\mathbf{y}$ is the compressive sensing data, and Φ is the sensing matrix. It should be noticed that the sensing matrix is known to the reconstruction algorithm. The signal x can be expressed as:

$$\mathbf{x} = \sum_{i=1}^{N} s_i \Psi_i \tag{2.47}$$

where $\mathbf{s}$ is the representation of the signal in the Ψ domain. The signal $\mathbf{x}$ is $K-$ *Sparse* if only K of the $\mathbf{s}$ coefficients are non-zero. The signal is compressible if it is $K -$ *Sparse*. The $\mathbf{y}$ can be written as:

$$\mathbf{y} = \Phi\Psi\mathbf{s} \tag{2.48}$$

2.5.2.2 Reconstruction Process

The signal reconstruction process is to use the **M** measurements in $\mathbf{y}$, the measurement matrix Φ, and the basis Ψ to reconstruct the signal $\mathbf{x}$, or equivalently, its sparse representation $\mathbf{s}$. Since $M \ll N$, the equation is underdetermined, which means there are infinite $\mathbf{x}$ (or $\mathbf{s}$) that satisfy the condition. Therefore, the signal reconstruction process is to find out the signal's sparse coefficient vector.

The classical approach is to find the vector in the translated null space with the smallest ℓ_2 norm by solving:

$$\hat{\mathbf{s}} = \text{argmin}||\mathbf{s}'||_2 \quad \text{such that} \quad \Phi\Psi\mathbf{s}' = \mathbf{y} \tag{2.49}$$

However, the ℓ_2 minimization usually has difficulty in finding a K-Sparse solution. ℓ_0 norm can recover a K-Sparse signal exactly with a high probability.

$$\hat{\mathbf{s}} = \text{argmin}||\mathbf{s}'||_0 \quad \text{such} \quad \text{that} \quad \Phi\Psi\mathbf{s}' = \mathbf{y} \tag{2.50}$$

Unfortunately, solving Eq. (2.50) is both numerically unstable and NP-complete. While ℓ_1 norm can exactly recover K-Sparse signal and closely approximate the signal with a high probability.

$$\hat{\mathbf{s}} = \text{argmin}||\mathbf{s}'||_1 \quad \text{such} \quad \text{that} \quad \Phi\Psi\mathbf{s}' = \mathbf{y} \tag{2.51}$$

This is a convex optimization problem and can be conveniently reduced to a basis pursuit problem, with a computational complexity about $O(N_3)$.

2.5.2.3 Reconstruction Evaluation Criteria

Several numerical derivations are used to evaluate the performance of individual reconstruction algorithms and dictionaries. The commonly used criteria include compression ratio and signal-to-noise and distortion ratio.

The Compression Ratio (CR) is defined as:

$$\text{CR} = \frac{N}{M} \tag{2.52}$$

The signal-to-noise and distortion ratio (SNDR) is defined as [137]:

$$\text{SNDR} = 20 \times \log \frac{||\mathbf{x}||_2}{||\mathbf{x} - \hat{\mathbf{x}}||_2}. \tag{2.53}$$

2.5.3 System Overview

The paradigm of the hypothetical chronic wireless neural signal acquisition system is illustrated in Fig. 2.40. The system has a dedicated implantable subsystem and a flexible external subsystem. The implantable subsystem contains the proposed compressed sensing neural recording SoC, an inductive charging module, and a super capacitor. The device will need to be sealed in a biocompatible package. The device can be placed under the skin, above the skull bone. The recording electrode can be placed in any brain area of interest. The external subsystem

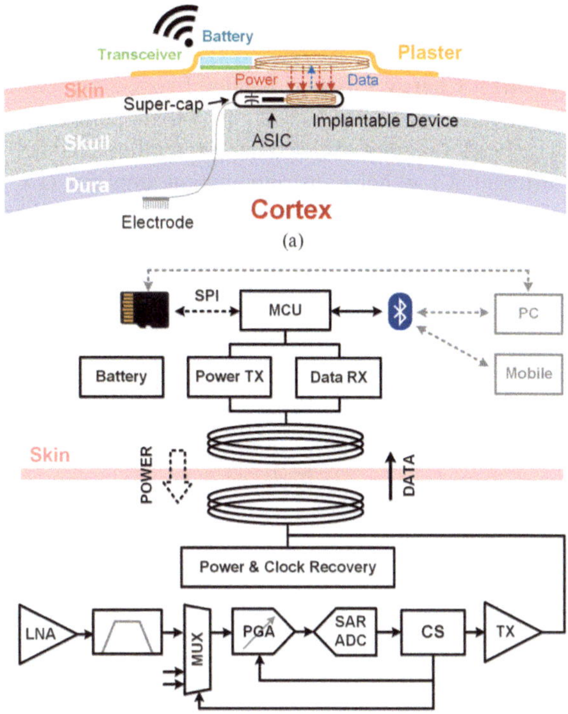

Fig. 2.40 (**a**) Illustration of the hypothetical chronic neural signal recording system using the fully integrated compressed sensing chip, and (**b**) the block diagram of the system

consists of a standard wireless transceiver, a rechargeable battery, and a coil. The external subsystem powers the implanted device and collects data back through back-scattering.

The advantages of the proposed system are threefolds: (1) the implanted wireless device leaves the skin intact, which reduces the risk of infection, (2) the battery is left externally so that the device's lifetime will not be limited by the battery's recharging cycles, and the toxicity associated with batteries will not be a potential danger to the subject, (3) the external transceiver makes the system flexible and versatile, for instance, different wireless solutions or flash memory can be used for different applications. Upgrading the system is also much easier, since the chronic implant can be used for years or even decades while the external digital and wireless electronics can be upgraded easily.

A single pair of coils is used for both power delivery and data read back. A carrier frequency of 13.56 MHz is chosen considering the trade-off between the power transfer efficiency and the data rate. Compressed sensing reduces the data rate of the wireless uplink, which is especially helpful for the multiple channel recordings.

2.5.4 *Circuit Implementation*

2.5.4.1 Energy Efficient Analog Front-End

The block diagram of one analog recording channel is shown in Fig. 2.41. A fully differential low-noise instrumentation amplifier (IA) is used to amplify the neural signal. The following Gm-C based high pass filter stage (HPF) conditions the

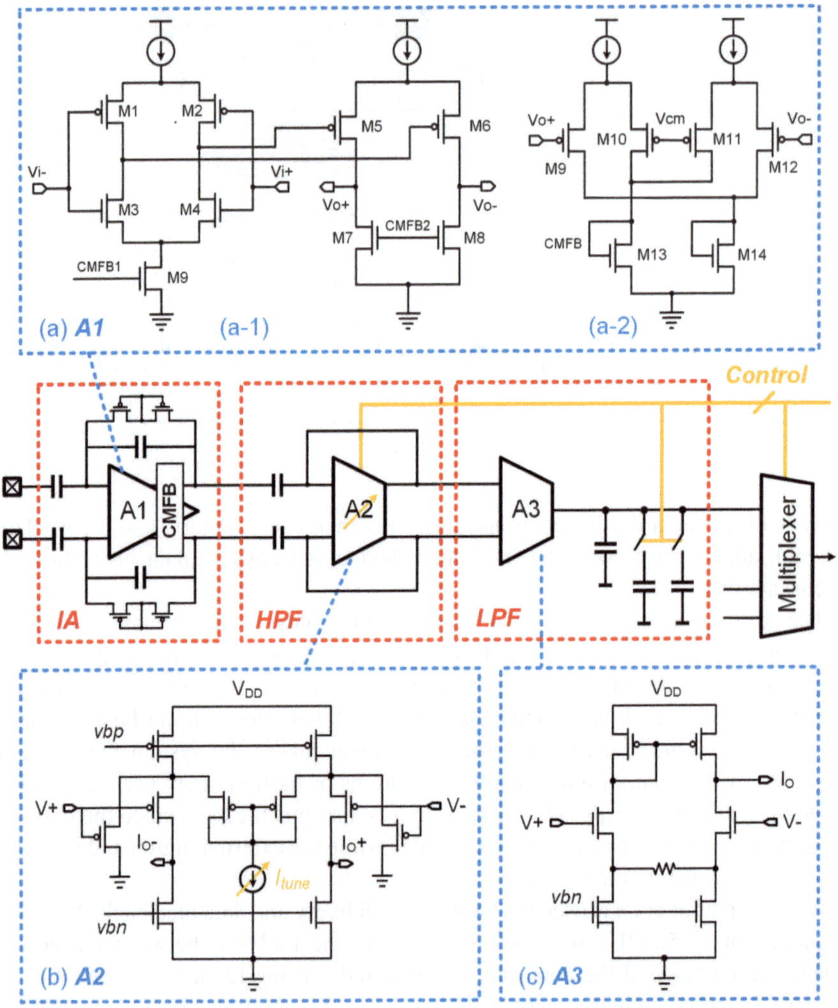

Fig. 2.41 The circuit schematic of one analog front-end channel of the proposed system (Part I). The signal chain includes: amplification, filtering, voltage-to-current conversion, multiplexing and digitization. The circuit schematic of (**a**) the low-noise neural amplifier, (**b**) the OTA with an extended linear range, and (**c**) the OTA with a programmable transconductance

signal with a tunable cut-off frequency. The next stage (LFP) is an operational transconductance amplifier (OTA) that converts the voltage signal into a current in a programmable low-pass frequency corner.

The IA in this work is a fully differential capacitor-coupled neural amplifier, which amplifies the weak neural signal in a wide frequency band. The input capacitors block the large electrode offset and half-cell potential from the interface, giving a maximum input range. The closed-loop differential gain is set to be 34 dB to relieve the noise requirement of the following stages. The core of the IA is a low-noise OTA, as shown in Fig. 2.41a-1. The OTA has been designed to maximize the noise and power efficiency. Compared with the design presented in Sect. 2.2.2, a two-stage topology is used to provide a sufficient open-loop gain. A complementary input stage (M1–M4) is used to increase the overall transconductance without increasing the quiescent current. The complementary input amplifier suffers from PVT variations [110], thus additional common-mode feedback circuit, as shown in Fig. 2.41a-2, is adopted to stabilize the DC output at half of the supply voltage. All input transistors are biased in the sub-threshold region to achieve a high energy efficiency. Since the complementary stage has a limited input range, a fully differential structure is chosen. The first stage dominates the noise, and the input-referred noise of the OTA can be expressed as:

$$\overline{v_{i,n,\mathrm{tot}}^2} = \frac{1}{(g_{m1}+g_{m3})^2}\left[8KT\gamma(g_{m1}+g_{m3}) + 2\left(\frac{K_N g_{m3}}{C_{\mathrm{ox},N} f W_N L_N} + \frac{K_P g_{m1}}{C_{\mathrm{ox},P} f W_P L_P}\right)\right]\Delta f \tag{2.54}$$

where g_{m1} ($=g_{m2}$) are the transconductance of the transistor M1 (M2), and g_{m3} ($=g_{m4}$) are the transconductance of the transistor M3 (M4). The flicker noise can be reduced by increasing the widths and lengths of the input transistors. A biasing current of 1 μA is used in the first stage as a trade-off between power and noise. A biasing current of 20 nA is used in the second stage. The dominant pole is set at the second stage, and the stability is guaranteed by adding an additional capacitive load.

An ultra low-power programmable bandpass filter is integrated into each channel for selecting the frequency band of interest. The first stage is a fully differential Gm-C highpass filter. The circuit schematic of the Gm block is shown as A2 in Fig. 2.41b. Current division and local feedback are used to achieve a low transconductance and an extended linear input range. The cut-off frequency can be programmed by tuning the transconductance. The second stage of the filter is a single-ended Gm-C based lowpass filter. The circuit schematic of the Gm block is shown as A3 in Fig. 2.41c. Source degeneration is used to achieve a high linearity. The differential voltage signal is converted into a single-end current signal. Since a standard current mirror load is used, no extra power is wasted for this conversion, but the single-ended operation reduces the capacitor array size by half, which is important for this design to be implemented at the channel level. The lowpass frequency can be programmed by selecting the load capacitor.

The shared part of the analog recording front-end is shown in Fig. 2.42. The current output from each channel is multiplexed and then converted to a voltage using a transimpedance amplifier (TIA) with a programmable gain. A single-to-differential (S2D) converter is used to drive the differential input ADC with an

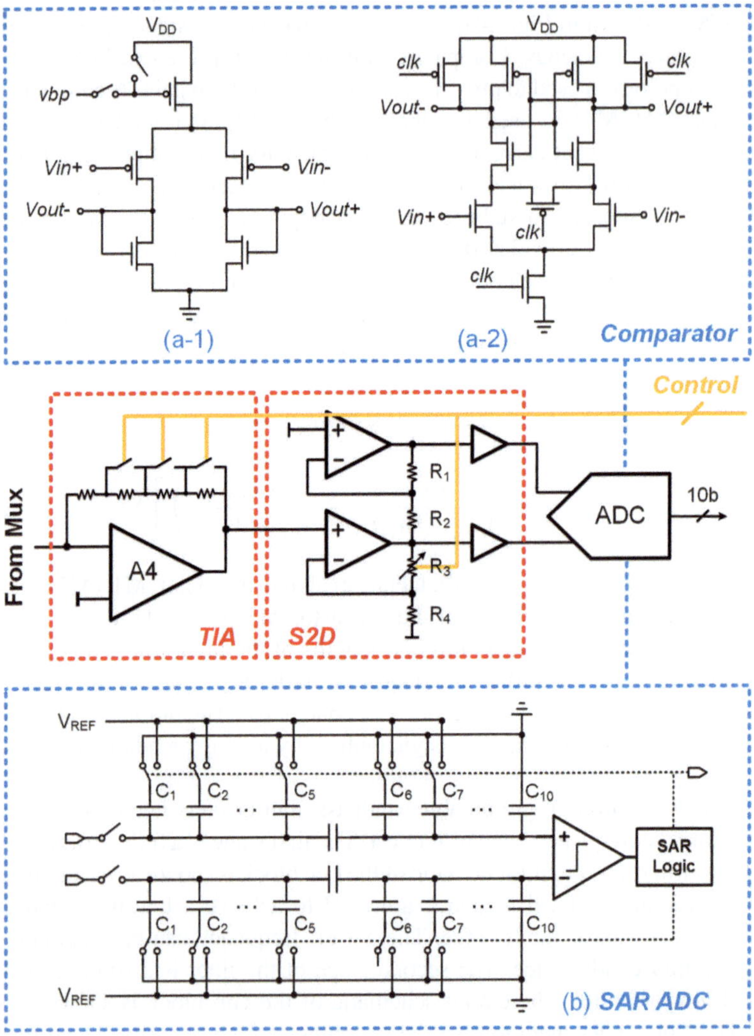

Fig. 2.42 The circuit schematic of the analog front-end of the proposed system (Part II). A current-to-voltage conversion with a programmable gain and a 10-bit SAR ADC is used to digitize the signal. The boxed windows show the circuit schematic of (**a**) the comparator and (**b**) the SAR ADC

additional programmable transimpedance. A 10-bit SAR ADC digitizes the signal. The design details of the ADC have been presented in Sect. 2.4.

The single-ended current output from the 16 channel is selected by a multiplexer. The single-ended signal reduces the effort in routing, and the R-I drop in the long routing trace doesn't corrupt the current signal, thus making it less susceptible to noise. The following TIA stage is used to convert the current signal back to a voltage in a programmable gain, as shown in Fig. 2.42. The gain can be set to be 5×, 6×, 7×, 8× by the compressed sensing digital processor. The gain of 2×, 4× can be easily achieved by shifting bits in the binary digital processor, and the 3× can be achieved from shifting the 6× signal by 1 bit.

2.5.4.2 Compressed Sensing Module

The compressed sensing processing is implemented in the digital domain. The digitized neural signal, x_{in}, of a single channel is fed into the digital processing module.

$$y = \Phi x_{\text{in}} \tag{2.55}$$

that can be written as,

$$\begin{bmatrix} y_0 \\ y_1 \\ \vdots \\ y_M \end{bmatrix} = \begin{bmatrix} \Phi_{11} & \Phi_{12} & \cdots & \Phi_{1N} \\ \Phi_{21} & \Phi_{22} & \cdots & \Phi_{2N} \\ \vdots & \vdots & \ddots & \vdots \\ \Phi_{M1} & \Phi_{M2} & \cdots & \Phi_{MN} \end{bmatrix} \begin{bmatrix} x_{\text{in}_0} \\ x_{\text{in}_1} \\ \vdots \\ x_{\text{in}_N} \end{bmatrix} \tag{2.56}$$

Equation (2.56) can be rewritten in the form of a sum of vector multiplications, as:

$$\begin{bmatrix} y_0 \\ y_1 \\ \vdots \\ y_M \end{bmatrix} = \sum_{i=1}^{M} \left(\begin{bmatrix} \Phi_{1i} \\ \Phi_{2i} \\ \vdots \\ \Phi_{Mi} \end{bmatrix} x_{\text{in}_{i-1}} \right) \tag{2.57}$$

There are two modes of operation. In the simple mode, the entries of the sampling matrix Φ are assigned to be 0, +1, or −1; in the high resolution mode, the entries can be assigned to be $0, \pm\frac{1}{8}, \pm\frac{2}{8}, \ldots, \pm\frac{7}{8}$.

In order to avoid a large on-chip storage for the sampling matrix, a shift register chain is used to preload the coefficients at the beginning of each sampling loop. Figure 2.43 shows the block diagram of the compressive sensing processing unit. Parallel output values from the ADC are fed into the digital model. A simple sign control is applied before sending the ADC output to the adder under the simple mode. Under the high resolution mode, the entries coefficients $\pm\frac{3}{8}$, $\pm\frac{5}{8}$ and $\pm\frac{7}{8}$ are

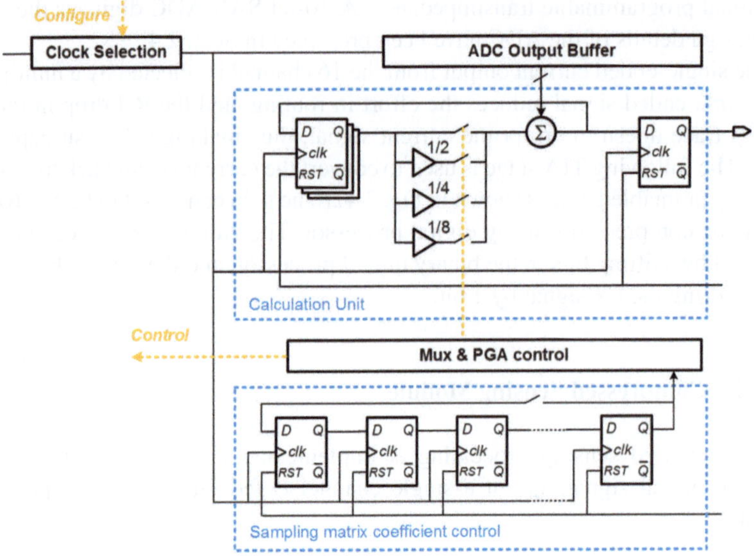

Fig. 2.43 The block diagram of the compressive sensing processing module. A linear congruential pseudo random number generator is used to generate all the entries of the sampling matrix

realized by configuring the gain of the analog amplifier to 3, 5, and 7, respectively, while shifting the ADC output by 3-bit before sending the ADC's output to the adder.

There are M vector multiplication units integrated in the system. The entries of Φ are randomly generated off-line and used for the logic control inside of each vector multiplication unit. The output measurement y is reset after every N iteration. The dimension of x_{in} is controlled by the iteration times. A parallel to serial convertor is integrated in the system for the readout of the measurements.

According to the CS theory, a dictionary for sparsifying neural signals is required for sparse recovery. In this section, neural data recording without compression is performed to generate a database for the algorithm analysis at the very beginning. The database is divided into two halves, where one half is used for training signal dependent dictionary **D** by an unsupervised dictionary learning algorithm [150] and another half is used for testing the recovery performance. In the proposed CS framework, we adopt an on-chip Bernoulli sensing matrix $\mathbf{\Phi}$ to compress the neural spikes or LFP $\mathbf{x}$ of length N into measurements y of length M, where normally $M \ll N$ and compression ratio is defined by $\frac{M}{N}$, as in Eq. (2.5.2.3). The recovery problem can be solved by Orthogonal Matching Pursuit [154],

$$\min_{\mathbf{a}} ||\mathbf{y} - \mathbf{\Phi}\mathbf{D}\mathbf{a}||_2^2 \quad s.t. \ ||\mathbf{a}||_0 \leq \Phi, \tag{2.58}$$

where **a** is the sparse coefficient vector and S indicates the sparse level. The recovered signal is defined as $\hat{\mathbf{x}} = \mathbf{Da}$ and the recovery quality is quantitatively evaluated by the SNDR, as defined in Eq. (2.5.2.3).

2.5.4.3 On-Chip Wireless Power and Data Link

A low-power backscatter based wireless transmitter is designed to communicate with an external transceiver. The backscatter transmitter consists of a PWM encoder and a buffered transistor for the antenna impedance modulation [48].

An active rectifier is used to achieve a high power efficiency [155]. Coupling coils are implemented off-chip. The system clock is recovered from the power waveform [86]. The circuitry of the clock recovery and division module is shown in Fig. 2.44. The module consists of a Schmitt trigger and several D flip-flops. The Schmitt trigger makes the circuit more resistant to the noise in the power waveform. Figure 2.45 shows the circuit schematic of the Schmitt trigger [156]. The D flip-flop guarantees the clock has a 50% duty cycle. Several different clocks can be divided from the following D flip-flops. The clock frequency selection is configured in a register.

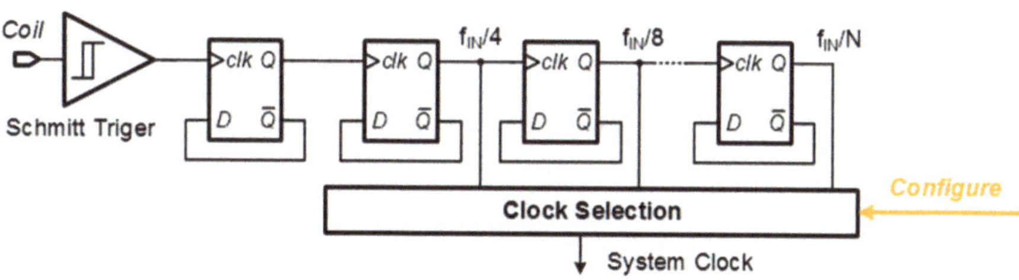

Fig. 2.44 The circuit schematic of the clock recovery and clock division module

Fig. 2.45 The circuit schematic of the CMOS Schmitt trigger

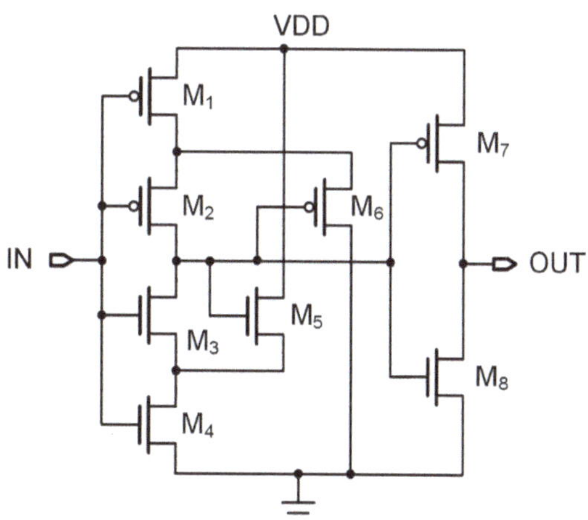

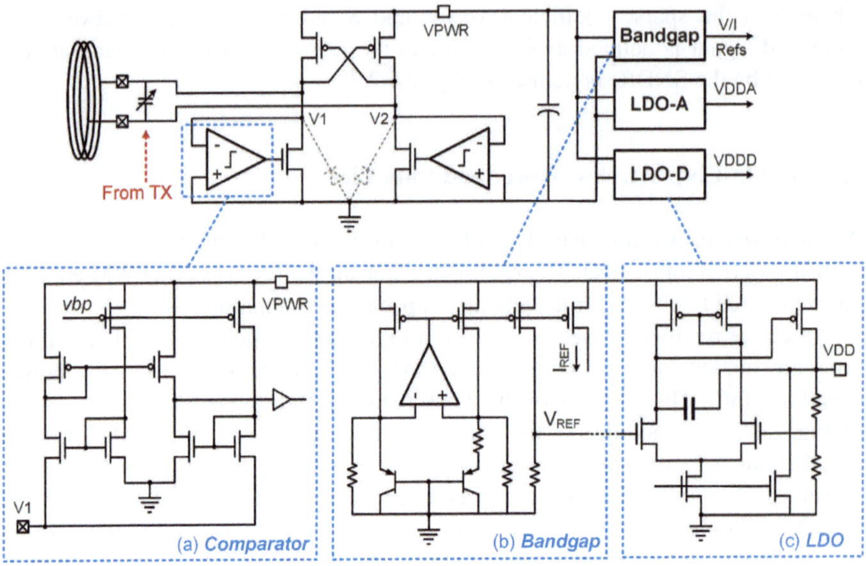

Fig. 2.46 Inductive power management module, including active rectifier and LDOs for analog and digital power supplies. (**a**) Circuit schematic of the comparator, (**b**) bandgap reference, and (**c**) LDO (start-up circuits are not shown)

Standard bandgap reference and low drop-out (LDO) circuits are used in the power management unit. The block diagram and the circuit schematics of the power management module are shown in Fig. 2.46. A push-pull comparator with source input is used to drive the active diodes. The design details of the active rectifier can be found in references [155, 157, 158].

2.5.4.4 External Wireless Relay Board

An external wireless relay board has also been designed to demonstrate the proposed paradigm. The external subsystem consists of a microcontroller with an integrated wireless transceiver, envelope detection circuits for reading the backscattered signal, power transmitter circuits, and a battery management module.

A 32-bit ARM Cortex-M0 based wireless transceiver (Nordic Semiconductor nRF51822) is used as the central processor and the wireless transceiver. It features a 2.4 GHz transceiver with an integrated Bluetooth 4.0 low-energy protocol framework, which provides an easy interface to the computer or mobile devices. A reliable wireless communication up to 5 meters was measured in normal indoor environment. A Serial Peripheral Interface (SPI) based microSD card interface is optional in the system to allow a long-term wireless recording without a limited receiver range.

A computer user interface has been developed in Matlab to configure the device and read back the data. The signal conditioning and off-line analyses can also be performed in the user interface.

2.5.5 Measurement Results

The proposed SoC design has been fabricated in IBM 180 nm standard CMOS technology, occupying a silicon area of 2.1 mm × 0.8 mm, excluding the IO pads. A microphotograph of the fabricated chip is shown in Fig. 2.47, with major building blocks highlighted.

Bench testing was conducted to verify the functions of the chip and the system. The measured frequency response of the low-noise amplifier is shown in Fig. 2.48. The frequency response was measured point by point using a function generator 33521A and an oscilloscope MSO7034B from Agilent. The phase shift was calculated in the oscilloscope. The measured midband gain is 34.1 dB. The measured CMRR and PSRR of the analog front-end in the frequency range of 0.5 Hz to 7 kHz are >80 dB and >67 dB, respectively.

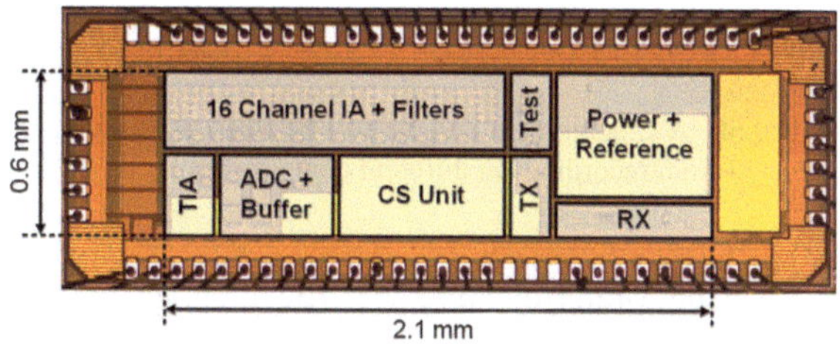

Fig. 2.47 The micrograph of the fabricated fully integrated compressed sensing neural recording front-end chip

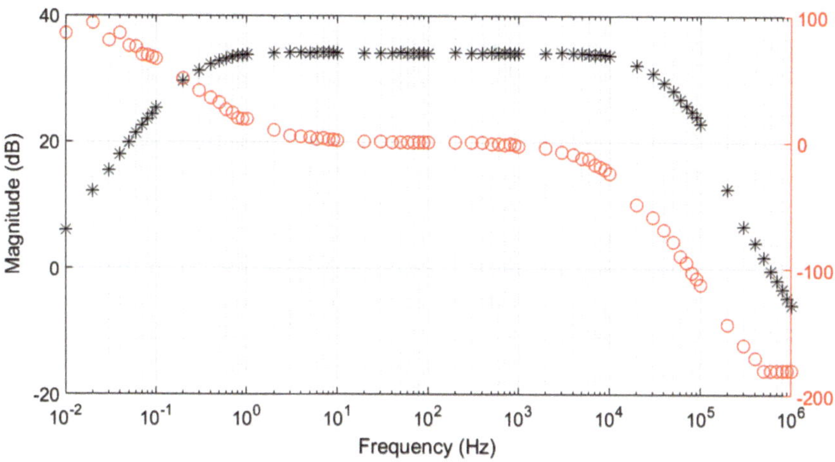

Fig. 2.48 The measured frequency response of the low-noise amplifier (without filtering stages)

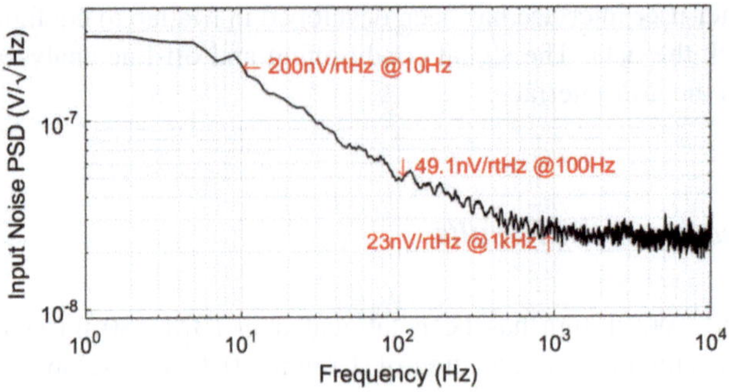

Fig. 2.49 The measured input-referred voltage noise spectrum

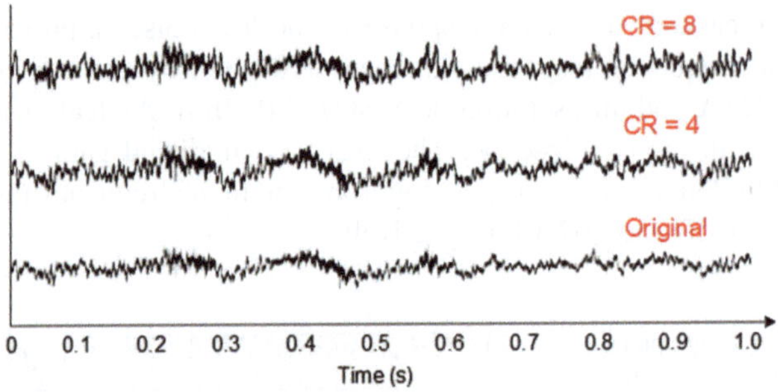

Fig. 2.50 A time-domain comparison between the uncompressed recording and the data reconstructed from recordings in different compression ratios (CR)

The input-referred noise spectrum is shown in Fig. 2.49. The noise was measured with the inputs shorted by an internal switch. The noise spectrum density is 200 nV/$\sqrt{\text{Hz}}$ at 10 Hz, 49.1 nV/$\sqrt{\text{Hz}}$ at 100 Hz, and 23 nV/$\sqrt{\text{Hz}}$ at 1 kHz. An integration under this curve from 1 Hz to 7 kHz yields an rms noise floor of 2.85 μV. The total harmonic distortion of the amplifier was measured to be −63 dB, with an input amplitude of 1 mV.

An invasive neural recording was performed in an anesthetized rat with a tungsten microelectrode placed in its motor cortex. Action potential data was extracted by configuring the filter with a passband of 300 Hz to 7 kHz. Different compression ratios from 2, to 4, to 8 and to 16 have been applied, respectively. Signal-to-noise distortion ratio (SNDR) of 3.60 dB, 9.78 dB, 30.60 dB, and 52.99 dB are achieved for compression ratios 16, 8, 4, and 2, respectively. Dual-threshold level-crossing action potential detection was used for both the uncompressed data and the restored data. A near-lossless action potential detection can be achieved while a compression ratio lower than 8 was applied.

Figure 2.50 compares the time-domain waveforms of the uncompressed and the reconstructed local field potential (LFP) sampling data sets. And Fig. 2.51 shows the comparison of the spectrums of the original uncompressed and reconstructed LFP sampling data sets. The LFP exhibited rhythmic bouts of broadband power

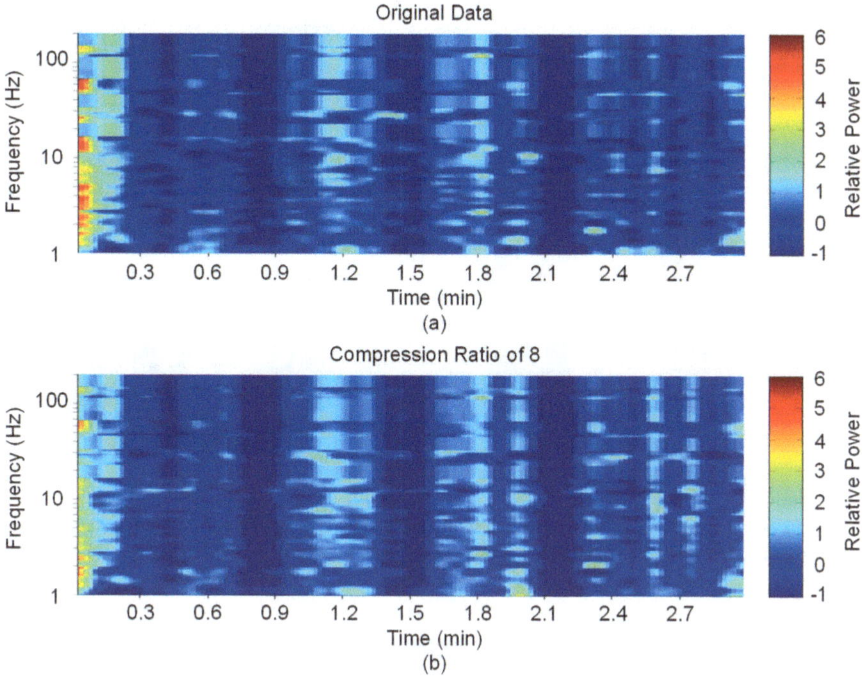

Fig. 2.51 A comparison of the spectrograms of (**a**) the uncompressed recording and (**b**) the data reconstructed from the recording with a CR of 8

interleaved with low power epochs. According to Fig. 2.51, the time-frequency content of the restored signal was very similar to the uncompressed LFP. Signal-to-noise distortion ratio (SNDR) of 9.04 dB, 4.85 dB, and 3.78 dB were achieved for compression ratios 4, 8, and 16, respectively.

A demonstration system was developed to show the proposed concept, as shown in Fig. 2.52. An open cavity plastic package was used for packaging the chip, and the size of the demonstration implantable system was limited by the package. Commercial coils were used for the inductive power and data transfer. An additional ceramic capacitor was used to improve the impedance matching. Two LEDs were used only for debugging purpose. A couple of programming and debugging pads are left (not shown). No other off-chip components were required.

In vivo evaluation of the device for a long-term operation was conducted in a rhesus macaque. An electrode was chronically implanted in the hippocampus. The recording device, including an external transceiver, was housed in a small chamber that was fixed to the skull. Figure 2.53 shows the spectrogram of a 24-h continuous recording while the monkey was freely behaving in his home cage. The recording shows the states of hippocampal activities throughout the day. Greater power at higher frequencies (>20 Hz) was associated with periods in which the animal was awake and freely moving about his home cage (hours 0–7.5 and 19–24). Greater power at low frequencies (<20 Hz) was associated with sleeping (hours 7.5–19). Individual sleep cycles can be seen. Some broadband chewing artifacts were also

Fig. 2.52 Photograph of an assembled demonstration system. (**a**) Power and data transmission testing setup across a 5 mm plastic cap, (**b**) the external transceiver board, and (**c**) the implantable device

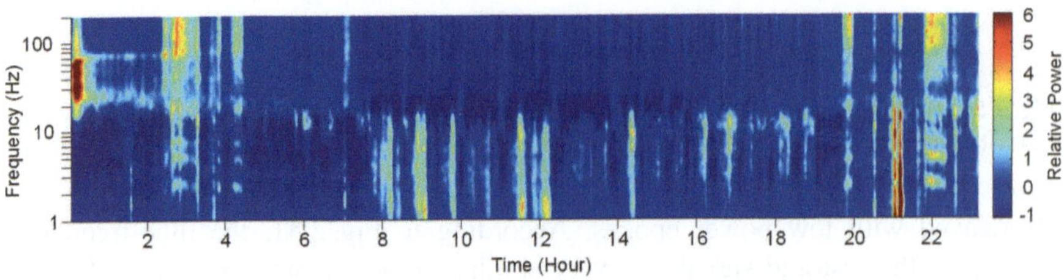

Fig. 2.53 A 24-h continuous recording in the hippocampus of a rhesus macaque during free behavior

Table 2.6 The chip specifications summary

Neural amplifier		*CS processor*	
Midband gain	34.1 dB	Input channel	Up to 16
Bandwidth	0.5 Hz–7 kHz	CS ratio	up to 8×
LNA noise	2.85 μVrms	Clock freq.	4 MHz
THD (1 mV)	−63 dB	*Wireless power and data*	
NEF/PEF	1.58/4.5	Carrier freq.	13.56 MHz
CMRR	>80 dB	Power efficiency	up to 73%
PSRR	>67 dB	Distance	up to 10 mm
SAR ADC		*Power*	
ENOB	9.1	Analog front-end	2.5 μW (per ch.)
Sampling rate	1 MSps	ADC	35 μW(@1 MSps)
INL (LSB)	+0.62/−0.85	CS processor	77 μW
DNL (LSB)	+0.69/−0.92	TX transmitter	27μW
FoM (fJ/step)	34.2	Total (avg.)	254 μW

Table 2.7 Comparison with state-of-the-art works

Reference	Deepu [106]	Gangopadhyay [159]	Zhang [160]	Biederman [56]	This work
Publication	2014 JSSC	2014 JSSC	2015 JNE	2015 JSSC	–
CMOS technology	0.35 μm	130 nm	180 nm	65 nm	180 nm
No. of channels	4	64	4	64	16
Signal type	ECG	ECG	Extracellular	Extracellular	LFP/extracellular
Input-referred noise	1.46 μV	<2 μV	3.1 μV	7.5 μV	2.8 μV
Sampling rate/ch	256/512 Hz	2 kHz	20 kHz	20 kHz	20 kHz
Front-end NEF/PEF	3.31/26.3	–	–	3.6/12.9	1.58/4.5
ADC ENOB	9.3	6.5	–	8.2	9.1
AFE + ADC power/ch	0.54 μW (512 Hz)	28 nW (2 kSps)	15 μW (20 kSps)	1.84 μW (20 kSps)	3.2 μW (20 kSps)
Compression method	Lin slope predict.	CS	CS	Spike Dect.	CS
Compression ratio	2.55×	up to 6×	8×–16×	8.3× (epochs)	8×–16×
Reconstruction SNR	–	–	<9 dB (16×)	–	9.78 dB (8×)
Wireless	–	–	–	–	Backscattering
In vivo experiment	Yes	–	Yes	Yes	Yes
System integration	–	–	–	Complete	Complete

present (around hours 3–4.5 and 20–22) corresponding to the times when the animal was fed. The overall activity pattern matches previous observations of sleep–wake changes in neural activity. The measured specifications of the chip are summarized in Table 2.6.

In this work, a fully integrated wireless neural signal acquisition system is presented. A high efficiency wireless neural signal recording SoC with integrated compressed sensing processor was designed and fabricated in 180 nm CMOS technology. An external wireless relay was used to power the implantable SoC, read back the data through backscattering, and transmit the data through a universal wireless link. The system features high energy efficiency, high flexibility, compatibility, upgradability without compromising the signal recording quality. By performing on-chip compressive sampling, the data rate is significantly reduced, which allows the system to support more recording channels without a power penalty. According to the experimental results, a compression ratio up to 8× will cause negligible loss of the data quality and/or information contained in the data. A pre-implantable system was assembled and successfully demonstrated the proposed paradigm. Bench testing and in vivo experimental results are presented. Table 2.7 compares the performance of the proposed work with prior published compressed neural signal recording front-end designs. The system shows a promising chronic neural signal recording paradigm for neuroscience research and BMI applications.

Chapter 3
Neural Feature Extraction

3.1 Introduction

Feature extraction, or feature learning, is an important technique to transform the raw data input to a representation that can be effectively understood [161]. Neural feature extraction allows one to acquire the qualitative and quantitative information from the neural signal. It has been widely used in the neuroscience and neuroprosthetic research for pattern recognition, numerical or symbolic regression, probability estimation, and dynamical system modeling [162]. Moreover, neural feature extraction provides inputs to a real–time decision support system, such as a brain–machine interface (BMI) system with machine learning for seizure on-site detection [107]. The implementation of real–time neural feature extraction is especially important for closed-loop BMI systems, where most processing is online. By applying feature extraction and machine learning techniques, the BMI devices have been successfully used in decoding motor function [163, 164], detecting epilepsy [107, 165], Parkinson's disease [166], depression [167], and so on.

The neural feature extraction can be performed in different domains, including: (1) time domain, (2) frequency domain, (3) wavelet domain, (4) statistics process, (5) information theory (e.g. entropy, mutual information), (6) fractal geometry, and so on [162]. However, the implementation of real–time feature extraction in BMI devices is usually limited by the hardware resources including the computation ability, the memory size, and the power consumption [168]. As a result, an energy efficient implementation is especially important. Some energy-efficient neural feature extraction techniques have been reported in the literature [107, 139, 169–172]. It should be noticed that the choice of a suitable set of neural features is also a challenging task. The brain signal contains a large number of simultaneous sources, and the information of interest might be overlapped with other sources in time or frequency. There are many existing theories for feature selection, including

X. Liu, J. Van der Spiegel, *Brain-Machine Interface*,
https://doi.org/10.1007/978-3-319-67940-2_3

principal component analysis (PCA) [173], independent component analysis (ICA) [174], genetic algorithm (GA) [175], sequential forward/backward selection (SFS) [176], and so on.

Although the use of neural features varies significantly in different applications, it is very helpful to identify the most commonly used features for real–time BMI systems. Both field potential and action potential features have been used in real–time, closed-loop BMI devices, as briefly summarized below:

- **Field potential (FP) features**
  - **Energy in multiple frequency bands**: spectral characteristics of neural field potentials have been used to identify, classify, and analyze brain activities [169, 177];
  - **Features of different brain states**: different features of field potentials can be used to identify different brain states [178, 179];
  - **Synchronization between electrodes**: synchronization of oscillations between different brain areas can be used to identify different brain activities [180, 181].
- **Action potential (AP) features**
  - **Action potential detection**: action potentials reflect activities of individual neurons, which have been widely used in BMI system for linking two brain sites, reinforcing motor activity, generating synaptic plasticity, and so on [182, 183];
  - **Action potential alignment and sorting**: on-line action potential sorting can be used to identify different neurons presented on a same electrode [182, 184];
  - **Action potential firing rate**: the action potential fire-rate presents the active-level of an individual neuron [185, 186].

This chapter presents the analysis and design of the neural feature extraction from three different perspectives, the field potential energy extraction, the action potential detection, and the phase-amplitude coupled feature extraction. Several novel techniques in the circuitry, algorithm, and system levels are proposed, with a focus on the energy efficient implementation for closed-loop BMI systems. The chapter is organized as follows. Section 3.1 introduces the neural features and neural feature extraction, and summarizes the commonly used features for real–time BMI systems. Section 3.2 describes the energy features in the LFP, and proposes a novel extraction circuit with frequency tuning in the natural logarithmic domain. Section 3.3 analyzes the real–time action potential detection and classification, followed by the design of a low-power current-mode action potential circuit module. Section 3.4 analyzes the matched filter with pre-whitening and its application in phase-amplitude coupled neural feature extraction.

3.2 Natural Logarithmic Domain Field Potential Energy Extraction

3.2.1 Introduction

The field potentials reflect the summed activity of thousands to millions of neurons. A large amount of important information regarding brain states, motor intent and behavioral processes can be inferred from the field potential recordings [187–189]. Oscillations are particularly prominent in field potential recordings which reflect synchronous, rhythmic changes in the activity across the neural network. As a result, field potential energy extraction is a commonly adopted method in neuroprosthetic applications. Some of the figures presented in this section were originally published in [57] ©IEEE. Reused, with permission.

A variety of distinct brain oscillations exist, with center frequencies spaced logarithmically [190], as illustrated in Fig. 3.1. Commonly used field potential frequency bands include: delta band (1–4 Hz), theta band (4–10 Hz), beta band (10–30 Hz), gamma band (30–80 Hz), and fast band (30–80 Hz). Neural oscillations associated with a certain cognitive state can be in a very narrow frequency band, especially in the low frequency range. For example, the first discovered and one of the best-known frequency bands is the alpha activity, which is 7.5–12.5 Hz [191]. This places a big design challenge in the energy extraction circuit. If the frequency tuning uses a fixed-step (i.e., linear tuning), a high-frequency resolution will have to be realized. Similarly, if a Fast Fourier Transform (FFT) analysis is used, it requires a large number of FFT points and a large memory size for buffering the data. In order to address this problem, a natural logarithmic domain tuning is proposed in this work, which provides a sufficient resolution for extracting the low-frequency brain oscillations, without increasing the number of tuning steps. Figure 3.1 illustrates the frequency bins of the conventional linear step and the proposed natural logarithmic domain step, when they have the same number of frequency bins.

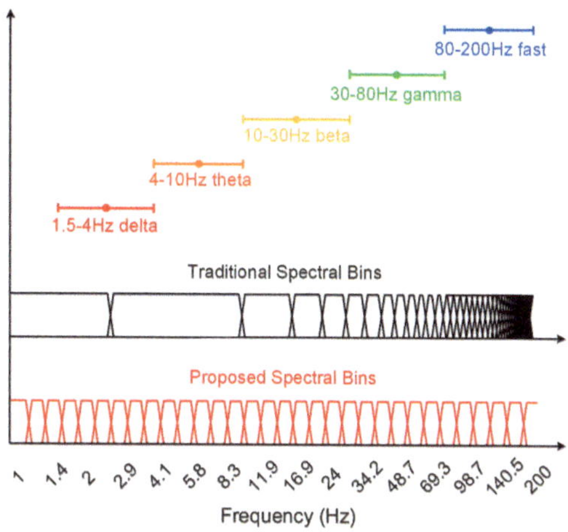

Fig. 3.1 Brain oscillation bands are shown in the natural logarithmic domain. The frequency tuning bins for energy extraction in traditional linear steps and in the proposed natural logarithmic domain are plotted for comparison. A total of 32 steps in a frequency range from 1 to 200 Hz is used in both cases for illustration

3.2.2 *System and Circuits Implementation*

The processing flow of the proposed LFP energy extraction circuit is shown in Fig. 3.2. A lowpass filter with a frequency corner of 300 Hz is first used to remove the high-frequency content in the signal. Then, two second-order stagger-tuned biquad filters are cascaded to bandpass the signal in a programmable center frequency and quality factor [170]. The filtered signal is then squared in a Gilbert multiplier to find the energy. Finally, the energy integral is produced by a leaky integrator with a tunable time-constant [192].

A prototype system that consists of 16 neural feature extraction channels is designed in this work. Each channel can be programmed independently. The feature extraction module in each channel can also be combined to perform spectrum analysis for one channel as a filter bank. Figure 3.3 illustrates the configuration of the system.

3.2.2.1 Design of the GmC Filter

Given the low frequency nature of the neural signal, filters with very large time-constant have to be integrated on-chip. There are several methods to implement such filters in CMOS circuits: (1) Op-amp based filters can achieve a high linearity and a

Fig. 3.2 The processing flow of the LFP energy extraction

Fig. 3.3 The diagram of the 16-channel programmable energy extraction module, and the configuration of the filter bank

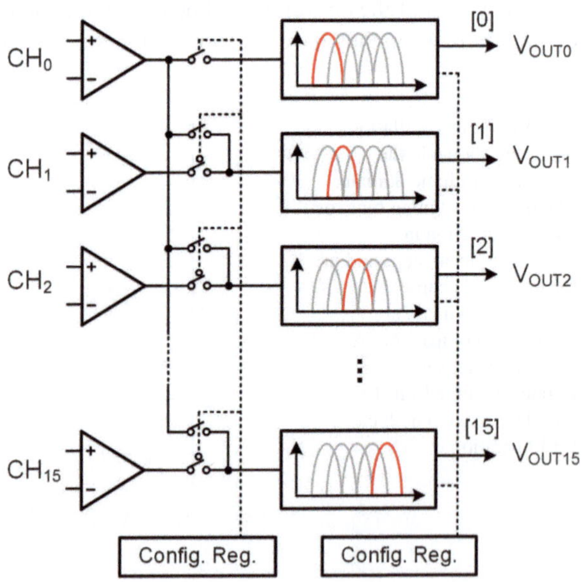

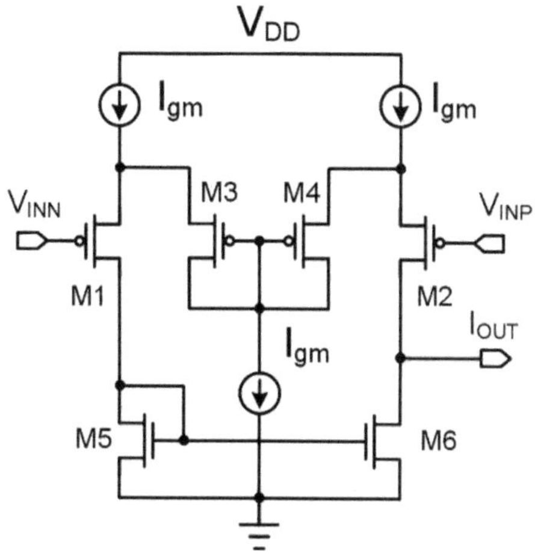

Fig. 3.4 The circuit schematic of a typical MOS Gm block with source degeneration

good signal-to-noise ratio (SNR), but suffer from a high power consumption, large passive components (non-linear if MOS resistors are used instead), and difficulties in tuning; (2) Switched capacitor filters can achieve a high linearity and a high frequency accuracy with tunability [169], but they have limitations and design challenges in the tunable range, the capacitor size, the non-idealities from the clocks and mismatches; (3) Gm-C filters can realize a large time-constant in an ultra-low power consumption and in a minimum silicon area. As a result, these filters have been widely used in biomedical applications [170]. But Gm-C filters also have limitations in linearity and the frequency corner accuracy. (4) Digital filters can achieve good filter characteristics, but they require a pre-digitization of the signal, a memory for buffering the data, and a dedicated DSP core [107]. A high-order digital filter requires accurate coefficients and a sufficient number of bits during the computation to prevent overflow. It should be mentioned that, with the development of advanced CMOS technology, digital filters may surpass analog filters in both accuracy and power consumption. However, traditional CMOS technology with thick oxide and large gate length is sometimes preferred in analog circuit design for a low noise profile and a low leakage current. In summary, design trade-offs need to be carefully considered when designing filters for different applications.

In this work, a Gm-C based filter topology is chosen. The filter is designed with a tunable transconductance in a range of two decades, with an extended linear input range to meet the specifications. The circuit schematic of a simple MOS Gm block is shown in Fig. 3.4. The input transistors are biased in the sub-threshold region [193]. In the sub-threshold region, the transconductance of the transistor has a linear relation with the biasing current [194], as expressed:

$$g_m = \frac{I_{\mathrm{DS}}}{\zeta U_T} \tag{3.1}$$

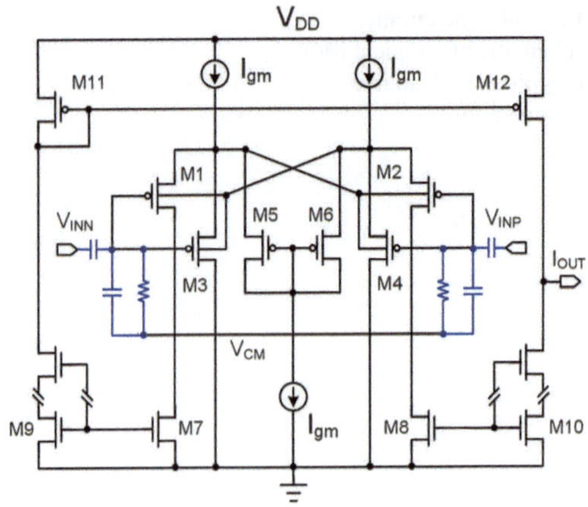

Fig. 3.5 The circuit schematic of the Gm block with a reduced transconductance and an extended linear input range. The biasing current is used to tune the transconductance

where ζ is a parameter that depends on the process, and $U_T = kT/q$. Thus, the transconductance of the Gm block can be directly tuned by the biasing current I_{DS}. A local feedback is used to reduce the distortion [115]. Instead of using resistors, dioded-connected transistors M_3 and M_4 are connected to the sources of the input transistors M_1 and M_2. The total resistance between the sources of the input transistors becomes $2/g_{m3}$. If a reduction factor n is defined as:

$$n = 1 + \frac{g_{m1}}{g_{m3}} \tag{3.2}$$

The third-order harmonic distortion HD_3 can be reduced by a factor of n^2 [195]. In this work, M_1 and M_3 are set to have the same dimension with the same biasing current to maximize the common mode input range [196].

Thick oxide transistors are used in this circuit for a lower transconductance and to reduce the leakage current. The current mirrors are biased in the strong inversion region with a large *vdsat* for a better matching. It should be noticed that there is a limitation in the minimum current that can be reliably copied in the circuit. To reduce the capacitor size in the Gm-C filter while keeping the time-constant of the filter, the transconductance needs to be further reduced. So, a modified version of the Gm block is implemented. The proposed circuit schematic is shown in Fig. 3.5. The input transistors (M_{1-6}) are also biased in the sub-threshold region. Current division is used at the input differential pair to reduce the transconductance. Besides the local feedback as in Fig. 3.4, a bulk degeneration [197] is used to further enhance the linear input range. In a testing version, capacitor attenuation [198] is used to further attenuate the input signal swing and reduce the effective overall transconductance. The attenuator is not used in the final integration considering the cost of the silicon area.

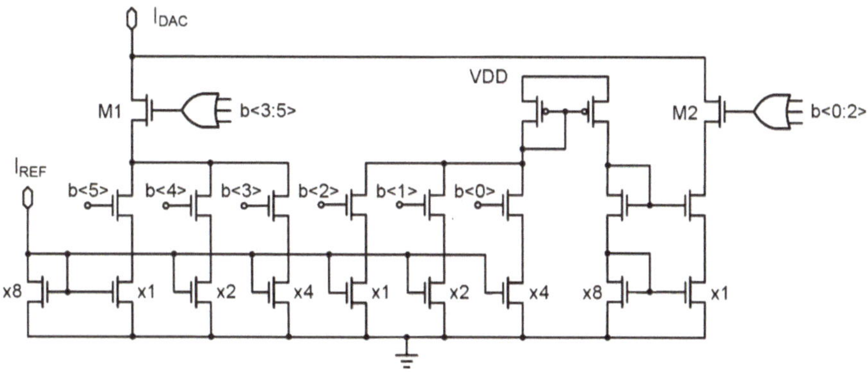

Fig. 3.6 The circuit schematic of the 6-bit current-mode DAC used for generating the biasing current for the Gm block

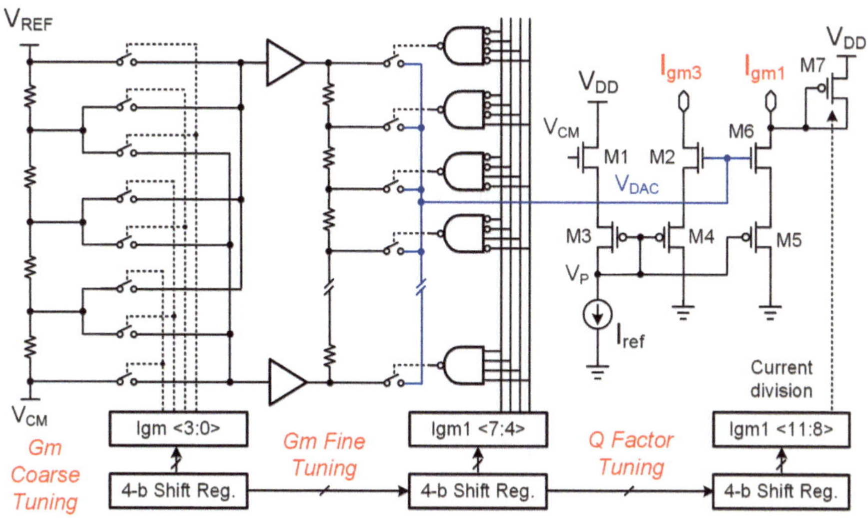

Fig. 3.7 The circuit schematic of the biasing current generation module. A 64-step natural exponentially spaced biasing current can be generated

3.2.2.2 Biasing Current Generation

In the first version, a linear 6-bit current-mode DAC is integrated for generating the biasing current for the Gm block. The circuit schematic is shown in Fig. 3.6. Thick oxide transistors are used to reduce the leakage current and to improve the matching in this CMOS process. The DAC has two segmentations with binary weighted transistors. Two gating transistors M_1 and M_2 are used to further reduce the current leakage and to allow a complete shutdown of the circuit.

The second version of the biasing current generation module is designed for the proposed natural logarithmic domain tuning. Figure 3.7 shows the programmable biasing current generation module. A two-step 6-bit resistor ladder based DAC is used to generate a 64-step linear tuning voltage between V_{cm} and V_{ref}. A custom designed current generation module (M_1–M_6) converts the linear voltage to a natural exponentially spaced current. In the current generation module, all transistors M_1–M_6 are biased in the sub-threshold region. Thick oxide devices are used. When the V_{DS} of a transistor biased in the sub-threshold region is higher than three or four times of the thermal voltage U_T, it is in the saturation region [112]. The equation of the sub-threshold current in the saturation region can be simplified to:

$$I_D = I_o \exp \frac{V_{GS}}{\zeta U_T} \tag{3.3}$$

In the current generation circuit, consider the transistors M_1–M_4. The equations for the currents can be written as:

$$I_{ref} = I_{on} \exp \frac{V_{GS1}}{\zeta U_T} = I_{op} \exp \frac{V_{GS3}}{\zeta U_T} \tag{3.4}$$

$$I_{gm} = I_{on} \exp \frac{V_{GS2}}{\zeta U_T} = I_{op} \exp \frac{V_{GS4}}{\zeta U_T} \tag{3.5}$$

where I_{ref} is generated by an on-chip bandgap reference, and is independent from the temperature and the supply voltage. Equations (3.4) and (3.5) lead to:

$$V_{GS1} - V_{GS3} = V_{GS2} - V_{GS4} = \zeta U_T \ln \left(\frac{I_{op}}{I_{on}} \right) \tag{3.6}$$

Also, from the circuit:

$$V_{CM} = V_{GS1} + V_{GS3} + V_P \tag{3.7}$$

$$V_{DAC} = V_{GS2} + V_{GS4} + V_P \tag{3.8}$$

Substituting Eqs. (3.4) and (3.5) into Eqs. (3.7) and (3.8) gives:

$$V_{GS1} = \frac{1}{2} \left(V_{CM} - V_P + \zeta U_T \ln \left(\frac{I_{op}}{I_{on}} \right) \right) \tag{3.9}$$

$$V_{GS2} = \frac{1}{2} \left(V_{DAC} - V_P + \zeta U_T \ln \left(\frac{I_{op}}{I_{on}} \right) \right) \tag{3.10}$$

The generated biasing current can be expressed as:

$$I_{gm} = I_{ref} \exp \frac{V_{DAC} - V_{CM}}{2 \zeta U_T} \tag{3.11}$$

Thus, the linear spaced voltage from the DAC is converted to an exponentially spaced biasing current. According to Eq. (3.1):

$$g_m \propto I_D \propto e^{(V_{DAC}-V_{CM})} \propto e^{code} \tag{3.12}$$

where code is the digital input of the DAC. Transistors with large gate area are used in the current generation module to minimize the mismatch. The process variation can be further calibrated by tuning the reference voltage V_{ref}.

The M_7 in Fig. 3.7 is a diode-connected transistor used to divide the generated current reference. The ratio of I_{gm1} and I_{gm3} can be programmed to tune the quality factor of the filter, which will be explained in the next section.

3.2.2.3 Design of the Biquad Filter

A staggered tuned fourth-order bandpass filter is implemented by cascading two biquad filters [199]. A biquad filter is a second order recursive linear filter containing two poles and two zeros [200]. Figure 3.8 shows the circuit schematic of the biquad filter implemented in this work. The biquad filter consists of four Gm blocks. The center frequency and the quality factor of each biquad are independently tunable. Only two capacitors with one terminal grounded are used in each biquad, resulting in a very compact layout. The transfer function is given by:

$$H(s) = \frac{s\frac{g_{m1}}{C_1}}{s^2 + s\frac{g_{m2}}{C_1} + \frac{g_{m3}g_{m4}}{C_1C_2}} \tag{3.13}$$

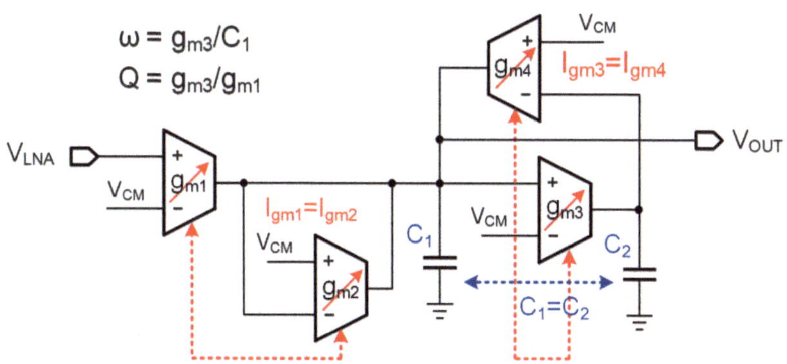

Fig. 3.8 The circuit schematic of the designed biquad filter. The center frequency and the quality factor of the filter can be tuned independently

The biasing currents for the Gm blocks are designed to be $I_{\mathrm{gm1}} = I_{\mathrm{gm2}}, I_{\mathrm{gm3}} = I_{\mathrm{gm4}}$, so that the transconductance of the Gm blocks are $g_{m1} = g_{m2}, g_{m3} = g_{m4}$. The capacitors are set to be $C_1 = C_2$. Thus,

$$\omega_C = \sqrt{\frac{g_{m1}g_{m2}}{C_1C_2}} = \frac{g_{m1}}{C_1} \tag{3.14}$$

If the C_1 is fixed, the center frequency is a function of g_{m1}. Also from Eqs. (3.12) and (3.14):

$$\omega_C \propto g_m \propto e^{\mathrm{code}} \tag{3.15}$$

Thus the center frequency of the biquad can be exponentially tuned by the digital code.

Also, the quality factor Q can be expressed as:

$$Q = \sqrt{\frac{C_1 g_{m3} g_{m4}}{C_2 g_{m2}^2}} = \frac{g_{m3}}{g_{m1}} \tag{3.16}$$

So the quality factor can be tuned by changing the ratio of I_{gm1} and I_{gm3}. As explained in Sect. 3.2.2.2, the M_7 in Fig. 3.7 is a diode-connected transistor with the same length as the current mirrors used for copying I_{gm} in the Gm block. The width of the M_7 can be programmed to divide the generated current reference, so the ratio of I_{gm1} and I_{gm3} can be programmed to tune the quality factor.

3.2.2.4 Multiplier and Integrator

A Gilbert multiplier is used to square the band-passed signal. The Gilbert multiplier, or Gilbert cell, is commonly used analog multiplier circuit introduced by B. Gilbert in 1963 [201]. A tutorial of analog multipliers design can be found in [202]. The circuit schematic of the Gilbert multiplier implemented in this work is shown in Fig. 3.9. The output current of the multiplier is determined by [203]:

$$I_O = \sqrt{2K_\alpha K_\beta}(V_{\mathrm{IN}} - V_{\mathrm{REF}})^2 \tag{3.17}$$

where K is the transconductance parameter, $K_\alpha = K_1 = K_2$, and $K_\beta = K_4 = K_5 = K_6$. So the gain can be tuned by the biasing current I_B.

The integral of the multiplier's output current is computed in a leaky Gm-C integrator [170]. The circuit schematic of the Gm-C based integrator is shown in Fig. 3.9. The integral window length can be changed by tuning the time constant of the integrator.

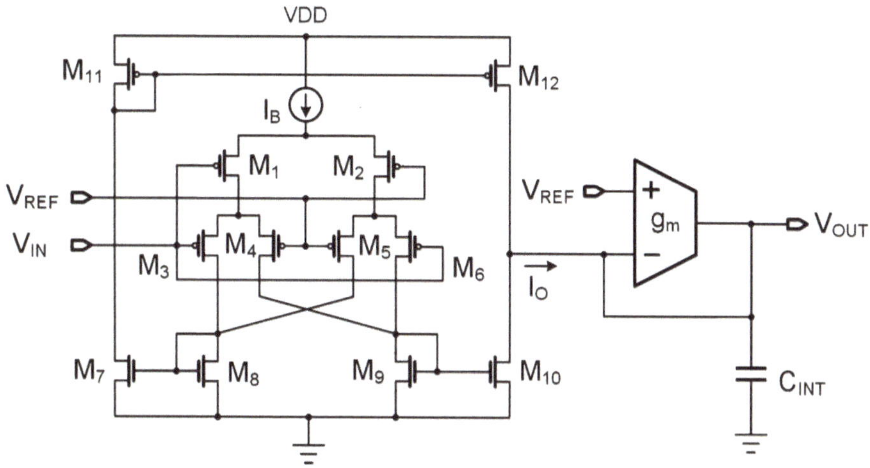

Fig. 3.9 The circuit schematic of the Gilbert multiplier and the integrator

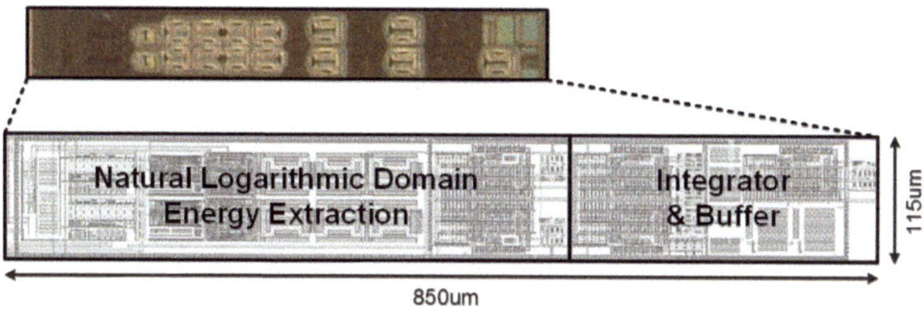

Fig. 3.10 The microphotograph and layout of one LFP energy extraction channel. The major building blocks are highlighted in the layout

3.2.3 *Measurement Results*

The design has been fabricated in 180 nm CMOS technology. Bench testing was conducted to verify the function and performance of the fabricated design. The microphotograph and layout of one LFP energy extraction channel are shown in Fig. 3.10. The occupied silicon area of the design is 850 μm$\times$115 μm.

Bench testing was conducted to verify the function and performance of the fabricated chip. The frequency response of the natural logarithmic tuning neural energy extraction module was measured. The measurement was conducted point by point using a function generator 33521A and an oscilloscope MSO7034B from Agilent. The reference voltage was calibrated to set the center frequency of the unit programming step. Figure 3.11 shows the measurements of every four steps out of the 64 possible steps, with a frequency ranging from 1 to 200 Hz. It should be noticed that the measurement step and *x*-axis in the figure are in the natural logarithmic domain.

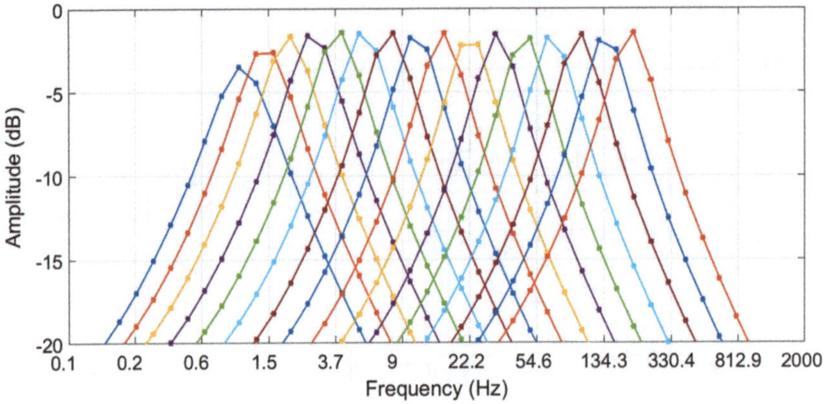

Fig. 3.11 The measured frequency response of the biquad filter tuning in the proposed natural logarithmic steps. A total of 16 steps were measured

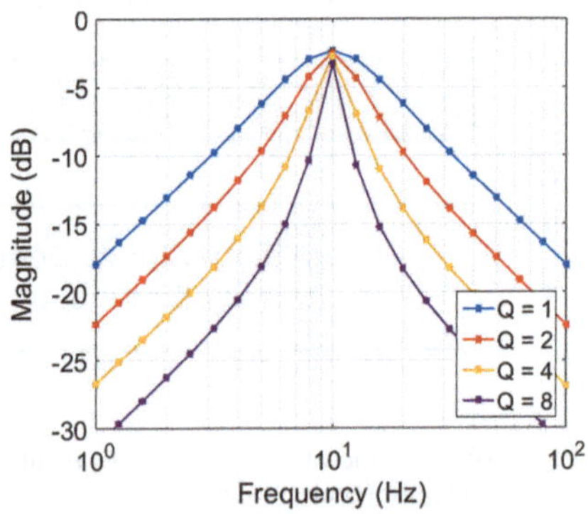

Fig. 3.12 The measured frequency response of the biquad filter with different quality factors. The center frequency is configured at 10 Hz

Similarly, the tuning of the quality factor was measured. Figure 3.12 shows the measurement result with the center frequency of the filter configured at 10 Hz. The quality factor can be configured at 1, 2, 4 and 8.

The biquad filter was also tested with a synthetic sinusoidal wave generated from a function generator. The sinusoidal wave has a constant amplitude, and the frequency was swept from 0.1 to 1 kHz logarithmically. The synthetic waveform and the output of the biquad filter are shown in Fig. 3.13. The frequency sweeping measurement verifies the response of the biquad filter in a straightforward manner.

The Gilbert multiplier was tested with an amplitude modulated 10 Hz sinusoidal wave generated from a function generator. The measurement result is shown in Fig. 3.14. The measured output of the Gilbert multiplier was plotted in comparison with the simulation result after a gain calibration. The measurement result matches the simulation closely.

Figure 3.15 shows the measured output of the leaky integrator with an amplitude modulated 40 Hz sinusoidal wave as the input. The measured output is compared with the theoretical computation of the signal's power. The measurement matches the computation closely.

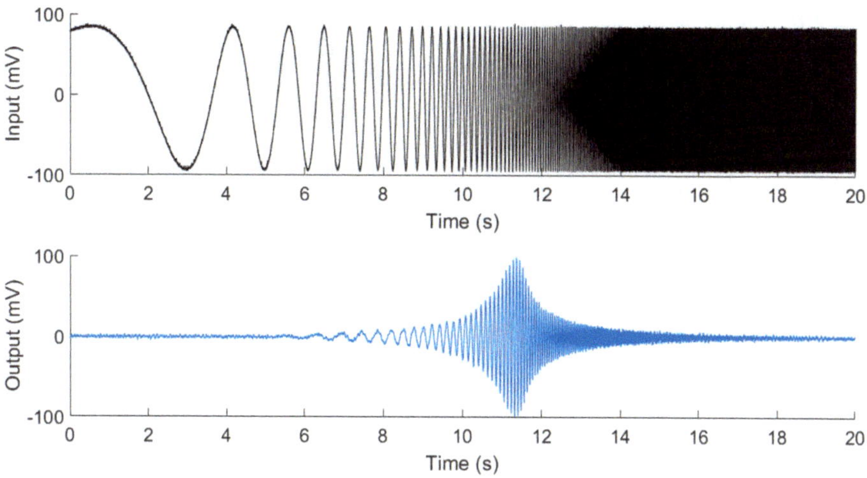

Fig. 3.13 The measured response of one biquad filter with a synthetic sine wave with frequency sweeping from 0.1 Hz to 1 kHz

Fig. 3.14 The measured output of the Gilbert multiplier with an amplitude modulated 10 Hz signal. The measurement result is plotted in comparison with the simulation (after a gain calibration)

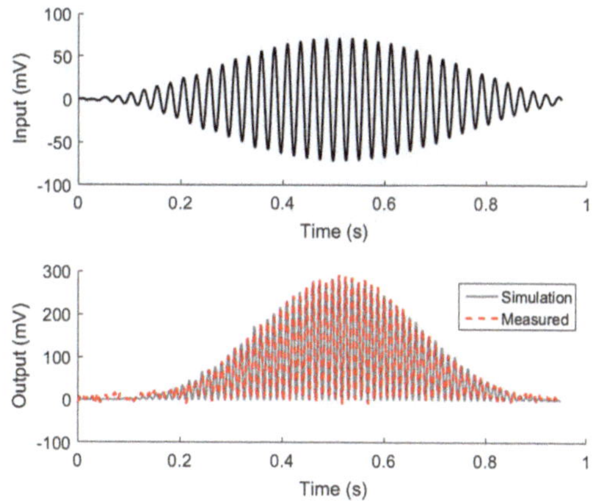

Figure 3.16 shows the power spectrum of a 6-h recording from a male rhesus macaque (*Macaca mulatta*) with electrodes implanted chronically in the left hippocampus. The recording presents an awake to asleep transition of the animal's state. The activities from different brain oscillation bands are clearly visible in the figure.

Figure 3.17 shows a 20-s segment of the LFP recording. The original recorded signal is shown in the top row. The energy in four commonly used frequency bands (solid lines) was extracted using the designed chip, including θ band (4–10 Hz), β band (10–30 Hz), γ band (30–80 Hz), and Fast band (80–200 Hz). The measured output is compared with the theoretical computation plotted in the dashed lines after a gain normalization. A close matching between the waveforms can be observed.

Fig. 3.15 The measured outputs of the multiplier and the LFP energy integrator (phase shift has been corrected)

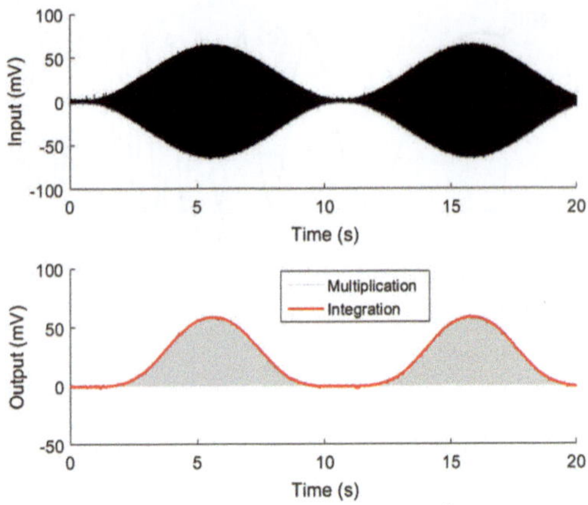

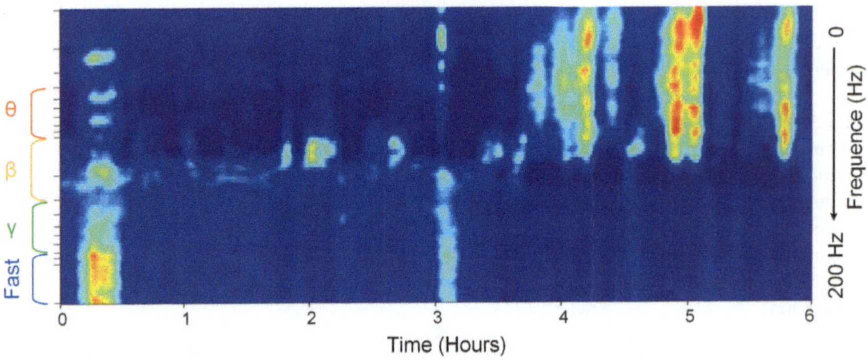

Fig. 3.16 The spectrum of a 6-h continuous recording using the prototype device. The animal's brain state changed from awake (high-frequency oscillation more active) to sleep (low-frequency oscillation more active) during the recording

3.3 Action Potential Discrimination

3.3.1 Introduction

Monitoring neuron activity is the basis for understanding the brain mechanisms [204]. When multiple neurons are close to one recording electrode, it is important to discriminate the action potentials corresponding to different neurons. Even nearby neurons have similar responses, it is important to distinguish them and observe their individual characteristics [205]. Given the distance and orientation relative to the recording electrode, different neurons present different waveforms. The action potentials can then be classified into different clusters, and this process is known as *spike sorting* [206].

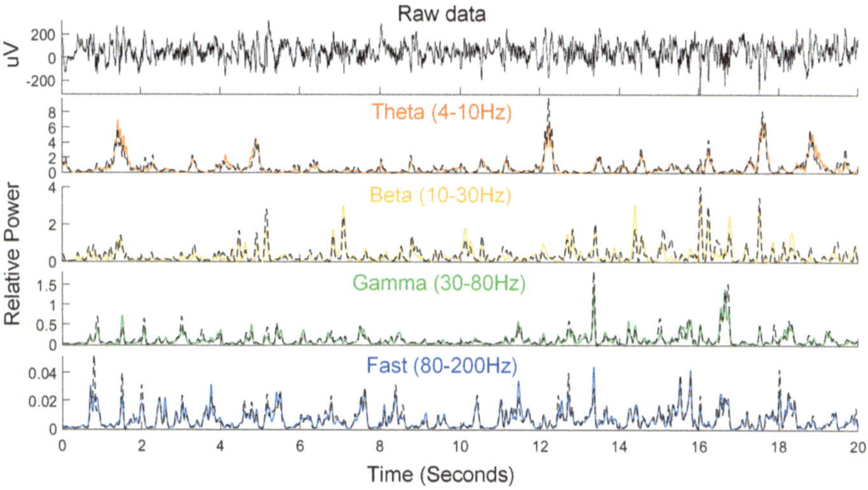

Fig. 3.17 In vivo recording of a Rhesus macaque using the designed chip. The extracted energy in four brain oscillation bands (Theta, Beta, Gamma, and Fast) compared with the theoretical computations (*dashed lines*)

In this section, a simplified action potential model is presented, followed by a review of detection and classification methods. The design of an energy efficient continuous-time current-mode action potential detection module is described. The circuit implementation and experimental results are presented. Some of the figures and tables presented in this section were originally published in [51, 57] ©IEEE. Reused, with permission.

3.3.1.1 Integrate and Fire Model

A good understanding of the action potential working principle is the basis for designing a good discrimination circuit. The Hodgkin-Huxley Model (HHM) is a well-known model which can approximate the generation of an action potential accurately [207]. The goal of this section is to implement the HHM Model for a better understanding of the integrate and fire process of a single neuron. The HHM is constructed by the membrane current as the sum of a leakage current, a delayed-rectified K^+ current, and a transient Na^+ current [208]:

$$i_m = \bar{g_L}(V - E_L) + \bar{g_K}n^4(V - E_K) + g_{\mathrm{Na}}m^3h(\bar{V} - E_{\mathrm{Na}}) \quad (3.18)$$

where n, m, and h are the gating variables. A channel controls the membrane conductance according to the gate variables. When the neuron is depolarized, the gate open probability increases; when the neuron is hyperpolarized, the gate open probability decreases. In general, *n*, *m*, *h* are variables of the voltage and time, and are within 0 and 1. They can be estimated by the following equations [208]:

$$\tau_n(V)\frac{dn}{dt} = n_\infty(V) - n \tag{3.19}$$

$$\tau_n(V) = \frac{1}{(\alpha_n V + \beta_n V)} \tag{3.20}$$

$$n_\infty(V) = \frac{\alpha_n(V)}{\alpha_n(V) + \beta_n(V)} \tag{3.21}$$

where α_n and β_n can be found by:

$$\alpha_n = \frac{0.01(V+55)}{(1 - e^{-0.1(V+55)})} \tag{3.22}$$

$$\beta_n = 0.125e^{-0.0125(V+65)} \tag{3.23}$$

where *m* and *h* can be calculated in:

$$\alpha_m = \frac{0.1(V+40)}{(1 - e^{-0.1(V+40)})} \tag{3.24}$$

$$\beta_m = \frac{1}{1 + e^{(-0.1(V+35))}} \tag{3.25}$$

$$\alpha_h = 0.07e^{-0.05(V+65)} \tag{3.26}$$

$$\beta_h = \frac{1}{(1 + e^{(-0.1(V+35))})} \tag{3.27}$$

The above HHM model was simulated in Matlab. Runge-Kutta method was used to find the arithmetic solution of the differential equations. The method requires initial conditions, which was taken from the reference [208]. The membrane potential from the simulation is shown in Fig. 3.18. HHM model can be used in the simulation and evaluation of action potential detection method.

3.3.1.2 Review of Action Potential Discrimination Methods

Real–time action potential detection and classification methods have been widely reported in the literature since the pioneering work in the 1920s [182]. Comprehensive reviews of detection and classification algorithms can be found in papers [182, 204, 205, 209]. In summary, an effective discrimination method relies on

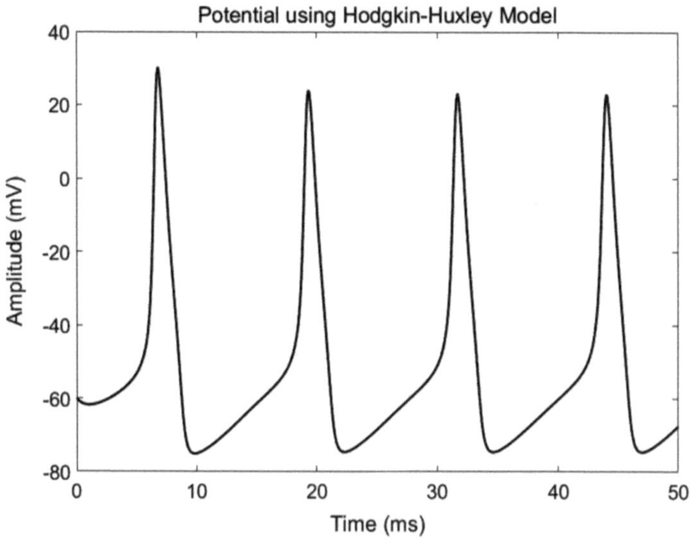

Fig. 3.18 Matlab simulation of the HHM model using Runge-Kutta method for solving the arithmetic solution

a good signal-to-noise ratio and a robust algorithm. The general steps for action potential discrimination are:

1. **Filtering**: filter the raw data between around 300 Hz to 6 kHz to remove low-frequency field potentials for the following processing;
2. **Detection**: detect the action potential, e.g. by applying an amplitude threshold on the filtered signal. Artifacts or noise might be detected as action potentials in this step;
3. **Extraction**: extract the relevant features of the action potential waveform;
4. **Classification**: apply classifier on the extracted features for action potential discrimination.

Both action potential detection and classification have been implemented on-chip [139, 171, 172]. A simple action potential detection can be easily performed in real–time. The detection results can be used to either reduce data transmission rate [69] or trigger pre-defined stimulation in a closed-loop BMI system [36, 38]. Commonly used action potential discrimination methods are summarized here:

1. **Absolute threshold detection**, which uses a predefined threshold for action potential detection [209]. The threshold can be manually set or using a weighted (3×–5×) root mean square value of the signal;
2. **Non-linear energy operator (NEO)**, which extracts the energy from the action potential signal to improve the detection integrity [138]. A modification of the NEO, named the multiresolution Teager energy operator (METO) combines the results of NEO in different resolution scales, also shows a good performance [210];

3. **Wavelet analysis**, which projects the signal to certain wavelets domain [172]. The wavelet transform can also be considered as a bank of matched filters. Different choices of mother wavelet have been reported for action potential detection [211, 212].

The performance of various action potential detection algorithms has been compared in [171, 209]. One of the conclusions is that for real–time action potential detection in systems with limited computational resources, applying an absolute threshold on the signal is just as effective for detecting action potentials as applying more elaborate energy-based nonlinear operators.

Commonly used features for action potential classification include: (1) maximum spike amplitude, (2) minimum spike amplitude, (3) spike width, and so on. It should be noticed that choosing the features manually sometimes yields a poor separation. One method of automatical feature choosing is principal component analysis (PCA) [182]. PCA can find an ordered set of orthogonal basis vectors that capture the data variation in the largest direction.

There are many methods for clustering [182], including K-means clustering, Bayesian clustering, support vector machine (SVM), and so on. For example, K-means clustering, or nearest-neighbor clustering is a hardware friendly classifier. K-means clustering defines the cluster location as the mean of the data within that cluster. An action potential signal will be classified to the cluster with a minimum Euclidean distance. The performance of different classifiers has been compared in [182, 204]. It should be noticed that there are other issues affecting the spike sorting algorithms, including electrode drifting, spike overlapping, neuron bursting, and so on. These issues should be taken into account when designing real–time BMI system.

3.3.2 Circuit Implementation

The action potential detection can be performed in analog domain [138, 213] or in digital domain [214]. The duration of an action potential is less than 2.5 ms [205]. For the accuracy of the classification, the digital sampling rate should be at least 10kSps, or an asynchronous sampling can be used [215]. Analog spike detection can achieve an ultra-low power consumption, while digital domain processing can achieve superior performance and classification accuracy.

In this work, a current-mode continuous-time action potential discrimination unit has been designed. Current-mode circuits present a signal as a current instead of a voltage, thus the dynamic range of the signal is not limited by the supply voltage. This can be very useful for implementing a large dynamic range signal processing in advanced CMOS technology, where a low supply voltage is often used. The block diagram of the designed current-mode action potential detection module is shown in Fig. 3.19. The overall system consists of a low-noise amplifier, a bandpass filter, and the action potential discrimination unit. The bandpass filter is second order

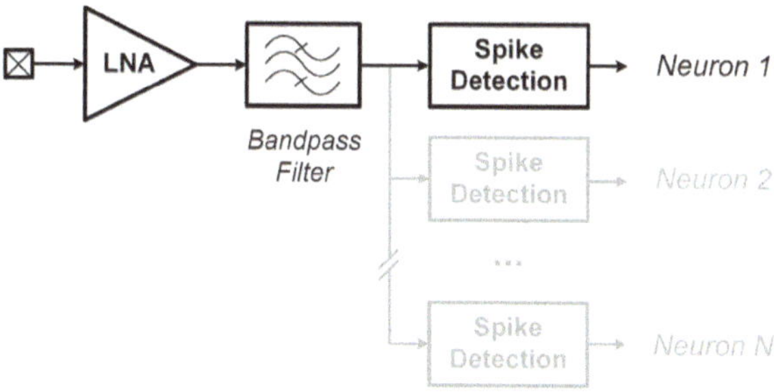

Fig. 3.19 The block diagram of the action potential detection module

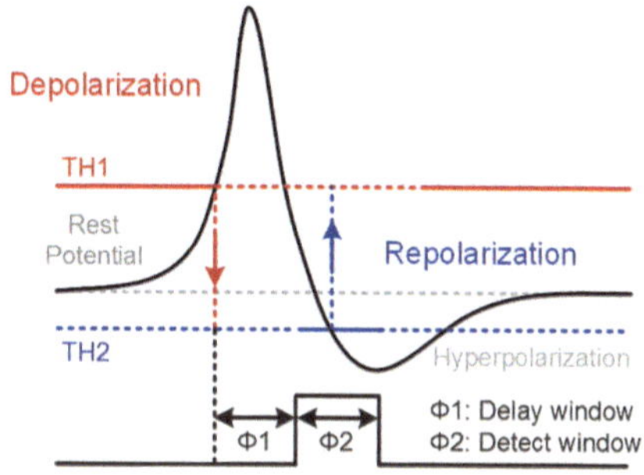

Fig. 3.20 Illustration of the window discriminator for action potential detection

with a passband from 300 Hz to 6 kHz. Multiple units can be connected together to discriminate more than one neuron per channel. In a multiple channel recording system, this can be achieved by designing a multiplexing module among channels, as shown in Fig. 3.19.

The working principle and timing of the action potential detection unit are shown in Fig. 3.20. Two amplitude thresholds and time windows are used to discriminate the action potentials from different neurons [51]. After a bandpass filter, the signal is compared with a depolarization threshold $TH1$. If the signal exceeds $TH1$, the comparator is then disabled for a period of $\Phi1$. After that, the signal is compared with a repolarization threshold $TH2$ for a period of $\Phi2$. If the signal crosses $TH2$ during $\Phi2$, an action potential is detected. The $TH1$ and $\Phi1$, $TH2$ and $\Phi2$ are programmable and will be set independently for each channel.

The circuit block diagram of the implemented action potential detection unit is shown in Fig. 3.21. The unit consists of a transconductance amplifier, a current-mode DAC, a current-mode comparator, and a digital timing and logic module. The transistor M1 is biased in the deep triode region, and its transconductance is set by the biasing voltage V_{Tune}. The threshold currents are generated by a 6-bit current-mode DAC. The circuit schematic of the DAC is shown in Fig. 3.22. The DAC uses binary weighted current mirrors. The current steering can be disabled by shorting I_{OUTp} to ground. Disabling the current steering reduces the power consumption with a lower settling speed, which may not be a problem as long as the threshold current can settle before the discrimination window starts. No additional calibration is implemented in this work. It should be noticed that a 6-bit resolution is usually more than sufficient for the window discrimination algorithm. A fine-tuning of the threshold values won't result in a better discrimination accuracy.

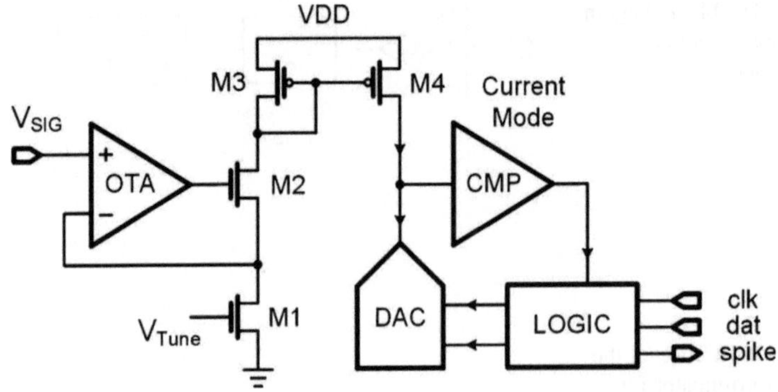

Fig. 3.21 The block diagram of the proposed current-mode action potential detection unit with an integrated programmable amplitude-window discriminator. The filtering stage is not shown in this figure

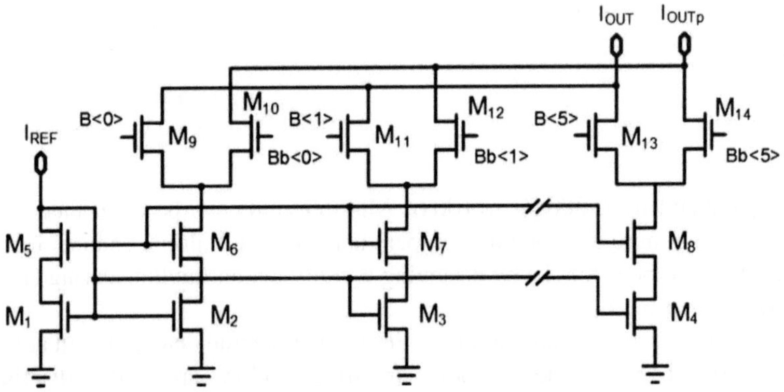

Fig. 3.22 The circuit schematic of the 6-bit current-mode DAC

A current-mode comparator is also designed. The circuit schematic is shown in Fig. 3.23. A current comparator usually is designed with a low input impedance with a relatively constant input node voltage [216]. However, a capacitive input stage can detect a low current amplitude with a much faster response [217]. But the input node voltage cannot be well controlled. As a result, a combination of capacitive and resistive feedback is implemented. The transistors $N3$ and $P3$ work as non-linear feedback resistors to set the input node voltage. When the input signal is small, the feedback loop is disabled and the comparator appears capacitive characteristic, which ensures a high resolution and speed. It is important to minimize the input capacitance, especially when the designed current amplitude is small. A differential pair is used in the second stage, followed by a current starved buffer stage to give rail-to-rail output.

The digital timing and logic module are designed with custom two wire interface (TWI) [57]. There are four registers for the two threshold amplitudes and time windows. Like many real–time spike detection algorithms, one drawback of this

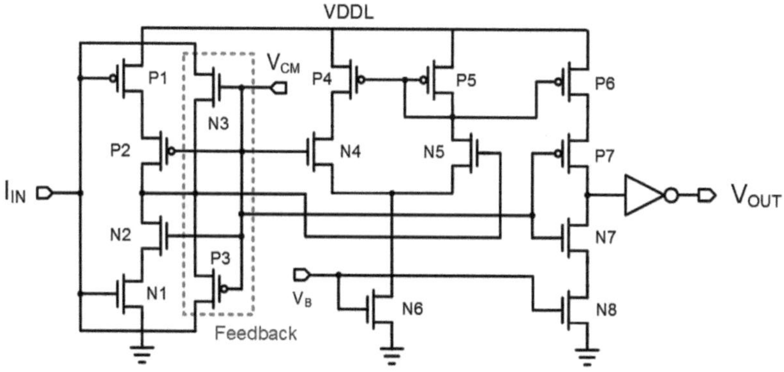

Fig. 3.23 The circuit schematic of the current-mode comparator

design is the false negative detection occurring right after an artifact. In this case, the spike detection won't be able to recover in time. A data buffer can be used to address this problem by the re-alignment of the input data. However, an analog buffer is difficult to implement. An alternative solution is to use multiple detection units in parallel. A digital logic will need to be designed to properly address the conflict by allowing one unit to process one action potential once a time.

Interestingly, with the current-mode comparator and DAC, the action potential discrimination unit can be extended to a current-mode ADC if a SAR logic is added. It might be beneficial to first detect the action potential in the analog domain and then digitize the signal for the further processing.

3.3.3 Experimental Results

The designed action potential detection unit has been fabricated in 180 nm CMOS technology. The design has a dimension of 125 μm × 25 μm. Figure 3.24 shows the microphotograph of the fabricated chip and the layout of one channel with major building blocks highlighted.

A couple of bench tests were conducted to verify the functions and evaluate the performance of the designed module. The experimental results are presented as follows. The DAC was measured with a worst INL and DNL less than 1 LSB. The ENOB is 5.6 bit. The average power consumption of the module is 4 μW with a supply voltage of 1 V. The clock frequency is set to be 100 kHz, which gives a maximum latency of 10 μs.

Synthetic neural signals with different SNRs were generated using an arbitrary function generator 33521A from Agilent to test the action potential detector. The original signal was recorded by a custom designed recorder from an anesthetized rat in the whisker motor cortex. A 2-min recording segment was used for testing. The first 10-s signal is shown in Fig. 3.25 for illustration. Figure 3.25a shows the original

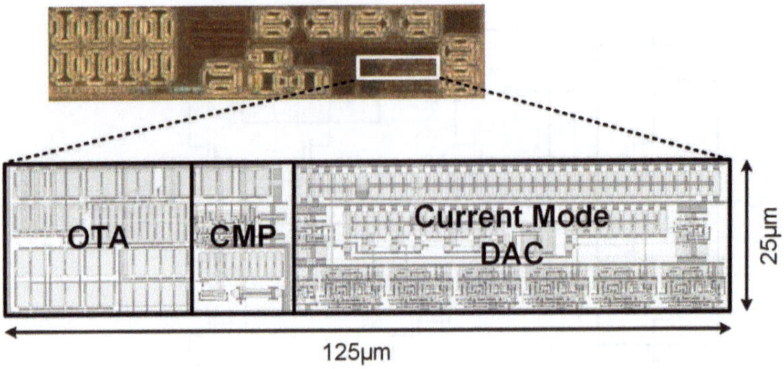

Fig. 3.24 The microphotograph and layout of the designed action potential discrimination module

signal recorded from the whisker motor cortex in an anesthetized rat. The sampling rate was 21 kSps. The recorded data has an SNR of 25.7 dB. The real action potentials are observed by experts and are marked by red triangles in the figure. Figure 3.25b and c shows the recorded signal with intentionally added white noise and artifacts resulting in a SNR of 20 dB and 15 dB, respectively. The artifacts are designed to mimic the motion and chewing effects, which commonly occur during the neural recording in freely behaving animals. There are N real action potentials in the recording, where $N = 117$ in this case. The true positive T_P is defined as the correct recognitions. The false negative F_N is defined as the wrong recognitions. The false positive F_P is the missed action potentials. The evaluated performance of the designed module is listed in Table 3.1. The ratio was calculated over the total number of $T_P + F_N + F_P$. The performance of the two-window discrimination is also compared with the detection using only one threshold.

The experimental results suggest that with a good SNR, both a simple thresholding and the window discrimination give excellent detection results. The simple thresholding method gives a slightly better result than the window discrimination, mainly because several action potentials fail to pass the second window due to distortion. The performance of the simple thresholding method drops significantly after adding artifacts. With an increased noise level, it is difficult to set the threshold voltage, and the window discrimination clearly rejects more false detections than the simple thresholding method. But the window discrimination also makes more mistakes in the higher noise environment. Some artifacts are mistaken as the real action potentials when the noises pass the second window.

A cluster analysis was performed in the microcontroller to study the real–time action potential classification. The algorithm was programmed in the C language. In this experiment, two neurons were captured simultaneously on the same recording electrode. The normalized maximum and minimum amplitudes in an action potential waveform were calculated and used as the features for the clustering analysis. The K-means clustering was used to separate the two neurons. Figure 3.26a illustrates the analysis result in the feature domain, which clearly shows the two

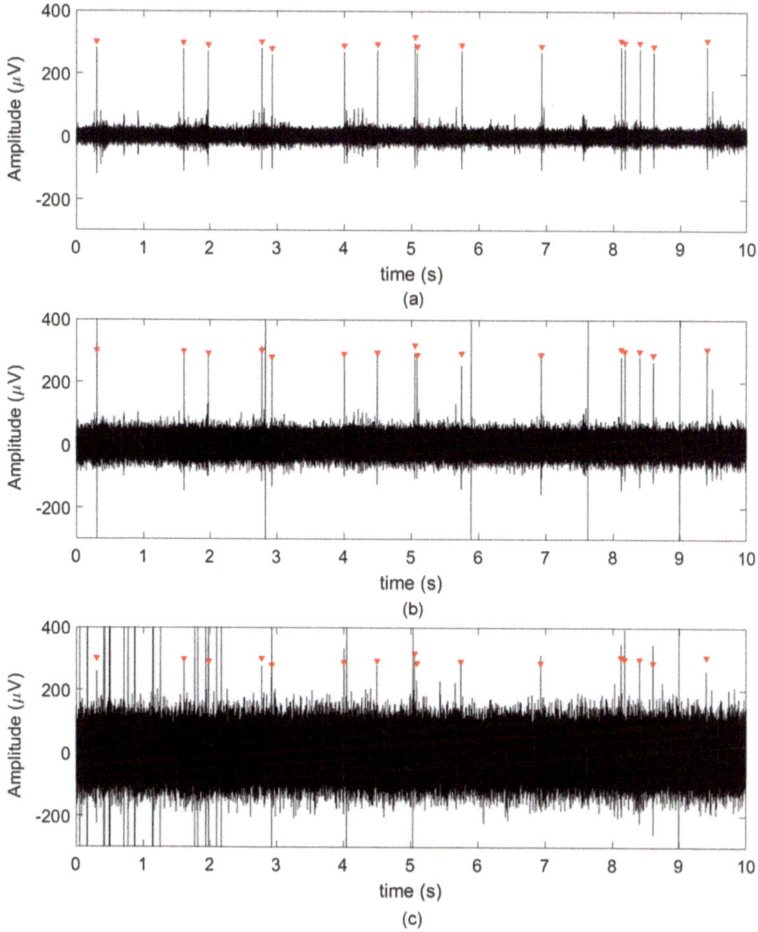

Fig. 3.25 The action potential signal used for testing the designed module. The real action potentials are marked by *triangle markers*. (**a**) The original signal with an SNR of 25.7 dB. (**b**) and (**c**) are synthesized testing signal with added white noise and artifacts, the SNR is 20 dB and 15 dB, respectively

clusters well separated. The action potentials are replotted with color coding based on the classification results, as shown Fig. 3.26b.

In this section, the design of a real–time current-mode action potential discrimination unit is presented. The design features low power consumption, robust detection, and small silicon area, which is suitable for an integration into a high channel count neural recording front-end or a closed-loop BMI system. A K-means classification was implemented in the microcontroller. The experimental results show that the system is capable of identifying multiple neurons from a single electrode in real–time.

Table 3.1 The measured AP detection accuracies from signal with different SNR

SNR	Algorithm	TP	FN	FP
25 dB	Win Discrim	95.8%	4.2%	0%
	Threshold	97.4%	2.6%	0%
20 dB	Win Discrim	93.2%	6.8%	0%
	Threshold	77.6%	2.0%	20.4%
15 dB	Win Discrim	83.8%	6.2%	10.0%
	Threshold	50.5%	8.0%	41.5%

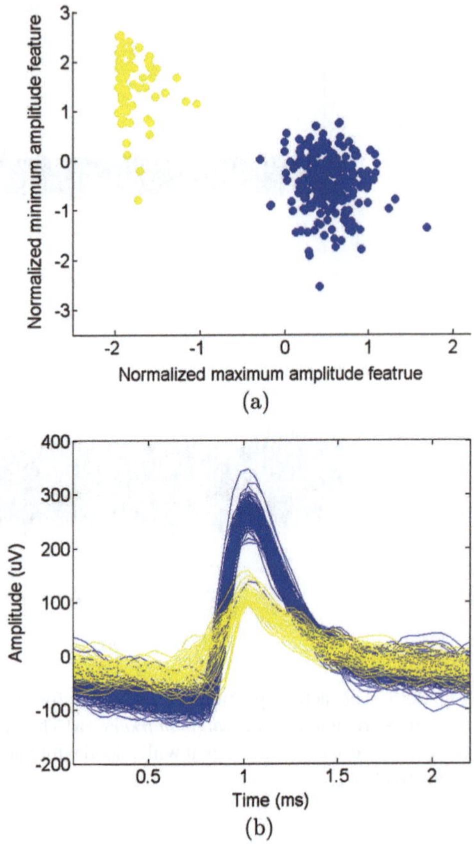

Fig. 3.26 A cluster analysis of the action potentials from two neurons. (**a**) Normalized maximum and minimum amplitudes are calculated and used as two features for the analysis. (**b**) The action potentials are labeled with different colors according to the classification results

3.4 Matched Filter for Neural Feature Extraction

3.4.1 Introduction

A matched filter is the optimal linear filter for maximizing the signal-to-noise ratio in presence of additive stochastic noise [218]. The matched filtering is performed by correlating a known template with the unknown input signal to detect the presence of the desired signal [219]. Matched filters are commonly used in wireless communications [220], radar and sonar [221], gravitational-wave astronomy [222], medical applications [223], and so on.

A number of studies propose to implement matched filters for action potential detection [224–226]. In addition, the matched filter can be applied to detect phase-amplitude coupled low-frequency neural rhythm [227]. For example, the cortical μ rhythm is an event-related desynchronization commonly used for BMI control. However, the μ rhythm's typical frequency band is 8–12 Hz, which overlaps with the virtual α rhythm (Sect. 3.2). Thus, an energy based feature extraction method often has difficulties in discriminating them. In this case, matched filters have the advantage in accurately modeling the phase-coupled rhythm.

Moreover, the performance of the matched filters can be optimized by pre-whitening the signal. This process can be achieved by implementing the pre-whitening filter proposed in Chap. 2. By combining the phase correction filter and matched filter, an energy efficient hardware implementation can be achieved. By programming the coefficients of the filter, it can be used in a wide range of applications, and is very suitable for an integration on a neural interface.

This section presents the analysis, design, and testing of a matched filter with pre-whitening for neural signal extraction. In the end of the section, a compressive domain matched filter is explored to further reduce the requested computation and hardware cost.

3.4.2 Matched Filter and Pre-whitening for Optimum Correlation Detection

The process of the matched filter is illustrated in Fig. 3.27. The input $x(t)$ consists of the signal $s(t)$ corrupted by a white noise $w(t)$, which can be expressed as:

$$x(t) = s(t) + w(t) \tag{3.28}$$

where $w(t)$ has a zero mean and power spectral density of $N_o/2$. If the filter is linear, the output is:

$$y(t) = s_o(t) + w_o(t) \tag{3.29}$$

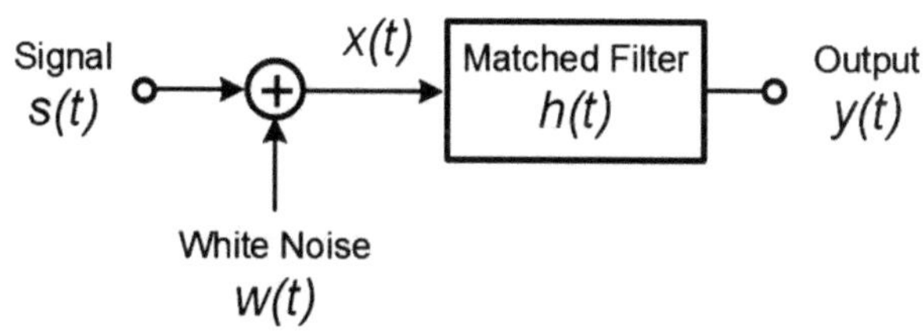

Fig. 3.27 The block diagram of the matched filter

The signal to noise ratio (SNR) is defined as:

$$\mathrm{SNR} = \frac{|s_o(T)|^2}{w_o^2(t)} \tag{3.30}$$

Assume the noise spectral density is $N_0/2$, and the transfer function is $H(f)$, Eq. (3.30) can be written as [228]:

$$\mathrm{SNR} = \frac{\left| \int_{-\infty}^{\infty} H(f) S(f) e^{j2\pi f t_d} df \right|^2}{\frac{N_0}{2} \int_{\infty}^{\infty} |H(f)|^2 df} \tag{3.31}$$

To find the maximum SNR, use the conclusion of Schwarz inequality [229]:

$$\left| \int_{-\infty}^{\infty} f_1(x) f_2(x) dx \right|^2 \leq \int_{-\infty}^{\infty} |f_1(x)|^2 df \int_{-\infty}^{\infty} |f_2(x)|^2 dx \tag{3.32}$$

only if

$$f_1(x) = k f_2^*(x) \tag{3.33}$$

Now set

$$f_1(x) = H(f) \quad \text{and} \quad f_2(x) = S(f) e^{j2\pi f t_d} \tag{3.34}$$

So Eq. (3.32) can be rewritten as:

$$\left| \int_{-\infty}^{\infty} H(f) S(f) e^{j2\pi f t_d} df \right|^2 \leq \int_{-\infty}^{\infty} |H(f)|^2 df \int_{-\infty}^{\infty} |S(f)|^2 df \tag{3.35}$$

And Eq. (3.31) can be rewritten as:

$$\begin{aligned} \mathrm{SNR} &= \frac{\left| \int_{-\infty}^{\infty} H(f) S(f) e^{j2\pi f t_d} df \right|^2}{\frac{N_0}{2} \int_{\infty}^{\infty} |H(f)|^2 df} \\ &\leq \frac{\int_{-\infty}^{\infty} |H(f)|^2 df \int_{-\infty}^{\infty} |S(f)|^2 df}{\frac{N_0}{2} \int_{-\infty}^{\infty} |H(f)|^2 df} = \frac{2}{N_0} \int_{-\infty}^{\infty} |S(f)|^2 df \end{aligned} \tag{3.36}$$

Thus the maximum SNR can be found using Eq. (3.36), when:

$$H(f) = k S^*(f) \tag{3.37}$$

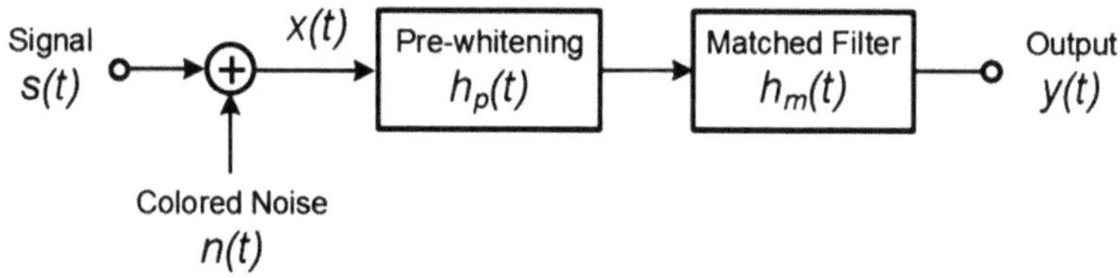

Fig. 3.28 The block diagram of the matched filter in combination with the pre-whitening filter for the correlation optimization

In the time domain,

$$h(t) = ks^*(t_d - t) \tag{3.38}$$

where k is an arbitrary constant. The matched filter $h(t)$ is just a time-reversed version of the signal with a gain factor.

The above analysis assumes that the noise has a white spectral density. However, the neural signal and electronics noise both have a frequency-dependent spectral density, as analyzed in Chap. 2. But if the noise and the background signal can be pre-whitened, the correlation detection can still be optimized by the matched filtering [230]. The process is illustrated in Fig. 3.28.

The generation of a rigorous pre-whitening filter requires a prior knowledge of the noise spectrum. However, this is usually not feasible in a real–time system. The pre-whitening filter proposed in Chap. 2 is a low-cost hardware solution to generate a pseudo pre-whitening filter to improve the performance of the matched filter. The following analysis and testing results verify the hypothesis.

3.4.3 Methodologies

3.4.3.1 Dataset

The dataset used in this study is a 5-min neural recording from an anesthetized rat. The sampling rate was 24.41 kHz and was down-sampled to 2441 Hz before the following processing. Figure 3.29a shows a 5-s recording segment from five different channels. Lots of 1-Hz oscillations (typically called an "up-down" state) can be observed in the recording. The frequency analysis in Fig. 3.29b shows the oscillation has an energy peak at 1 Hz. The oscillation also has a strong amplitude-phase coupling [231], as shown in Fig. 3.29c.

A cycle-triggered average analysis was applied to find out these 1-Hz oscillations: (1) a total of 310 segments were detected in the recording, which were used as the data bank in the following study; (2) these segments were all aligned on the "down-state" peaks; (3) the average of these segments was used as the target neural feature waveform, and is referred to as *template* in the following study. Figure 3.30a

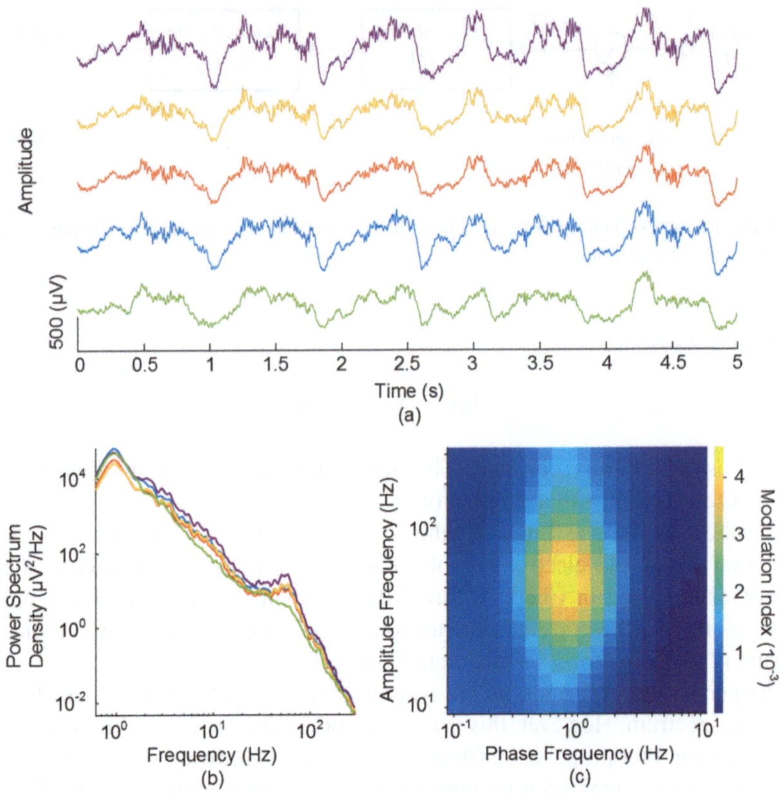

Fig. 3.29 (**a**) A 5-s recording segment from an anesthetized rat. (**b**) The power spectrum density of the recording. (**c**) The phase-amplitude coupling analysis of the signal

shows the neural feature waveform (template). Clearly, the waveform has more time in the "up-state" than the "down-state," so it is not a single frequency sinusoid wave. Figure 3.30b shows the frequency analysis of the template. The dominant frequency components are from 0.6 to 1.5 Hz.

3.4.3.2 Bandpass Filter

Bandpass filters are used for comparing the detection performance with the proposed matched filter. Several bandpass filters have been implemented, including Butterworth filters, Chebyshev filters, Biquad filters, and different types of FIR filters. Figure 3.31 shows a comparison of these filters' frequency responses. The hardware cost for a FIR filter to achieve such a narrow frequency band is significantly higher than the IIR filters, especially when the sampling rate is high. A second order Butterworth filter is used as the bandpass filter in the following study for the comparison purpose. It can be implemented in either analog or digital circuits. The cut-off frequencies were chosen to be 0.6 and 1.5 Hz.

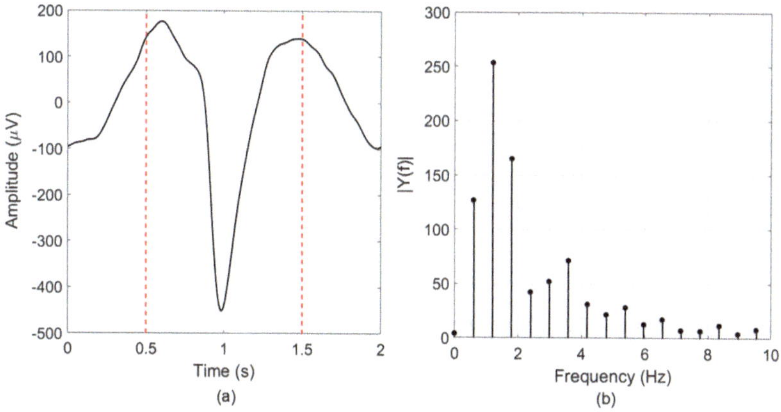

Fig. 3.30 (**a**) The neural feature waveform (template). The waveform has more time in the "up-state" than the "down-state," so it is not a single frequency sinusoid wave. (**b**) The frequency analysis of the neural feature template

Fig. 3.31 A comparison of different filters' frequency responses. These filters are used in this work for extracting the slow oscillation

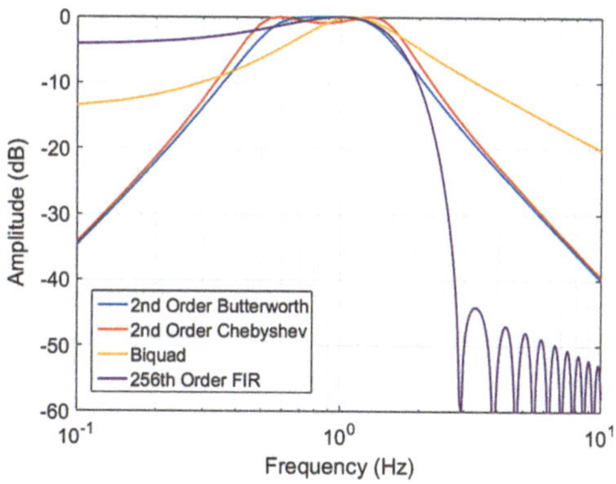

3.4.3.3 Matched Filter with Pre-whitening

As derived in Eq. (3.38), the matched filtering was performed by the convolution of the signal and the time reversed version of the template:

$$y(t) = x(t) * h_m(t) \tag{3.39}$$

where $x(t)$ is the input signal, $h_m(t)$ is the matched filter, and $y(t)$ is the output signal. The pre-whitening filtering was implemented by a first order highpass filter, with a corner frequency of 100 Hz. This mimics the actual hardware circuit implementation of the pre-whitening filter proposed in Chap. 2. The template was also pre-whitened by the same filter to compensate the phase distortion.

$$y(t) = (x(t) * h_w(t)) * (h_m(t) * h_w(t)) \tag{3.40}$$

where $h_w(t)$ is the pre-whitening filter. It should be noticed that the $h_m(t) * h_w(t)$ can be pre-computed to save the online computation cost.

After the filtering, the output signal was squared to find the energy. A moving average filter with a window size of 1-s is applied to find out the envelope, and a threshold is used to detect the event (onsite of the oscillation).

3.4.4 Experimental Results

3.4.4.1 Detection of Synthesized Signal

This section presents the testing performance in detecting synthesized signal with different SNR. Take the 2-s template signal (noiseless) $s(t)$ and add random generated pink noise $n(t)$ as the test signal $x(t) = s(t) + n(t)$. Since the signal power is known, the SNR can be controlled by changing the energy of the pink noise.

$$\text{SNR} = \frac{P_{\text{Signal}}}{P_{\text{Noise}}} \tag{3.41}$$

Figure 3.32 shows an example set of 16 synthesized testing signals with SNR ranging from 0.2 to 6. Each of the 2-s synthesized signals was superimposed on a 100-s pink noise signal. Different filters were applied to the 100-s data for detecting the target oscillation. The detection accuracy is defined as:

$$\text{Accuracy} = \frac{T_P}{T_P + F_N + F_P} \tag{3.42}$$

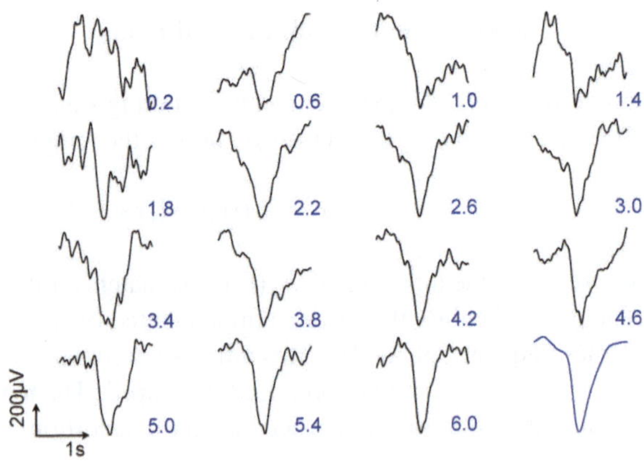

Fig. 3.32 Examples of the synthesized test signals with SNR ranging from 0.2 to 6. The last one is the template

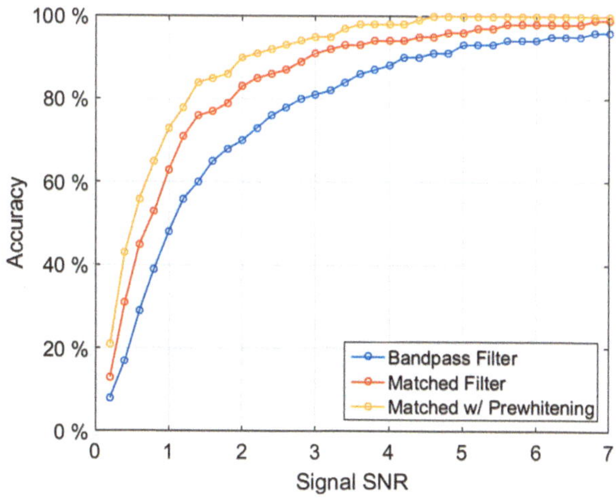

Fig. 3.33 Different filters' accuracies in detecting signals with different SNR (0.2–7 in a step of 0.2). The accuracy calculated for each SNR step was an average of 100 trials with random pink noise. A total of 3500 trials were tested for each filter in this experiment

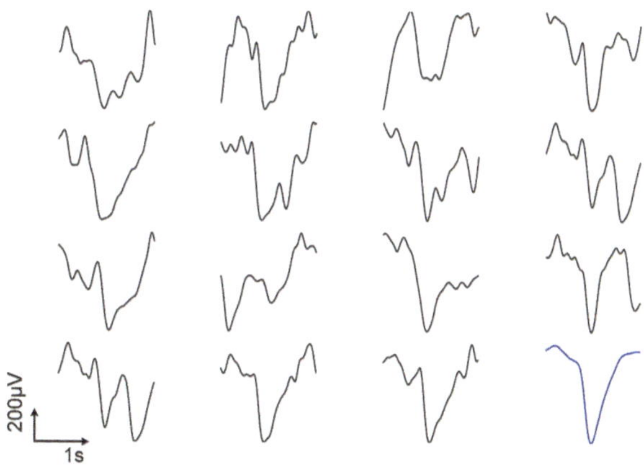

Fig. 3.34 Examples of the randomly selected real neural signal segment from the data bank. The last one is the template

where the true positive T_P is for the correct detection, the truth negative F_N is for the wrong detection, the false positive F_P is for the missed ones.

Testing signals with SNR ranging from 0.2 to 7 with a step of 0.2 were generated. 100 trials were generated for each SNR step with random pink noise. So a total of 3500 trials of 100-s testing signals were used for evaluating the performance of each filter. Figure 3.33 shows the testing results. The result shows that the matched filter has a better accuracy than the bandpass filter for detecting this neural feature. And the pre-whitening filter further improves the detection accuracy of the matched filter. The experimental results verify the hypothesis.

3.4.4.2 Detection of Recorded Neural Signal

This section evaluates the detection accuracy of the real neural signals from the data bank. 100 real neural signal segments were randomly selected from the data bank. Figure 3.34 shows 15 examples of the segments together with the template.

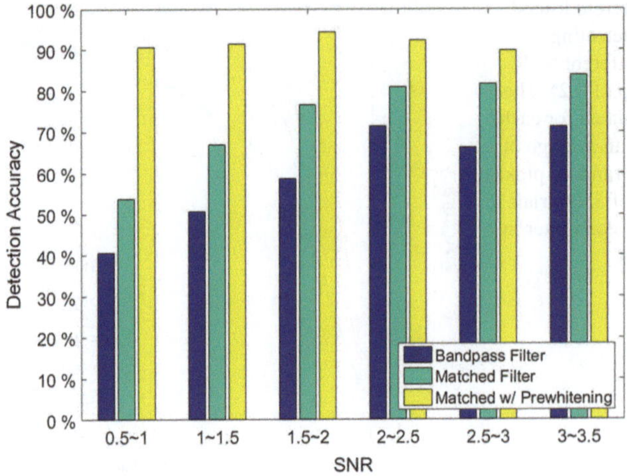

Fig. 3.35 Accuracies of different filters in detecting 100 signal segments from randomly generated pink noise. A total of 10,000 trials were tested for each filter in this experiment. The detection result shows that the pre-whitening filter improves the matched filter's performance, especially in low SNR cases

100 segments of pink noise were generated, and the 100 neural signal segments were superimposed on these pink noise segments. So a total of 10,000 testing signals were synthesized. The SNR for each trial was calculated. The detection results using bandpass filter, matched filter, and matched filter with pre-whitening are plotted in Fig. 3.35. The detection result shows that the pre-whitening filter improves the matched filter's performance, especially in the low SNR cases. The average detection accuracy is lower than the first experiment, which is mainly due to the existence of more than one oscillation in a 2-s data segment (only one true oscillation is assumed in each trial). This experiment verifies the hypothesis that matched filter with pre-whitening can achieve superior performance in detecting phase-amplitude coupled neural feature.

3.4.4.3 Compressive Domain Matched Filtering

One drawback of FIR filter and matched filter is the large coefficients and the required computation. However, most neural feature has slow oscillation period which potentially can be used to reduce the computation. This section shows the detection results by applying matched filter in the compressive domain. The experiment setup is the same as in previous sections. The template used for the matched filter is randomly sampled to create a sparse coefficient vector, and the incoming signal is sampled in the same way.

$$y''(t) = ((x(t) * h_w(t)) * V) * ((h_m(t) * h_w(t)) * V) \tag{3.43}$$

where V is a sparse vector contains only 0 and 1. The number of ones over the total length of the vector is the compression ratio. $(h_m(t) * h_w(t)) * V$ can be pre-computed to save the computational cost.

The experimental results are shown in Fig. 3.36. The detection accuracies of different filters with SNR ranging from 0.2 to 3 in a step of 0.4 are compared. The results show that the matched filter with pre-whitening has the best performance, while the conventional bandpass filter is the worst. Applying compressive matched filter doesn't compromise the detection accuracy up to a compression ratio of 16×. However, even with a compression ratio of 64×, the matched filter with pre-whitening still has a better detection accuracy than the matched filters without pre-whitening. Finally, the 64× compressive matched filter with pre-whitening achieves a detection accuracy over 90% given an SNR of 3 dB, and over 98% given an SNR of 6 dB. It should be noticed that all of the pre-whitening used in these experiments is simply a highpass filtering as proposed in Chap. 2, which can be easily implemented in hardware.

In summary, this section presents the design and testing of the matched filter for neural feature extraction. A pre-whitening filter is used to further improve the detection accuracy. In addition, compressive matched filter is introduced to reduce the computational cost. The experiment was based on a dataset of a recording in an anesthetized rat. A 1-Hz up-down state oscillation was used as the target neural feature. The experimental results suggest that: (1) The performance of the matched filter is better than the conventional bandpass filters in detecting this kind of features; (2) The pre-whitening processing further improves the performance of the matched

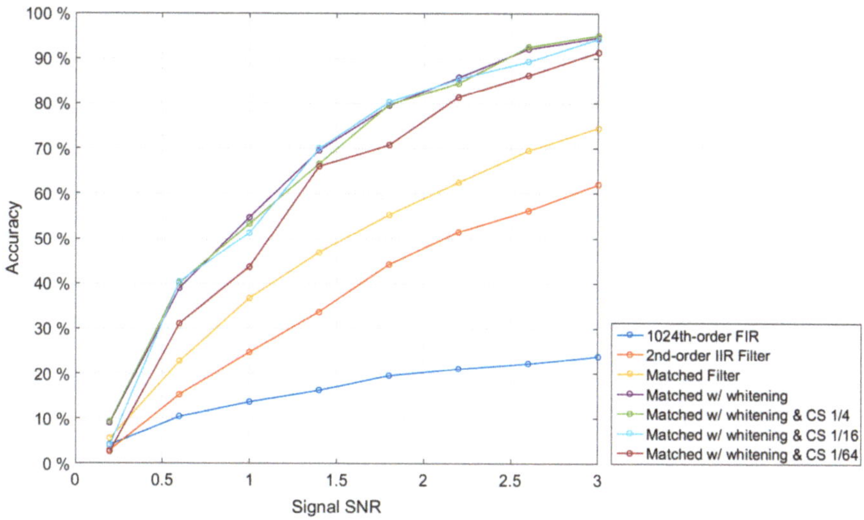

Fig. 3.36 Comparison of different filters' detection accuracies. The experimental results show that the matched filter with pre-whitening has the best performance. The compressive matched filter with pre-whitening and a compression ratio of 64× still has a better performance than the matched filter without pre-whitening

filter in correlation detection, especially in low SNR cases; (3) A compression of coefficients up to 64× can be applied to the matched filter without sacrificing too much accuracy. In conclusion, the matched filter with pre-whitening is a powerful and promising tool for extracting amplitude-phase coupled neural features. It can be used for a wide range of applications, and is especially useful for the integration in a real–time closed-loop BMI system.

Chapter 4
Neural Stimulator Design

4.1 Introduction

The electrical stimulation of excitable neurons is one of the most prevalent functions performed in implantable biomedical devices [232]. The first electrical brain stimulation was pioneered by researchers Luigi Rolando and Pierre Flourens in the early nineteenth-century [233], and the development of the medical stimulator began with the early pacemaker design in the 1930s [234]. The development of electronics engineering, especially integrated circuit technology, enables the design of accurate, reliable, and miniature stimulators for neuroscience research and clinical treatment. Nowadays, electrical stimulators have been widely used for deep brain stimulation (DBS), functional electrical stimulation (FES), spinal cord stimulation (SCS), visual and auditory neural stimulation, brain–machine interface (BMI), neuroprosthetics, and clinical therapeutic treatments [232, 235].

The clinical adoption of the electrical stimulation requires the neural stimulator device to be designed with a high-level of safety, reliability, and programmability, as well as a minimum-level of power consumption and heat dissipation. In addition, a sufficient channel-count and bandwidth, a flexible configuration, a device dimension suitable for an implant, and a wireless communication capability are all essential features. Lots of circuit techniques have been developed to address the challenges in the electrical neural stimulator development.

This chapter presents the analysis and design of high efficiency electrical neural stimulators. The design of a general–purpose neural stimulator is first reviewed and described, and a novel stimulation strategy is proposed to address a practical problem from the electrode–electrolyte interface. The chapter is organized as follows. Section 4.1 introduces the background of neural stimulation and the physicochemical properties of the electrode–electrolyte interface. Section 4.2 gives an overview of the electrical stimulator design. The key design requirements are summarized, and previous state-of-the-art techniques are reviewed. Section 4.3 presents a general–purpose programmable neural stimulator design. Section 4.4

X. Liu, J. Van der Spiegel, *Brain-Machine Interface*,
https://doi.org/10.1007/978-3-319-67940-2_4

describes the novel net-zero charge neural stimulator design. The circuit implementation and the experimental results are presented. All experimental procedures used in the studies presented in this chapter were approved by the institutional animal care and use committee (IACUC) of the University of Pennsylvania.

4.1.1 Background of Neurostimulation

Neurostimulation is a method for modulating the nervous system's activity using non-invasive or invasive means [236]. The controlled electrical, magnetic [237, 238], chemical [239, 240], or optical stimulation (optogenetic modulation [241, 242]) of central or peripheral nervous systems is usually referred to as *neuromodulation* in the medical literature [243]. The focus of this work is on the electrical stimulation.

The mechanism of electrical neural stimulation is a consequence of the depolarization and hyper-polarization of excitable cell membranes from the applied electrical currents. However, other mechanisms including thermal and neurohumoral effects may also involve with the process. The neuron membrane acts as a capacitor by separating the charges lying along its interior and exterior surfaces. The membrane conductance depends on the densities and types of the ion channels. The channels are highly selective, allowing only a single type of ion to pass, while maintaining a concentration difference of ions inside and outside of the neuron. When the membrane potential of a neuron reaches a certain threshold, it will typically fire an action potential. The relation of ion current and neuron potential is accurately described in the Hodgkin-Huxley model (HHM), proposed by A.L. Hodgkin and A.F. Huxley in 1952 [207].

4.1.2 Electrode and Electrolyte Interface

The essential process during an electrical stimulation includes the charge transfer and redistribution across the electrode and electrolyte interface. It should be noticed that in the electrode and the electrical circuits, the charges are carried by the electrons while in the physiological medium, the charges are carried by the ions, mainly including sodium, potassium, and chloride. Figure 4.1a illustrates the two primary mechanisms [244]:

- **Faradaic charge transfer**, or non-polarizable mechanism, where electrons transferred between the electrode and electrolyte interface cause reduction and oxidation reactions. It should be noticed that the Faradaic reaction may be reversible or irreversible.

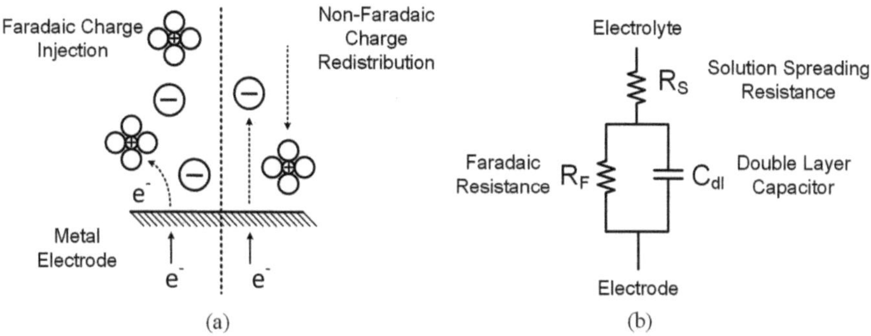

Fig. 4.1 Illustration of the electrode and electrolyte interface. (**a**) The physical representation, and (**b**) a simplified electrical circuit model

- **Non-Faradaic charge redistribution**, or polarizable mechanism, where a double layer capacitor C_{dl} is formed on the surface of the electrode, and the stimulation process involves charging and discharging the C_{dl} without direct electrons transfer [244, 245].

Modeling the electrode impedance accurately, however, is a rather complicated task [246]. A simplified linear model, modified from [244], is adopted in the analysis of this work. The model has been widely used in neural interface studies, and proves to be effective in estimating the electrode characteristics during neural recording and stimulation [78, 247, 248].

To verify the model, the impedance of two types of low-cost tungsten electrodes commonly used in this research was measured. Ten electrodes of each type were measured in 0.9 g/100 mil Sodium Chloride. The measurement results are shown in Fig. 4.2. The simplified linear model is used to fit the measurement results. The electrode with a diameter of 75 μm has an average C_{dl} of 55 nF, R_F of 7 MΩ, and a spreading resistance of 12kΩ. The electrode with a diameter of 50 μm has an average C_{dl} of 18 nF, R_F of 19 MΩ, and a spreading resistance of 20 kΩ. In general, the electrode with the smaller contact area gives a higher impedance. This measurement result gives a good insight of the electrode characteristics, and is used in the following studies.

4.2 Overview of Electrical Stimulator Design

A lot of electrical simulation techniques have been developed to produce the charges needed to recruit a neural response. An ideal stimulator triggers the desired neural response with minimum injected charges, and leaves no residue charge. However, the ideal stimulation effect is not always achievable. When it comes to the electronics design, the safety, the power efficiency, and the circuit function and performance all need to be taken into consideration. The design trade-off becomes

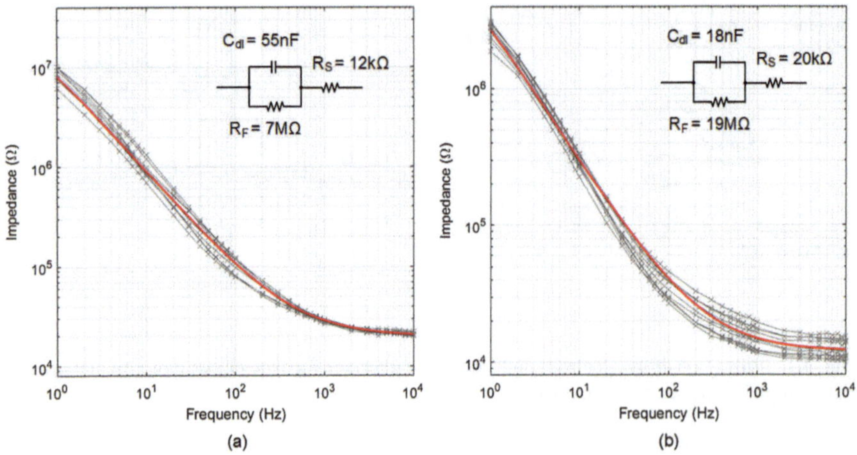

Fig. 4.2 The measured impedances of two types of custom-made tungsten electrodes. (**a**) and (**b**) shows the electrode with a diameter of 75 μm and 50 μm, respectively. Each figure shows an overlay of the measurements of ten electrodes in *gray*, and a fitting curve in *red*. The parameters of the fitting models are given

Fig. 4.3 The design considerations and trade-offs of an electrical neural stimulator. Safety, performance, and efficiency are the three main considerations

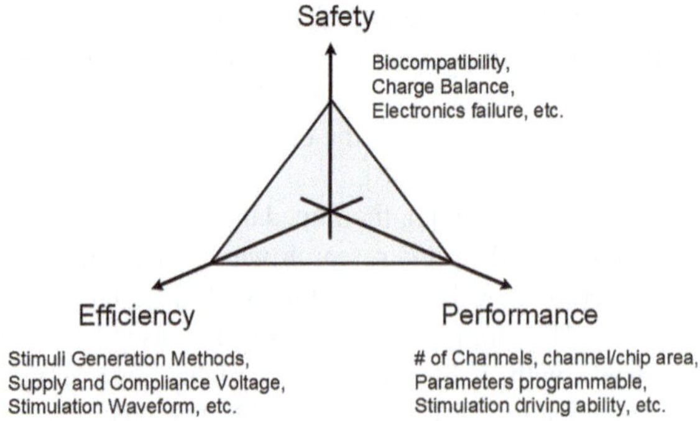

more difficult for an implantable device which requires a high channel-count, a minimum chip area, and a low power density. Figure 4.3 highlights the trade-offs in the neural stimulator design.

Firstly, a stimulator design should take the safety as the top priority. A safe long-term stimulation requires the stimulator to give a charge balanced stimulation with minimum direct current injection. Prior studies show that a direct current leakage of 100 nA will cause permanent damage to the tissue [112]. Secondly, the performance requirements of the stimulator mainly include the channel-count, the occupied silicon area, the programmability of the stimulation parameters, the stimulation current driving ability, and so on. The third consideration is the efficiency of the stimulator. The overall efficiency should consider both the power efficiency of the stimulator for generating the stimuli, and the efficiency of the generated stimuli in triggering the desired neural response. However, the latter is usually much more difficult to be quantized. Among all methods for generating the stimuli, voltage, current and charge regulation all have been reported with pros and cons. Besides, different electrode types, stimulation waveforms and parameters will all result in different efficiencies.

In summary, a very high power efficiency neural stimulator design may not give the best charge-balancing performance, a sophisticated charge matching technique may not be suitable for a high channel-count integration, while a high channel-count design may not allow all parameters to be programmable. An aggressive optimization on one dimension might cause drawbacks in the other two dimensions, and eventually, makes the overall system design non-practical. There is no best universal stimulator design, but only good designs for certain applications.

4.2.1 Methods of Stimuli Generation

In general, the neural stimuli is generated from the electronics by regulating the voltage, the current, or the total amount of charges. Essentially, the charges disturb the membrane equilibrium and evoke the neural response [244]. However, different generating methods give different control levels of the charges. Of course, a high controllability usually comes with a cost in circuit complexity and power consumption. The pros and cons of each method are summarized as below.

4.2.1.1 Voltage-Regulated Stimulation

In a voltage-regulated stimulation, a certain stimulus voltage is applied between two electrodes (or between one electrode and the common tissue ground). Current passes through the electrodes depending on the tissue and electrode impedance. Since the circuitry has no control over the total amount of injected charges, it is difficult to achieve a charge-balanced stimulation. For clinical use, the physicians will assign a proper stimulating voltage to achieve the desired neurophysiologic response in the safe range.

A voltage-regulated stimulator usually gives a high overall efficiency with simple circuitry. However, it is usually poor in the controllability in the charge injection and thus lacking safety. It has been used in applications requiring high-density integration like retinal implants, and power hungry clinical uses including pacemaker and deep brain stimulator. The circuit implementation of the voltage-regulated stimulation, and techniques for improving its safety have been reported in literature [30, 249, 250].

4.2.1.2 Current-Regulated Stimulation

In a current-regulated stimulation, a certain stimulus current is passed between two electrodes (or between one electrode and the common tissue ground). The compliance voltage between the two electrodes depends on the tissue and the electrode impedance, and is limited by the system supply voltage and the circuit topology. The total amount of injected charges can be controlled by the stimulus current and the stimulating time, and a charge balanced stimulation can be well achieved. Monophasic current stimulation [34] and biphasic current stimulation [38, 251, 252] have been reported in literature.

The current-regulated stimulation is the most widely used method in electrical stimulator designs, given its high controllability of charge injection and the high safety. However, current-regulated stimulators usually have poor efficiency. The circuit implementation of the current-regulated stimulators, and techniques for improving its efficiency have been reported in literature [34, 38, 57, 251–253].

4.2.1.3 Charges-Regulated Stimulation

In a charges-regulated stimulation, a capacitor tank is connected to one electrode and discharged to a reference electrode. The discharging current is used to excite the tissue. The circuit implementation of charges-regulated stimulator has been reported in literature [254, 255].

The charges-regulated stimulation potentially can achieve a high power efficiency as well as a good controllability of the total amount of injected charges. However, the discharge time constant is not well controlled depending on the load impedance, and the implementation of storage capacitors takes a large silicon area or has to be implemented off-chip.

4.2.2 Stimulation Waveform and Electrode Configuration

Various stimulation waveforms have been proposed and used in both research and clinical treatment. Among them, the biphasic stimulation is the most commonly used method. A typical biphasic stimulation mainly consists of a cathodic phase and an anodic phase. The cathodic phase elicits the desired physiological effect such as initiation of an action potential, and the anodic phase reverses the electrochemical processes occurring during the cathodic phase to minimize the damage. A cathodic-first order is usually preferred because the electrons move in the opposite direction of the current. Thus pulling a cathodic current is, in fact, pushing the electrons into the tissue.

A typical stimulation waveform is shown in Fig. 4.4, with key parameters labeled. A constant current-regulated method is used here for illustration, but all other methods share similar parameters. I_S and I_R are the amplitudes for the stimulation

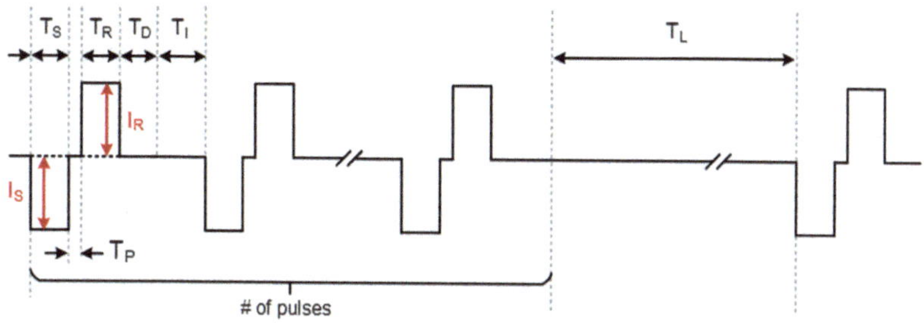

Fig. 4.4 Illustration of a typical biphasic stimulation waveform with the parameters marked. I_S: stimulation current, I_R: reversal current, T_S: stimulation phase time, T_R: reversal phase time, T_D: discharging phase time, T_P: phase interval, T_I: pulse interval, T_L: pulse group interval

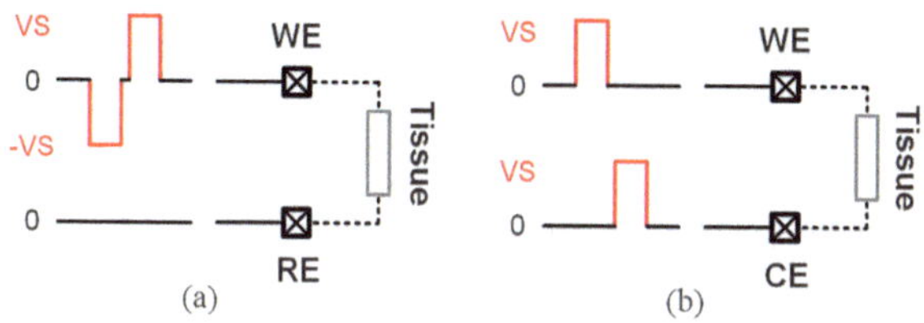

Fig. 4.5 Illustration of (**a**) monopolar and (**b**) bipolar stimulation methods. A voltage-regulated stimulation is used for illustration, but a current-regulated stimulation can be applied in the same way

and the reversal phase, respectively. If the same amplitude is used for both phases, it is referred to as a symmetrical biphasic stimulation. In some cases, a lower amplitude and longer time are preferred in the reversal phase to reduce the damage to the tissue, resulting in an asymmetrical stimulation. Both symmetrical [251, 252] and asymmetrical [38] biphasic current-regulated stimulation have been reported. T_S and T_R are the times for the stimulation and reversal phase, respectively. T_P is the interphasic delay between the stimulation and reversal phases. The interphasic delay is intentionally added for achieving a better stimulation effect. A discharging phase T_D usually follows the reversal phase to remove the residue charges. In some cases, the anodic phase is replaced by the discharging phase. In addition, T_I is the pulse interval, and T_L is the interval between the pulse groups. The terminologies are used consistently in the following study.

It should be noticed that monophasic/biphasic stimulation can be confused with monopolar/bipolar stimulation. However, they are different terminologies and are not directly related. In a monopolar stimulation, both cathodic and anodic phases are generated from a single electrode. In a bipolar stimulation, the cathodic and anodic phases are generated from a pair of electrodes. Both monopolar and bipolar configurations can be used to perform monophasic and biphasic stimulation. Figure 4.5 illustrates the typical electrode waveforms of the two configurations in generating a biphasic stimulation.

The monopolar stimulation is widely used in a high-density electrode array, in which case it delivers stimulus with respect to a common reference electrode with low impedance. Bipolar stimulation has a better-guided stimulus orientation than the monopolar stimulation, at the cost of a more complicated channel selection. The bipolar stimulation also favors a single-supply system design since it doubles the equivalent compliance voltage range [253]. Ideally, both monopolar and bipolar configurations can achieve a charge-balanced stimulation.

It should be noticed that even though we have been discussing rectangular stimulation waveforms, non-rectangular waveforms have also been proposed in literature [255–257]. Using a non-rectangular waveform, like exponential current stimuli, may achieve a better stimulation effect or a better power efficiency. Of course, these designs usually come with a cost of the circuit and control complexity. More importantly, it may cause difficulties in achieving a charge balance, which will be discussed in the following section.

4.2.3 Methods for Charge Balancing

The importance of charge balancing cannot be overemphasized. The building-up of the excess charges, even slowly, might cause toxic effects and lead to a permanent damage. One of the most commonly used traditional methods is to place a blocking capacitor in series with the stimulating electrode. The blocking capacitor limits the total charges, preventing a direct current injection. However, the capacitor cannot be too small, otherwise it will limit the output compliance voltage. For a typical functional electrical stimulation, these capacitors are in the order of tens of nanofarads to a few microfarads [256, 258–260]. The physical dimension of these capacitors is usually prohibitively big to be integrated on a silicon chip, especially in a high channel-count design. Various techniques have been developed to achieve the charge balance with and without the blocking capacitors. This section reviews the pros and cons of these techniques.

4.2.3.1 Matching and Calibration

Ideally, a charge balance can be achieved if the amount of charges of the stimulation and the reversal phases are the same. In a typical current-regulated stimulation, the amount of charges depends on the stimulation current amplitude and pulse width. The mismatch between the anodic and cathodic currents without calibration can be 2% even with careful matching in the layout [261]. This current mismatch might lead to a significant charge error in a heavy-duty stimulation. For that reason, a lot of research has been conducted with a focus on matching the stimulation and reversal currents, in order to achieve the net-zero change. Several important matching techniques are reviewed as follows.

J. Sit et al. from Massachusetts Institute of Technology proposed blocking capacitor free change-balanced stimulator design in 2007 [248]. The design used a dynamic current balancing method to achieve a current balancing. The work paid a special attention to the switch leakage and the loop stability in the dynamic current mirror. The reported DC current error was 6 nA.

K. Song et al. from Korea Advanced Institute of Science and Technology proposed a DC-balanced adaptive stimulator in 2012 [262]. The design proposed a current source mismatch compensation method. A precise current balance was achieved by sampling the mismatch current, and making a compensation accordingly. The challenge in this work was the requirement of a large time-constant sample and hold (S/H) circuit. The S/H circuit would need to hold the mismatch current ΔI for up to 0.5 s without variation. The reported current mismatch was less than 10 nA.

M. Monge et al. from California Institute of Technology proposed a high-density self-calibration epiretinal prosthesis design in 2013 [263]. This work used a fully digital calibration technique to match the biphasic currents during the stimulation. A multi-point calibration scheme was proposed, which included a 5-point calibration process for each driving site. With the help of the full-range calibration, the stimulation can perform arbitrary waveform stimulation. The reported mismatch was 2.24%.

In summary, the matching technique can be implemented in either analog or digital domains. In the analog domain, it requires a feedback or a large time-constant storage unit. In the digital domain, it requires an on-chip memory to save the mismatch parameter. The analog matching is attractive if only a single-point matching is needed. If a full-scale calibration is desirable, the digital calibration is more suitable.

4.2.3.2 Passive and Active Discharge

If the current matching is not sufficient in achieving the net-zero charge requirement, an additional discharge phase is used to remove the residue charges. The discharge can be as simple as shorting the stimulation electrode to a common or a reference electrode, which is referred to as passive discharging. However, the disadvantage of passive discharging is that the discharge current depends on the load and electrode impedance, which cannot be well controlled. If the impedance is too low, the discharge current might be too large, which might damage the tissue. Thus additional current limit circuit is required [252]. On the other hand, if the impedance is too high, the time for discharging might be too long, and the residue charges might not be able to clear before the next stimulus cycle. Residue charges will accumulate in this case.

K. Sooksood et al. from the University of Ulm proposed an active charge balance method in 2010 [264]. In the active approach, the residue charges or the electrode potential is monitored by active circuits, and additional discharge circuits are used to maintain the potential within a safe range in a closed-loop manner. One method to reduce the residue charges is by inserting short pulses after the reversal phase [264, 265]. In addition, E. Noorsal et al. from the same group proposed to regulate the residue charges by using DC biasing current [266]. In this work, a safe window is defined approximately as 100 mV (for a Pt black electrode). The electrode potential is compared with the safe windows right after the stimulation, and the biasing current sources can be adjusted accordingly.

In summary, passive and active discharge and charge cancellation techniques can be used to remove the residue charges after the stimulation. Passive discharging is simple but has limited control of the discharging current and time. Active discharging by continuously monitoring the residue charges can be more effective, but takes dedicated circuit design and silicon area.

4.3 Design of a General–Purpose Stimulator

This section describes the design of a 16-channel general–purpose neural stimulator. The motivation of this work is to design a fully programmable stimulator for various applications. The stimulator can perform monopolar or bipolar, monophasic or biphasic, symmetrical or asymmetrical constant-current, charge balanced stimulation. All of the parameters for the stimulator are programmable. The output current is from 0 to $\pm 255\,\mu$A in the low-current mode, and 0 to ± 2 mA in the high-current mode. The design has been fabricated in IBM 180 nm technology, and occupies a silicon area of $810\,\mu\text{m} \times 290\,\mu\text{m}$, excluding the IO pads. The system architecture, circuit implementation, and experimental results are presented in this section. Some of the figures and tables presented in this section were originally published in [57] ©IEEE. Reused, with permission.

4.3.1 Architecture of the Stimulator

The overall block diagram of the stimulator is shown in Fig. 4.6. The stimulator includes an analog part designed in 1.8/5 V and a digital part designed in 1.8 V. The stimulator integrates four independent driving sites. Each site includes: (1) a DAC generating a reference for the output current; (2) a current driver consisting of current sink and source output stages with high output impedance; (3) a 1:4 demultiplexer to support four channels and provide near-simultaneous stimulation; and (4) level-shifters to interface the low-voltage digital control signal with high voltage switches.

The analog part can be shut down when no stimulation is to be delivered to save the power consumption. This is especially important for low-power design since the stimulator output stage has high voltage and current. The digital part

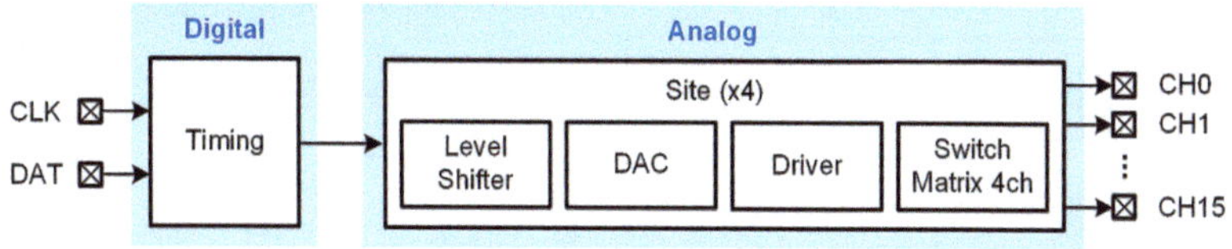

Fig. 4.6 The block diagram of the neural stimulator. The stimulator consists of an analog part and a digital part. The stimulator integrates four independent driving sites, and each site demultiplexes to four channels

can be configured via a custom designed two wire protocol. The parameters of the timing generation module in each stimulator site can be programmed individually. In addition to the regular mission modes, the stimulator can be configured to output a continuous current in order to test the DAC and the output stage.

4.3.2 Circuit Implementation

The circuit schematic of one stimulator driving site is shown in Fig. 4.7. A 6-bit current mode DAC is used to generate the reference for the stimulation current. A typical binary weighted current source array is used in the DAC [267]. The transistors are sized for a 6-bit accuracy [195]. Common-gate transistors are used to improve the matching and increase the output impedance. The DAC is designed in thin-oxide devices and powered at 1.8 V. A thick oxide transistor is cascaded in the output current path to reduce the overdrive voltage stress from the following stage. Thick oxide transistors also have a low leakage current which allows a complete shutdown of the DAC.

The output stage is designed using thick-oxide devices with a supply voltage of 5 V. Regulating amplifiers are used to achieve a high output impedance. A PMOS input folded-cascode amplifier is used in the current sink, and an NMOS input folded-cascode amplifier is used in the current source. These amplifiers are disabled when the stimulator is in the idle mode to reduce the power dissipation.

Level shifters are used to convert the low-voltage digital control signal to high voltage switches. The circuit schematic of the level shifter is shown in Fig. 4.8. It should be noticed that even some dynamic level shifter can achieve a high switching frequency and a lower power consumption, the risk of undetermined state may cause direct stimulation current leakage to the tissue, thus is not used in this design.

The command and parameter registers of the stimulator are listed in Table 4.1. An 8-byte register is used for saving the configuration. The definition of the parameters has been introduced in Sect. 4.2.2 (Fig. 4.4). A header of 0x57 is used to verify the start of the command in case of data loss or corruption.

To perform a monopolar stimulation, one electrode is activated at a time, and the stimulation current is passed between the selected electrode and the ground electrode. The timing of the monopolar stimulation and the corresponding control

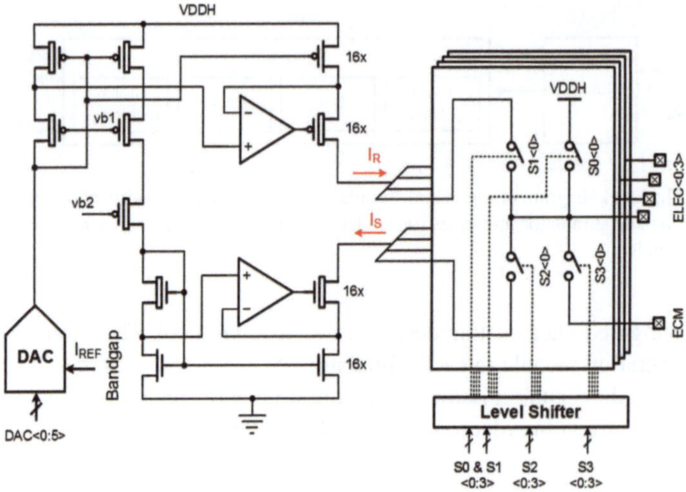

Fig. 4.7 The circuit schematic of the proposed multi-mode stimulator site. Each site consists of: (1) a current-mode DAC which generates a reference for the output current, (2) a current driver including current sink and source output stages with high output impedance, and (3) high voltage switches with level-shifters. Each site demultiplexes to four channels, and provides near-simultaneous stimulation

Fig. 4.8 The circuit schematic of the level shifter

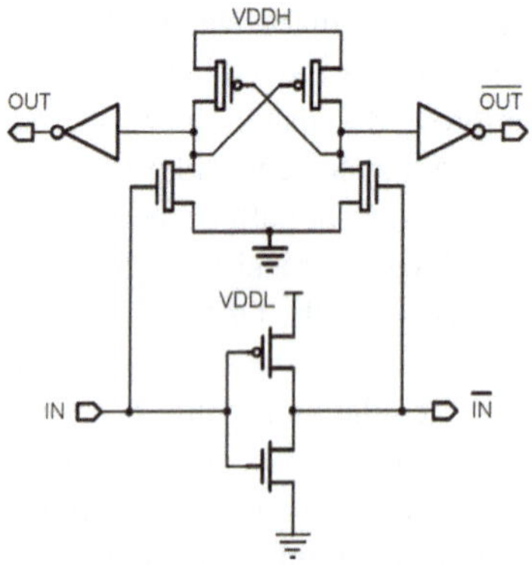

switches are shown in Fig. 4.9. The timing parameters are defined as in Fig. 4.4, and the switching signals are defined as in the circuit schematic in Fig. 4.7. XXX means the DAC can be any value, allowing the DAC to generate the reference value for another channel.

To perform a bipolar stimulation, two electrodes selected from the 16 channels are configured to work as the cathodic and anodic electrodes, and the stimulation current is passed between them. The timing of the bipolar stimulation and the corresponding control switches are shown in Fig. 4.10. The parameter definitions are the same as in Fig. 4.9. It should be noticed that the electrode X and Y can be in the same driving site or two different driving sites. The DAC # and DAC $ will be the same DAC if X and Y are in the same site.

Table 4.1 Stimulator command and parameter registers

CMD Addr.	Description	Value
Word 00 [0:1]	Header	0x57
Word 00 [2:3]	Active module position	0x00–0x11
Word 00 [4:5]	Stimu mode	00—biphasic, monopolar 01—biphasic, bipolar 10—monophase
Word 00 [6]	Stim. power	0—low (FS: ± 255 μA) 1—high (FS: ± 2047 μA)
Word 00 [7]	Stim. on/off	0—stimulator OFF 1—stimulator ON
Word 01 [0:1]	Stim. position	0x00–0x11
Word 01 [2:7]	Stim. magnitude (I_S)	0x01–0x3F
Word 02 [0:1]	Counter position	0x00–0x11
Word 02 [2:7]	Counter magnitude (I_R)	0x01–0x3F
Word 03 [0:7]	Stim. phase (T_S)	1–256 μs
Word 04 [0:7]	Reserval phase (T_R)	1–256 μs
Word 05 [0:3]	Phase interval (T_P)	1–16 μs
Word 05 [4:7]	Discharge phase (T_D)	16–1024 μs
Word 06 [0:3]	# of pulses	1–16
Word 06 [4:7]	Pulse interval (T_I)	640 μs–10 ms
Word 07 [0:7]	Pulse group interval (T_L)	8 ms–2 s

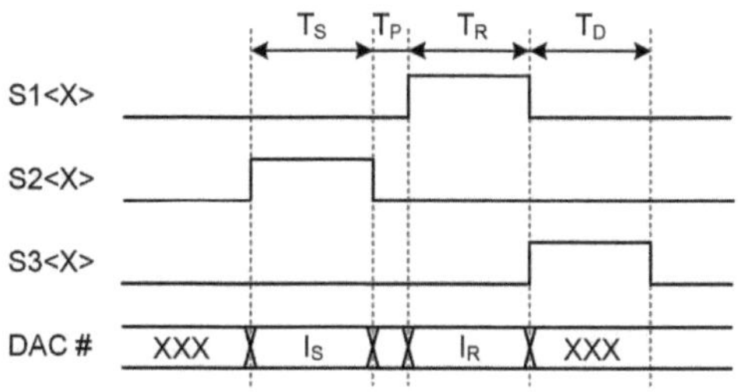

Fig. 4.9 The timing for generating a monopolar stimulation at the electrode X. The DAC # and the electrode X locate in the same driving site (I_S: stimulation current, I_R: reversal current, T_S: stimulation phase time, T_R: reversal phase time, T_D: discharging phase time, T_P: phase interval. The DAC can be any value in the XXX state)

4.3.3 Measurement Results

The design has been fabricated in IBM 180 nm CMOS technology. The occupied silicon area is 810 μm × 290μm. The microphotograph and the layout are shown in Fig. 4.11. The major building blocks are highlighted in the figure.

Several bench tests have been conducted to verify the function and evaluate the performance of the designed stimulator. Both static and dynamic performance are important in a stimulator design. Figure 4.12 shows the measured output currents

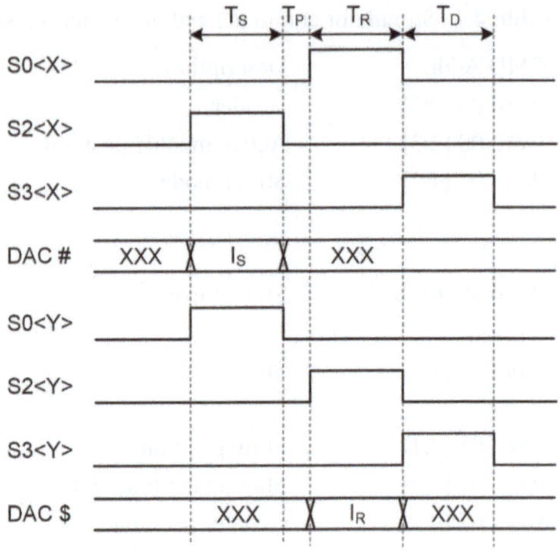

Fig. 4.10 The timing for generating a bipolar stimulation between the electrode X and Y. The DAC # and DAC $ are in the site where the electrode X and Y locates, respectively

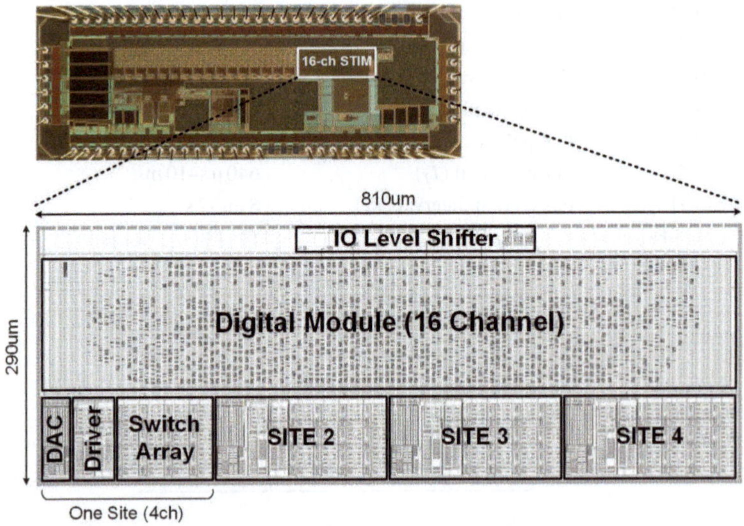

Fig. 4.11 The microphotograph and the layout of the neural stimulator. The major building blocks are highlighted in the layout

from the current sink and current source, with several digital input codes. The results show a large compliance voltage range with a circuit overhead less than 264 mV, corresponding to 5.28% of the supply voltage.

Figure 4.13 shows the measured output currents from the current source and sink of the stimulator output stage. The non-linearity of the source and sink current is 0.31% and 0.37%, respectively. The result shows a good matching between the source and sink, which is 1.29% without calibration. No additional analog calibration is used in this work, but a digital calibration in the digital code can be implemented for a better matching. The discharging phase should always be used to avoid charge accumulation in this case.

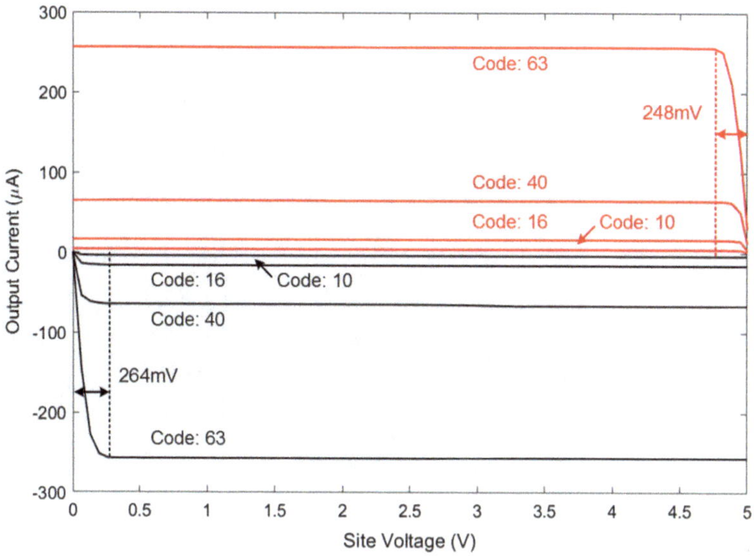

Fig. 4.12 The measured stimulator output current versus output voltage

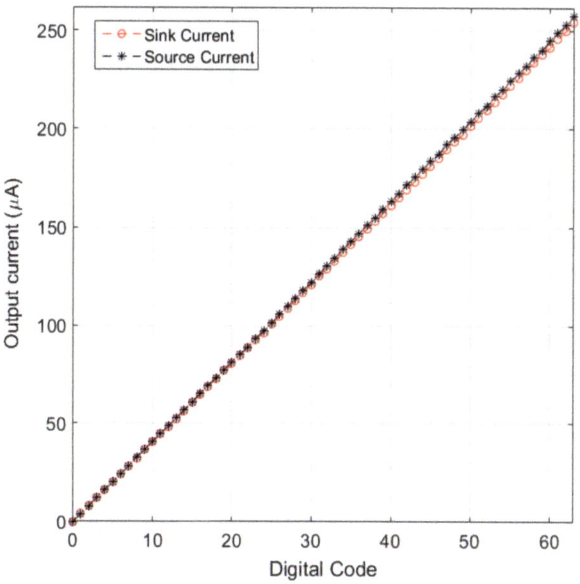

Fig. 4.13 The measured output currents from the current source and current sink of the stimulator output stage. The non-linearity of the source and sink current is 0.31% and 0.37%, respectively

The stimulation process was measured in 0.9 g/100 mil Sodium Chloride. The measured simultaneous stimulation outputs from four independent channels are shown in Fig. 4.14. Different pulse train interval times were intentionally used for each channel, showing the ability for this chip to drive multiple-channel simultaneous stimulation with different parameters.

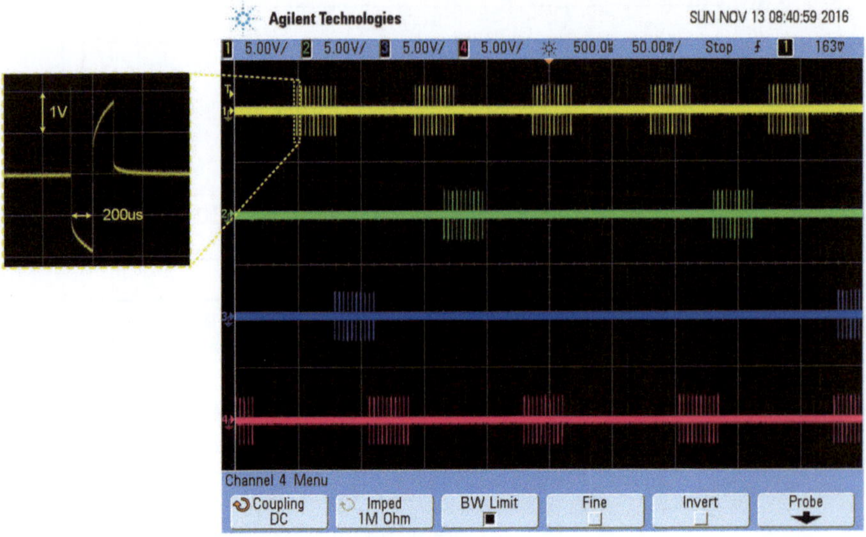

Fig. 4.14 The measured simultaneous stimulation output from four independent channels. The boxed window shows the measurement of a single pulse in high resolution

4.4 An Energy Efficient Net-Zero Charge Neural Stimulator

4.4.1 Introduction

As discussed in Sect. 4.1.2, two primary mechanisms occur at the interface between the electrode and the physiological medium during an electrical stimulation: the direct Faradaic charge transfer and the capacitive charge redistribution [244]. The Faradaic charge transfer usually involves reduction and oxidation processes, which may create damaging chemical species and dissolve the electrodes. So it is critical to avoid the onset of these reactions. A reversal phase is commonly used after the stimulation phase to reverse the electrochemical processes. However, it is not always possible to avoid the irreversible charge injection, resulting in a certain amount of unrecoverable charges during the stimulation [244].

In order to achieve an overall net-zero charge, a lot of techniques have been developed, as reviewed in Sect. 4.2.3. However, previous works have been exclusively focusing on matching the stimulation and reversal currents and charges, ignoring the unrecoverable charge injection during the stimulating phase. This work proposes a new stimulation strategy to achieve the net-zero charge by monitoring the residue charges directly on an inserted capacitor. Using the proposed method, over-reversal can be avoided. In addition, a perfect matching between the stimulation and reversal currents is not required, and an arbitrary stimulation waveform can potentially be performed without calibration.

As reviewed in Sect. 4.2.1, voltage-regulated [268], charges-regulated [269], and current-regulated [248, 262, 263] stimulation methods have been reported in literature. In summary, the voltage-regulated stimulation method has the highest efficiency, but it is difficult to control the amount of injected charges [251]. The charge-regulated stimulation limits the total amount of charges by discharging a capacitor tank, but the capacitors cost a large silicon area, and the discharging time cannot be precisely controlled. The current-regulated stimulation has a high controllability of the charge injection, thus it is the most widely used method. However, the traditional current-regulated method suffers from a low power efficiency [262]. In this work, an adaptive driving voltage is designed with a feedback control scheme to improve the power efficiency. The design also enables a constant low supply voltage operation for all active circuits except the driving voltage.

The remaining of this section is organized as follows. Section 4.4.2 highlights the innovations proposed in this work. Section 4.4.3 describes the system architecture and the circuit implementation, with emphasis on the stimulator output stage and the feed-forward compensation comparator design. Experimental results are presented in Sect. 4.4.4. Some of the figures and tables presented in this section were originally published in [253] ©IEEE. Reused, with permission.

4.4.2 Motivation and Innovation

4.4.2.1 Net-Zero Charges Stimulation

As discussed in Sect. 4.2.2, biphasic stimulation is the most commonly used stimulation waveform. During a biphasic stimulation, a stimulation phase first elicits the desired physiological effect (e.g., initiation of an action potential), and after an optional interphase delay, a reversal phase is used to reverse electrochemical processes [244]. The threshold current required to initiate the neural response decreases with an increasing stimulation pulse width. The threshold and pulse width relation, which can be experimentally quantified, is usually presented as a strength-duration curve [270]. Although it is not a physiological requirement to design the reversal current equals the simulation current, a same amplitude is commonly adopted in the circuit design for an easier matching. Although high current matching accuracies have been reported in literature [248, 262], these methods often ignore the fact that inevitable charge diffusion. The process is illustrated in Fig. 4.15. The traditional designs assume the ideal charge curve plotted in the dashed line, but the practical charge curve deviates from the ideal curve due to the irreversible reaction and the chemical products diffusion. So if taking the irreversible reaction and the chemical products diffusion into account, even perfect matched cathodic and anodic currents will still leave residual charges. These residual charges will accumulate in a simulation pulse train, resulting in a more serious damage if a discharge procedure is not properly assigned. This work addresses this problem by monitoring the residue charges on an inserted blocking capacitor. The reversal phase terminates when a net-zero charge point is reached, as illustrated in Fig. 4.15b. In

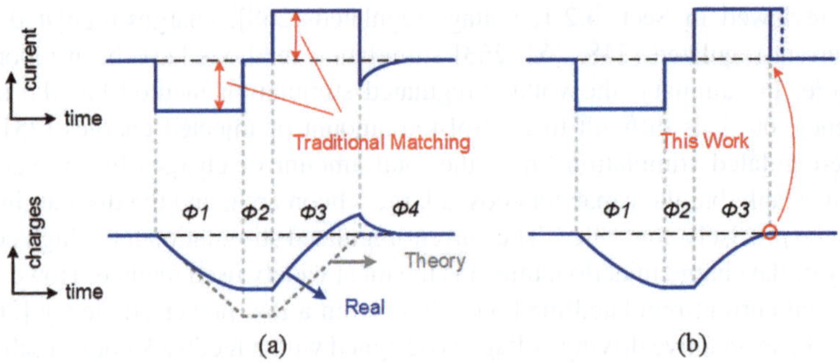

Fig. 4.15 (**a**) The traditional charge-balancing method matches the stimulation and reversal currents. The ideal charge curve on the electrode is plotted in a *dashed line*. The practical charge curve deviates from the ideal curve due to the irreversible reaction and the chemical products diffusion. (**b**) This work terminates the reversal phase based on the monitoring the net-zero charge crossing point. $\phi 1$: stimulating phase, $\phi 2$: interval phase, $\phi 3$: reversal phase, $\phi 4$: discharge phase

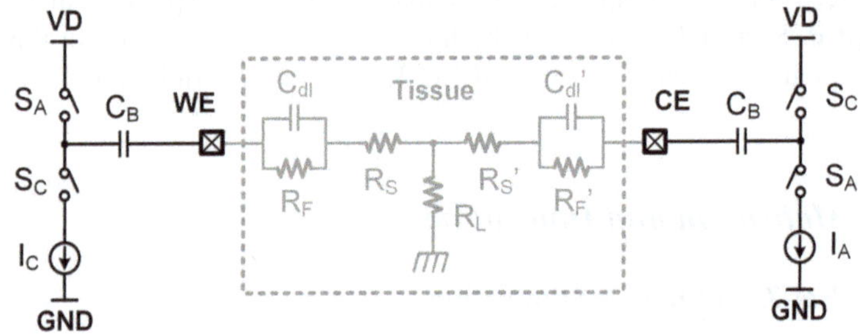

Fig. 4.16 The model for simulating the effects of non-ideal charge diffusion

this way, the systematic over-reversal in traditional methods can be avoided, and a perfect matching between the stimulation and reversal currents is no longer required.

To further illustrate the effects, a simplified linear simulation model is established using ideal components, including the current sources. A typical single-supply bipolar stimulation configuration is used in this simulation [256, 271]. The circuit schematic of the simulation model is shown in Fig. 4.16. A resistor R_L is used to mimic the impedance between the stimulation location and the tissue ground. The electrode parameters measured in vitro (Fig. 4.2) are used in this simulation. 500 kΩ, 1 MΩ, and 5 MΩ resistors are used as R_L. It should be noticed that electrochemical reactions are involved in practice, thus the charge reduction is more complicated and is not linearly depending on the electrical potential. Figure 4.17 shows the simulated voltage across the blocking capacitor C_B. As R_L decreases, the over-reversal is increasingly notable. Figure 4.18 shows the same simulation setup but in a pulse train without discharging. Even with the ideal current sources with equal amplitude and pulse width, the charges still build up on the electrode quickly. The residual charges may cause permanent damage to the tissue in a long term.

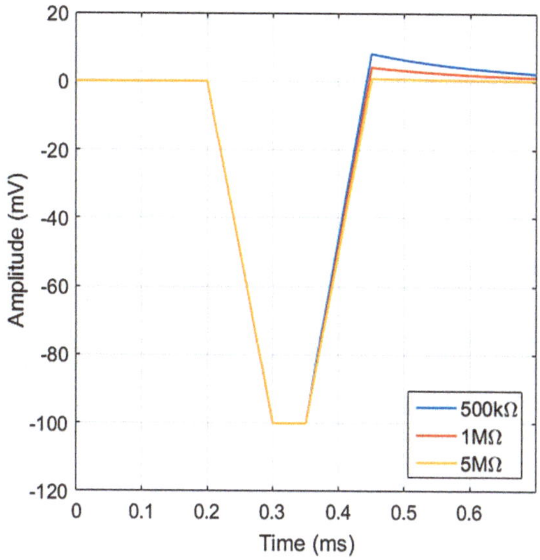

Fig. 4.17 The simulation of the voltage across the blocking capacitor C_B (Fig. 4.16) during a symmetrical single pulse. Ideal current sources with equal stimulation current amplitude and time are used. 500 kΩ, 1 MΩ, and 5 MΩ resistors are used to mimic the impedance between the stimulation location to the tissue ground

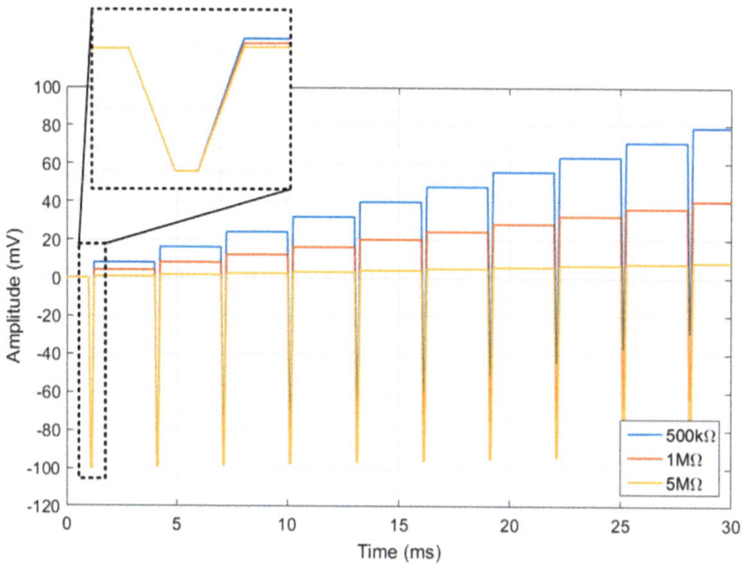

Fig. 4.18 The simulation of the voltage across the blocking capacitor of a stimulation pulse train without discharging. 500 kΩ, 1 MΩ, and 5 MΩ resistors are used to mimic the impedance between the stimulation location to the tissue ground. The charges build up even using the ideal current sources with equal amplitude and pulse width

It should be noticed that a blocking capacitor is commonly used to ensure the safety by preventing a direct current injection as well as limiting the maximum net charges. Thus this work doesn't require a major extra configuration change from the conventional stimulator designs.

4.4.2.2 Adaptive Driving Voltage

In the simplified linear model, the compliance voltage for a charge balanced biphasic stimulation can be expressed as:

$$V_C = 2R_S I_S + \frac{I_S T_S}{C_{\mathrm{dl}}} \tag{4.1}$$

In conventional designs, the supply voltage of the stimulator needs to be higher than the peak compliance voltage with headroom to avoid cut-off. However, the uncertainty and the drifting of the electrode impedance makes it difficult to predict the peak compliance voltage. So the supply voltage usually needs to be over-designed to guarantee a sufficient compliance voltage. As a result, a lot of power is wasted in the circuitry overhead instead of on the load tissue. The overall efficiency of the system can be expressed as:

$$\eta = \frac{P_{\mathrm{load}}}{P_{\mathrm{load}} + P_{\mathrm{circuits}}} = \frac{I_{\mathrm{stim}}^2 Z_{\mathrm{Tissue}}}{(I_{\mathrm{stim}} + I_{\mathrm{circuits}})V_{\mathrm{supply}}} \tag{4.2}$$

In this work, an adaptive driving voltage instead of a constant high supply voltage is used for improving the power efficiency. In contrast to the conventional output stage designs which include both current sink and source, this design only uses a current sink. The working electrode (WE) sinks the current, and the counter electrode (CE) only needs to generate a potential difference with respect to the WE [256, 271]. This operation is illustrated in Fig. 4.19a. As a result, all circuits for the WE can be designed in a low supply voltage, while only the driving voltage of the CE needs to be boosted. In the simplified linear model, the required counter electrode voltage V_{CE} can be expressed as:

$$V_{\mathrm{CE}} = V_{\mathrm{WE}} + 2\frac{I_{\mathrm{stim}}\Delta t}{C_{\mathrm{dl}}} + 2\frac{I_{\mathrm{stim}}\Delta t}{C_{\mathrm{bk}}} + I_{\mathrm{stim}} R_{\mathrm{tissue}} \tag{4.3}$$

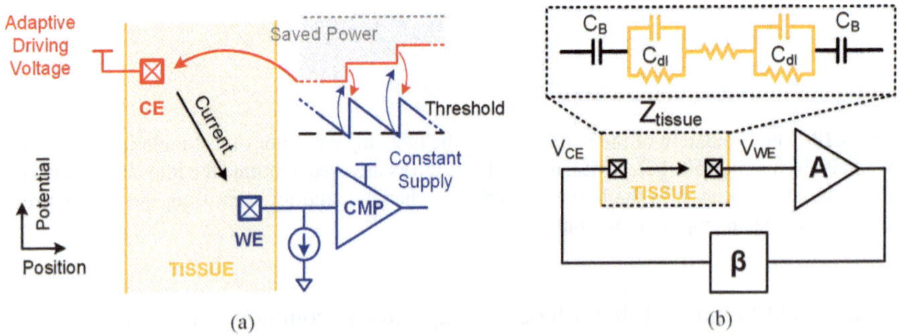

Fig. 4.19 Illustration of the adaptive driving voltage stimulation

where R_{tissue} is unknown and varies from site to site. Assume the minimum voltage for the current sink is V_{sink}, so the WE's potential should always be higher than V_{sink}. This gives an opportunity to design a feedback control scheme for the driving voltage by monitoring the WE's potential. A boosting converter for generating the driving voltage can be designed with continuous tuning or with several discrete output levels. When V_{WE} is lower than a pre-defined threshold V_{th} ($V_{th} > V_{sink}$), a 1-bit digital signal is generated to let the boosting converter's output increase in one step. The process can also be understood as a typical feedback control system, as shown in Fig. 4.19b. It should be noticed that the system is always stable if the boosting converter's output is set to be the minimum value at the beginning, and only changes in one direction (increasing).

4.4.2.3 Arbitrary Channel Configuration

In order to achieve the best stimulation performance, the ability to perform stimulation in an arbitrary location and direction from the implanted electrode array is very helpful. It can fully take advantage of the high-density electrode array. However, conventional stimulator design with the current source and sink matching technique can hardly realize an arbitrary channel configuration, since the matching is usually designed to perform between pre-defined electrode pairs. This is illustrated in Fig. 4.20a. With the proposed stimulation technique, a perfect matching is no longer required. Thus an arbitrary channel configuration of the working and counter electrodes is feasible, and more precision stimulation pattern can be generated from a limited number of electrodes, as illustrated in Fig. 4.20b.

4.4.3 Circuit Implementation

4.4.3.1 System Architecture

The block diagram of the proposed stimulator system is shown in Fig. 4.21. The system contains six driving sites. Each site can be configured as the working electrode driver (WE mode) or the counter electrode driver (CE mode). The

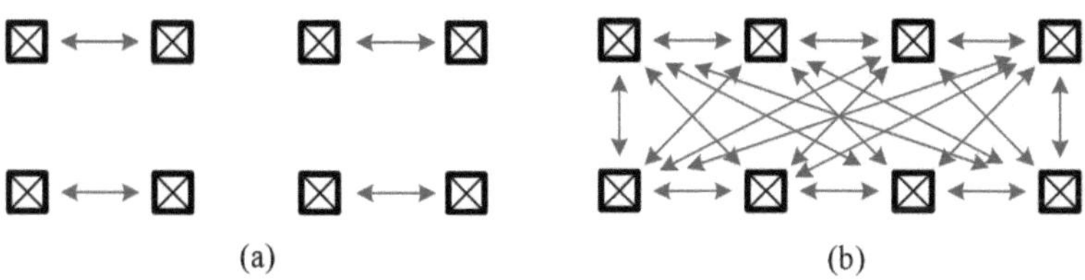

Fig. 4.20 The channel configuration of (**a**) the traditional stimulation, and (**b**) the proposed work. Arbitrary channel configuration is feasible in the proposed work without pre-calibration

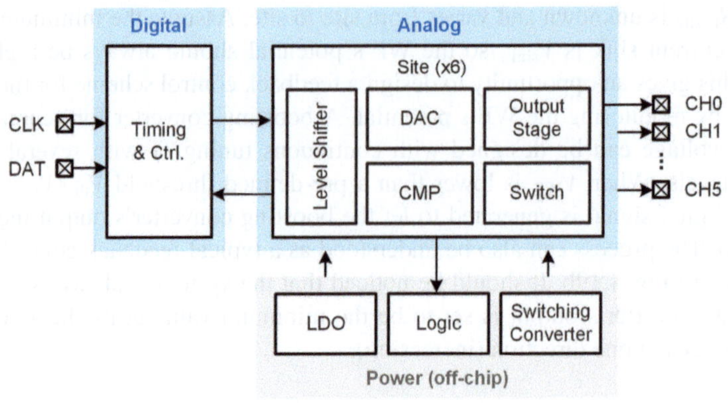

Fig. 4.21 The block diagram of the net-zero charge neural stimulation system. The system consists of an analog core, a digital module, and off-chip power management units

stimulation and reversal currents are generated by reversing the current path between the WE and CE. Each site contains a current sink and a 6-bit current mode DAC. Two comparators with different specifications are integrated into the driving site. The low-speed comparator is used to monitor the electrode voltage. The high-speed comparator is used to detect the zero-net charge crossing point, and to terminate the reversal phase. The functions of the digital module include: (1) output mode selection, (2) output stage switch control, (3) DAC and comparator configuration, (4) supply voltage adjusting request generation, and (5) feed-forward comparator calibration.

The working flow of the proposed stimulation strategy is shown in Fig. 4.22. After all stimulation parameters are received, the system first enables the selected WE and CE driving sites, and the DACs of the corresponding output stages are configured. The stimulation phase starts first, with a timer controls the stimulation time. The low-speed comparator monitors the compliance voltage and generates driving voltage adjustment signal accordingly. The stimulation phase terminates by the timer, and then the interphase timer starts. After the interphase, the reversal phase starts. The comparators are used to detect the net-zero charge point. The low-speed comparator first performs a coarse detection, and it triggers a high-accuracy high-speed comparator when the voltage gets close. The reversal phase stops by the output of the high-speed comparator. After the reversal phase, an optional discharge phase can be used to clear the capacitor and amplify the residue voltage across the blocking capacitor. This phase is usually used in a training mode. If the residue voltage is beyond the safe range, a calibration value is adjusted and stored in the register.

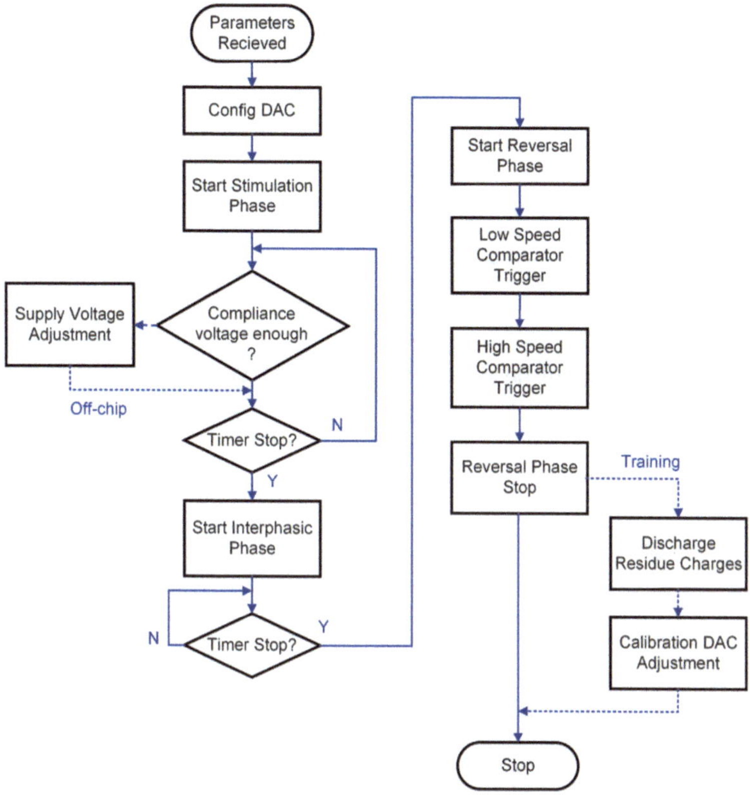

Fig. 4.22 The working flow chart of the proposed stimulation strategy

4.4.3.2 Output Stage with Dynamic Element Matching

An output stage with a high voltage compliance and a high output impedance is critical for a neural stimulator. A transconductance amplifier (OTA) with series–series feedback can make a current generator with a high output impedance, as shown in Fig. 4.23a. The output current can be controlled by adjusting the reference voltage or the tail resistor. The resistor can be implemented using a transistor biased in the linear region [271], or a batch of transistors biased in the deep triode region [256], as shown in Fig. 4.23b and c, respectively. A voltage mode DAC is used to bias the transistor to control the output current. However, the threshold voltage variation appears to be a problem. Matching is especially important when the circuits are implemented for driving a micro-electrode array contains hundreds of channels. In this work, a current-mode DAC with dynamic element matching is used to address this problem.

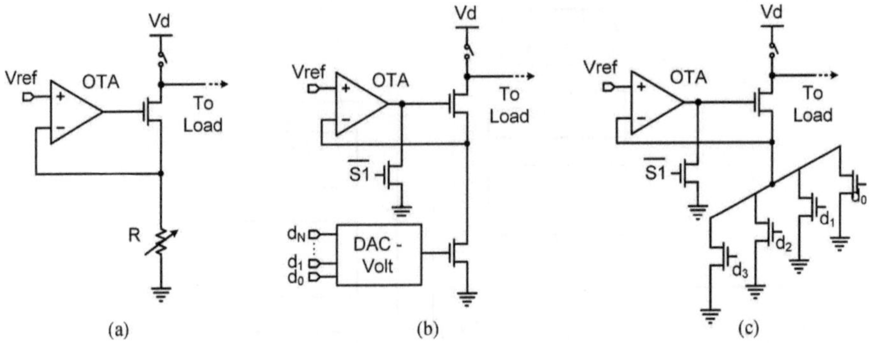

Fig. 4.23 Circuit schematic of (**a**) basic current generator, (**b**) output stage with voltage-controlled transistor, modified from [271], (**c**) output stage with digital-set DAC, modified from [256]

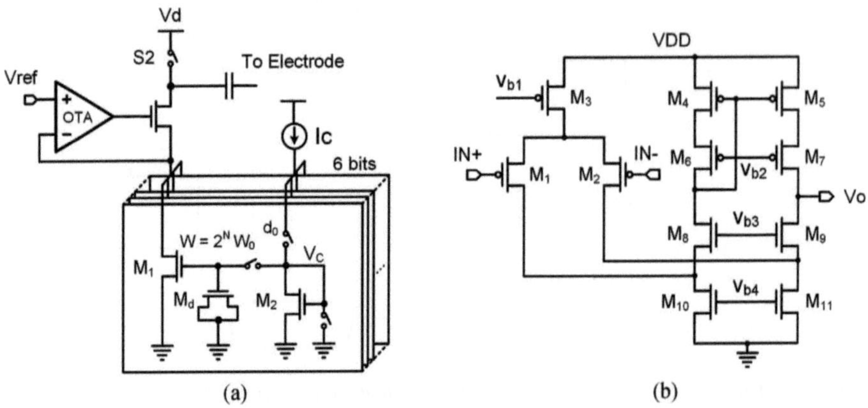

Fig. 4.24 (**a**) The proposed output stage with current set current-mode DAC. (**b**) The OTA used in the work

Figure 4.24 shows the simplified circuit of the proposed output stage. A 6-bit binary weighted DAC is used to generate the output current. The transistor M1 is one-bit of the DAC, a dummy cell M2 is set on the side of M1. Instead of using digital signal or voltage-mode DAC to bias M1, the gate voltage of M1 is generated by charging M2 using a reference current I_C. Thus, the gate voltage of the M2 is given by:

$$V_C = \sqrt{\frac{2I_C L_2}{\mu C_{\text{ox}} W_2}} \tag{4.4}$$

And the drain current of M1 can be calculated as:

$$
\begin{aligned}
I_{d0} &= \frac{\mu_n C_{\mathrm{ox}}}{2}\frac{W_1}{L_1}\left[2(V_C - V_{\mathrm{th1}})V_{\mathrm{ref}} - V_{\mathrm{ref}}^2\right] \\
&= \frac{\mu_n C_{\mathrm{ox}}}{2}\frac{W_1}{L_1}\left[2\left(\sqrt{\frac{2I_C L_2}{\mu C_{\mathrm{ox}} W_2}} + (V_{\mathrm{th2}} - V_{\mathrm{th1}})\right)V_{\mathrm{ref}} - V_{\mathrm{ref}}^2\right] \\
&\approx \frac{\mu_n C_{\mathrm{ox}}}{2}\frac{W_1}{L_1}\left[2\sqrt{\frac{2I_C L_2}{\mu C_{\mathrm{ox}} W_2}}V_{\mathrm{ref}} - V_{\mathrm{ref}}^2\right]
\end{aligned}
\tag{4.5}
$$

where V_{ref} is the reference voltage set by the OTA, V_{th1} and V_{th2} are the threshold voltages of M1 and M2, respectively. Thus the output current only depends on the threshold difference in the local area, which is much smaller than the variation itself. The output current of the DAC is:

$$
\begin{aligned}
I_{\mathrm{dac}} &= I_{d0} + I_{d1} + I_{d2} + I_{d3} + I_{d4} + I_{d5} \\
&= \sum_{n=0}^{5} 2^n d_n (A\sqrt{I_C} + B)
\end{aligned}
\tag{4.6}
$$

where d_n's are input digital codes, and A, B are constants controlled by design parameters and by V_{ref}. A 100-run Monte-Carlo simulation including both process corners and mismatch of the different output structures are shown in Fig. 4.25. The result shows that even with the worst variation, the proposed current-set dynamic element matching method reduces the output variation significantly. It should be noticed that the Monte-Carlo simulation doesn't take the adjacent layout placement into account, so the variation of the proposed method should be better in practice.

4.4.3.3 Feed-Forward Error Compensation Comparator

In the output stage, two comparators are connected to the blocking capacitor. A low-power, low-speed op-amp based continuous time comparator (LS comparator), and a high-speed high accuracy comparator (HS comparator). Both comparators have a shut-down option for saving the power consumption.

The HS comparator is designed with a three-stage preamplifier and a dynamic latch. The offset of the comparator originates from the imperfect symmetrical layout and the variation during the fabrication. The CMOS latches implemented with small devices have larger offsets compared to the pre-amplifiers. The output offset auto-zeroing circuits are employed for the three-stage pre-amplifier to suppress the offsets in this work. The sources of the comparator delay include the charging time of the blocking capacitor, the converging of the pre-amplifier and the latch, as illustrated in Fig. 4.26.

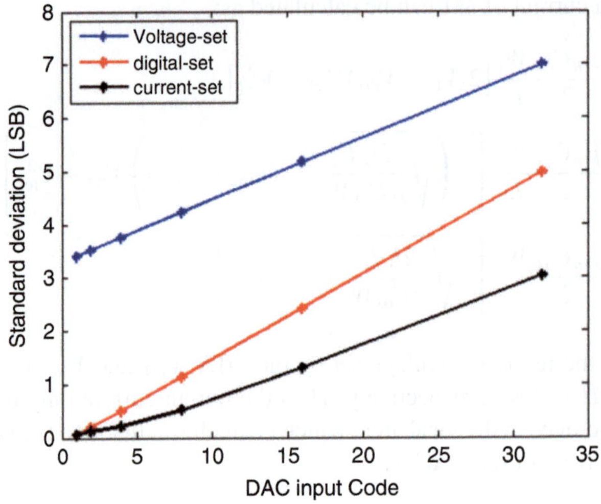

Fig. 4.25 100 runs Monte-Carlo simulation of the different output stage architectures with mismatch and process variation

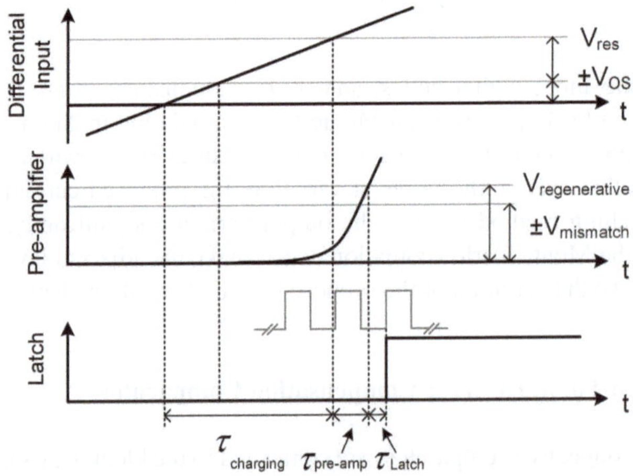

Fig. 4.26 Analysis of the comparator delay for determining the zero-crossing point of the blocking capacitor

The total time error for this comparator can be expressed as:

$$\tau_{total} = \tau_{charging} + \tau_{pre\text{-}amp} + \tau_{latch} \tag{4.7}$$

where $\tau_{charging}$ is the time it takes for the differential input voltage to meet the resolving voltage of the comparator, which depends on the stimulating current I_{stim} and the size of the blocking capacitor C_B.

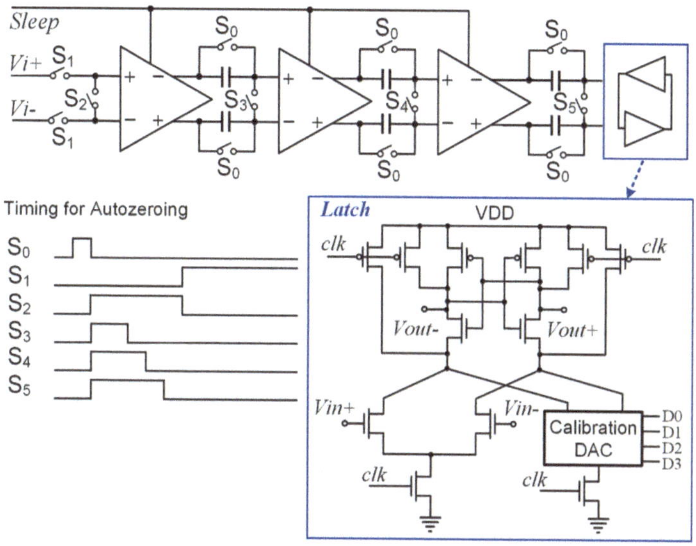

Fig. 4.27 The circuit schematic of the comparator consisting of a three-stage pre-amplifier and latch. The pre-amplifier has auto-zero calibration, and the latch has a 4-bit DAC for calibration

$$\tau_{\text{charging}} = \frac{(\pm\Delta V_{\text{OS}} + V_{\text{res}})C_B}{I_{\text{stim}}} \tag{4.8}$$

With a typical size of the blocking capacitor and a typical stimulation current, the τ_{charging} will be in tens of nanoseconds to several microseconds, which might dominate the total time error. This, unfortunately, causes a systematic delay, which is highly undesirable.

In order to resolve this problem, a feed-forward error compensation mechanism is introduced. The circuit schematic for the comparator is shown in Fig. 4.27. The comparator consists of a three-stage preamplifier with output auto-zeroing, and a dynamic latch with a 4-bit current DAC for calibration. Neural stimulation usually consists of stimulation pulse trains with the same amplitude and pulse width but varies in frequency (time interval between pulses). According to Eq. 4.8, the $\tau_{charging}$ will be the same for a train of pulses. So the delay of the comparator can be learned during the first few pulses and used to compensate the foregoing stimulation. Two four-bit DACs are used to calibrate the dynamic latch. The error of the comparison is learned from the residue charges after a stimulation pulse. In the first few stimulation pulses, an additional discharge phase is triggered after reversal phase to clear the charges on the capacitor. A switched capacitor amplifier is used to amplify the residue charges, and a dual threshold comparator is used to decide whether the residue charge is within the safe range or not. The schematic of the discharge and amplification circuit is shown in Fig. 4.28. The calibration DAC's value is changed according to the comparison result. The calibration DAC is designed to change 1

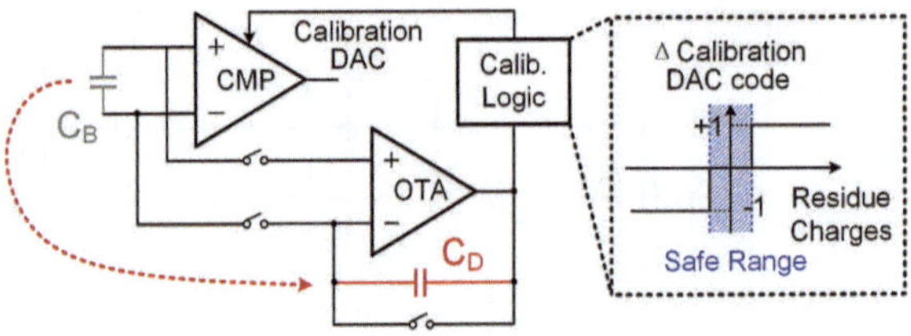

Fig. 4.28 The circuit schematic of the switched capacitor circuit used to discharge the blocking capacitor. The circuit is also used to amplify the residue voltage for the calibration purpose. The amplified residue voltage is compared with two pre-defined safe voltage window. If the residue charge is out of the safe window, the calibration DAC of the comparator will be changed accordingly

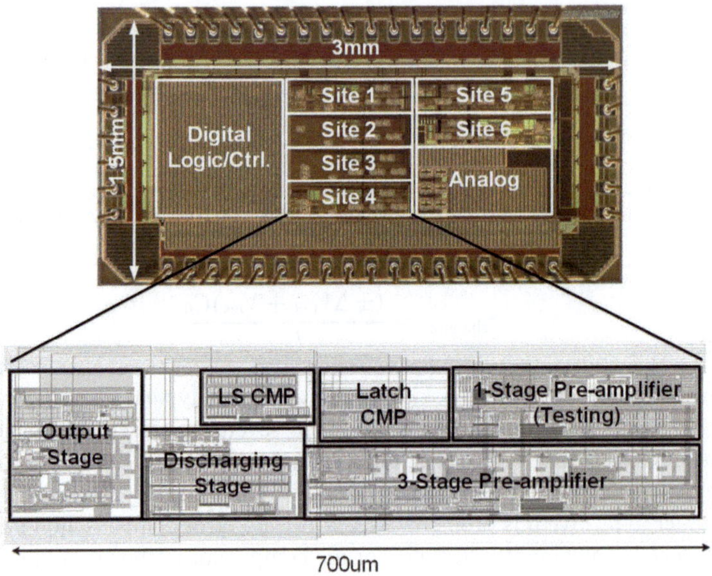

Fig. 4.29 The micrograph of the fabricated stimulator chip. The occupied silicon area is $3 \times 1.5\,\text{mm}^2$

LSB each time for stability and simplicity. So in the worst case, it takes 16 cycles to change from 0 to the full range of the DAC, which will be finished in one or two pulse groups.

4.4.4 *Experimental Results*

The design has been fabricated in IBM 180 nm CMOS technology. The occupied silicon area of the full chip is $3 \times 1.5\,\text{mm}^2$, including IO pads. One driving site features a dimension of $700\,\mu\text{m} \times 150\,\mu\text{m}$. The micrograph of the chip and the layout of one channel are shown in Fig. 4.29.

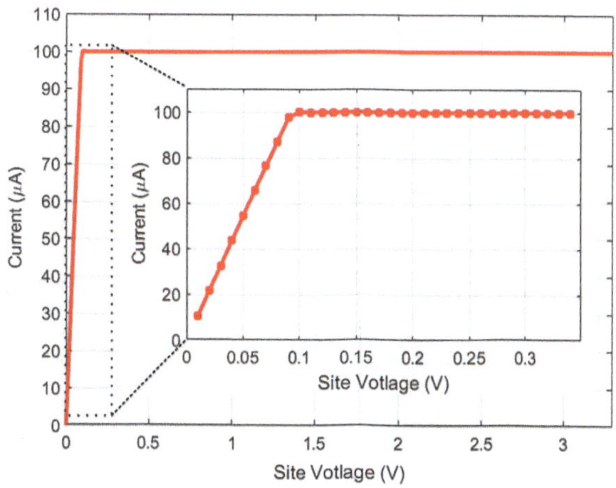

Fig. 4.30 The measured current from the output stage versus the output voltage. The embedded figure shows a zoom-in plot from 0 to 0.3 V

Bench testing was conducted to verify the functions of the chip. The measured currents from the output stage versus the output voltage are shown in Fig. 4.30. The measurement result shows a full compliance voltage range of 3.2 V out of the 3.3 V supply voltage at the current amplitude of 100 μA, which corresponds to 97% of the full voltage range. This is much higher than the result achieved in the general–purpose design presented in Sect. 4.3.3.

Both traditional digital-set method and the proposed current-set method have been implemented in the chip for a comparison purpose. The measured INL/DNL of the DAC using traditional digital-set method are 0.37/0.34 LSB, and are improved to 0.19/0.17 LSB using the proposed current-set method with the dynamic threshold variation cancellation technique. With the new technique, the charge error during a typical 100 μA and 200 μs is less than 0.05%.

The measured generated stimulation waveform and the finite state machine are shown in Fig. 4.31. The stimulation was measured with a high resistor load. The driving site was disconnected from the electrode when not activated to prevent leakage.

Another bench test was used to verify the function of the adaptive driving. Figure 4.32a shows a measurement of the electrode voltages during the driving voltage adjustment. Stimulation currents were measured under a load of two 10 nF capacitors and a 10 kΩ resistor in series. A blocking capacitor of 100 nF was applied. The probe placement is highlighted in the boxed figure. The boosting converter was implemented off-chip. Figure 4.32b shows the measured current across the load. The current maintains constant during the stimulation phase with the driving voltage adjustment.

In vitro tests are conducted using a 75 μm tungsten electrode in 0.9 g/100 mil Sodium Chloride. Figure 4.33 shows a comparison of the measured voltages over a 5-min continuous stimulation using a traditional method and the proposed method. It was measured at the same driving site under a different configuration. Given the same mismatch in the current sources, a drifting of the electrode voltage when

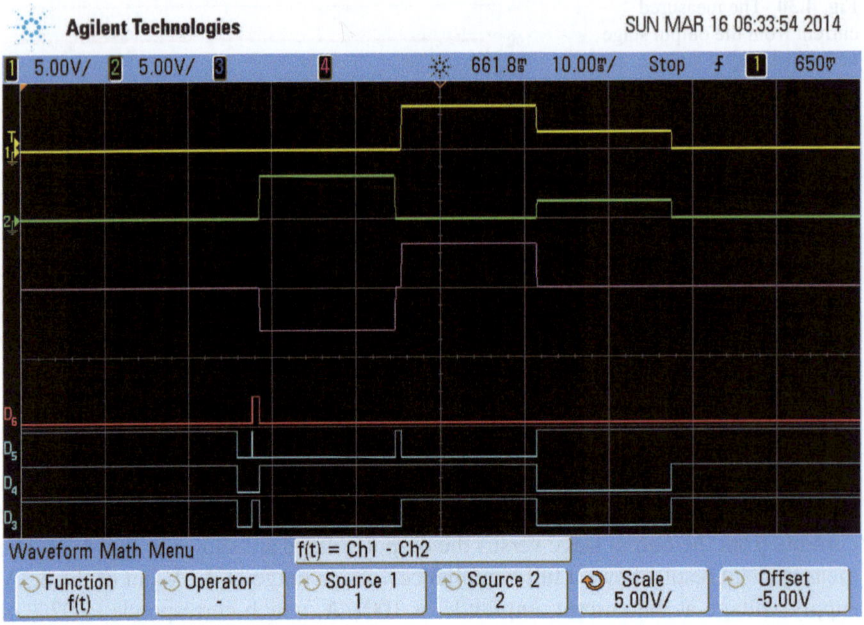

Fig. 4.31 The measured generated stimulation waveforms with a high impedance load. Channel 1 and channel 2 measure the output of the WE and CE, respectively. The Math channel calculates the difference between the two channels. Channels D3–D6 show the states of the digital module

Fig. 4.32 (**a**) The measurement of a stimulation pulse during the driving voltage adjustment. The load model is given with the measurement points highlighted. (**b**) The measured stimulating and reversal currents during the driving voltage adjustment

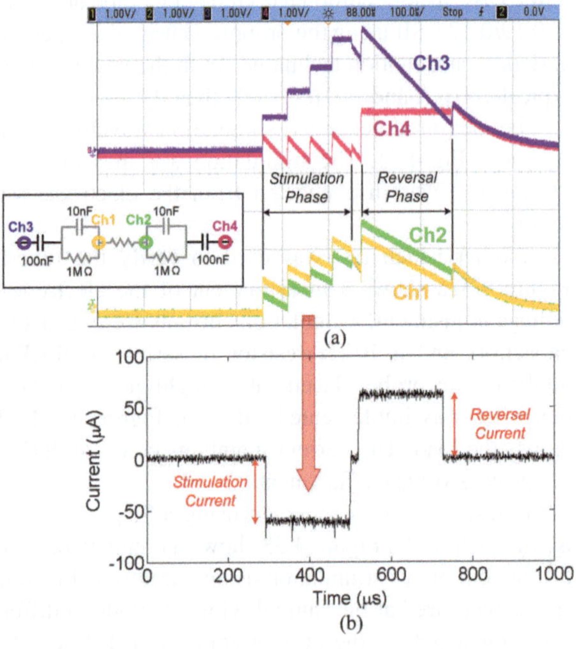

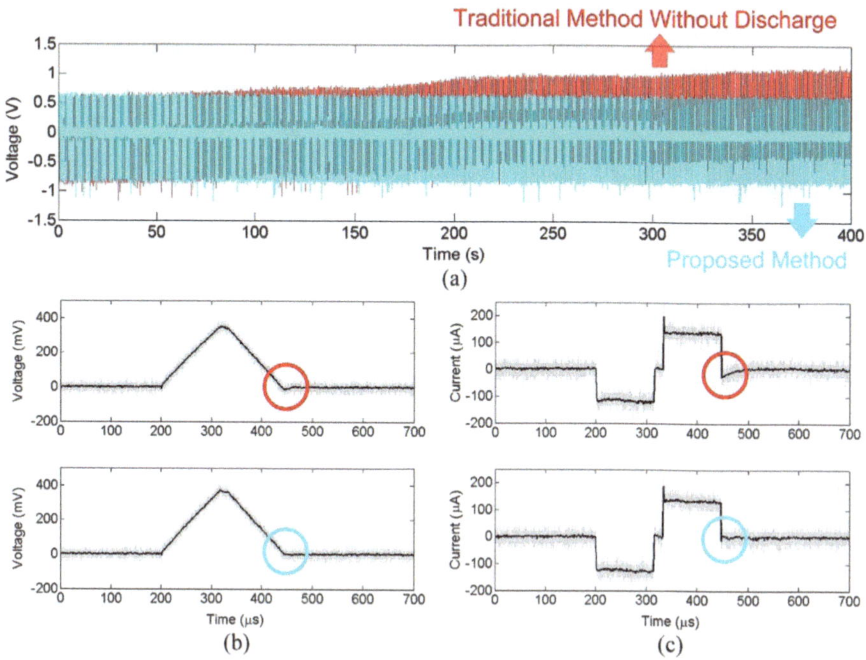

Fig. 4.33 The measurement of biphasic stimulation in the saline solution using the traditional method (*red*) and the proposed method (*blue*). (**a**) shows a 5-min continuous stimulation without discharge. (**b**) and (**c**) show an overlay of 20 measurements of the voltage across the blocking capacitor and the derived stimulation current

using the traditional digitally set DAC without discharge is shown in Fig. 4.33a, while the proposed method successfully resolves this problem. The measurements of the blocking capacitor's voltage in 20 trails are overlaid in Fig. 4.33b, and the derived currents are plotted in Fig. 4.33c. The over-reversal in the traditional method are shown from the test in the saline solution. The measurement results verify the theoretical analysis and the simulation.

To demonstrate that the stimulator is capable of evoking physiological activity, an in vivo experiment was performed in a sedated rat. Trains of biphasic stimulus pulses (10 pulses, 5 ms interpulse interval, 0.3 ms/phase) were delivered through a pair of insulated tungsten microwires, with a 50 μm diameter, implanted near the intrinsic muscles that protract the mystacial vibrissae. Whisker movements, as measured by an optical micrometer, were reliably evoked as shown in Fig. 4.34a. The stimulator chip was programmed by a microcontroller with a wireless transceiver. Whisker displacements were a function of current intensity as shown in Fig. 4.34b. This experiment can be further used to implement facial reanimation for patients who suffer from facial paralysis.

The measured performance of the chip is summarized in Table 4.2.

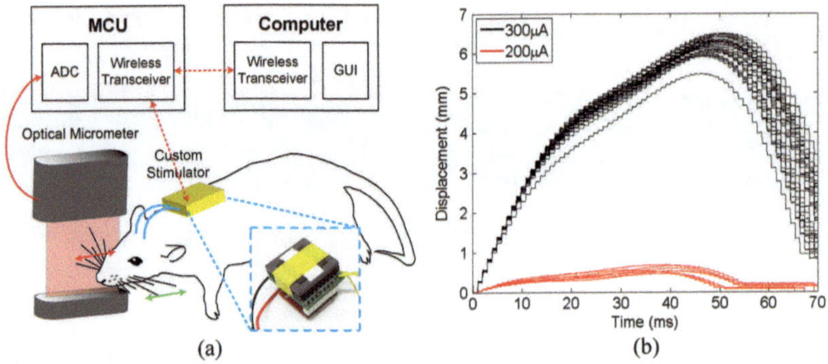

Fig. 4.34 In vivo experiment performed on a sedated rat. (**a**) Whisker movements, as measured by an optical micrometer, were reliably evoked by stimulation. (**b**) Whisker displacements were a function of current intensity

Table 4.2 Chip specification summary

Driving site	# of sites	6
	Area per site	0.1 mm^2
	Driving voltage	3.3 V
	Compliance range	97% (typical)
	Stim current	<2 mA
	Stim freq.	1–500 Hz
	Charge error	<0.05% (typical)
DAC	Resolution	6-bit
	INL	0.19 LSB
	DNL	0.17 LSB
	Full scale std	<0.7%
Comparator	Resolution	40 μV
	Calibration	Auto-zero/4-bit DAC
	Speed	40 MHz
Power	per site	136 μW
	Coin battery	1.2 V regulator off-chip
	Efficiency	81%

4.4.5 *Conclusion*

In this section, a high efficiency, tissue-friendly net-zero charge stimulator is proposed. The net-zero charge stimulation is achieved by controlling the timing of the reversal phase based on monitoring the residual charge. An arbitrary channel configuration is achieved without a pre- or on-the-fly calibration, which enables a more dedicated stimulation position and pattern. Feedback control of the adaptive driving voltage is further proposed to improve stimulation efficiency. A novel current-mode DAC is implemented to suppress the process variation across the driving site array. A digital feed-forward error compensation is used to calibrate the zero-crossing detection comparator in a continuous stimulation pulse train. Both in vitro and in vivo experiment results are presented.

Chapter 5
Bidirectional Neural Interface and Closed-Loop Control

5.1 Introduction

A bidirectional electronic neural interface enables simultaneous recording and stimulation of the neural system, establishing a two-way direct communication link between the brain and the external world [272]. The importance of a bidirectional closed-loop neural interface can be understood from several aspects: (1) In the development of prosthetic devices, an electrical neural stimulation can provide an artificial sensory feedback to the user, allowing the user to perceive the movement and the haptic interaction with external objects [273]. This is especially important for paralyzed patients to fully control a prosthetic; (2) in the treatment of the neural disorders, including Parkinson's disease, the mechanism underlying the deep brain stimulation remains not clear [18, 19]. Research shows that the application of the closed-loop stimulation has a greater effect than the conventional open-loop stimulation paradigms [19, 48]. (3) In the study of electrophysiology, the brain's response to an external stimulus is a result of various activities triggered by the sensory stimulus itself and the brain's internal state. Thus, the study needs more than statistical descriptions [22], and an open-loop approach cannot fully capture the characteristics of the brain's response. As a result, a closed-loop approach is critical in these studies [274].

Although the importance of the bidirectional closed-loop neural interface has already been recognized [275], it has not been widely used in neuroscience research. The electronics design of a bidirectional closed-loop neural interface is one of the bottlenecks. There are two primary design challenges in a bidirectional closed-loop neural interface system: (1) The artifacts caused by stimulating and recording simultaneously, namely the stimulation artifacts [247, 276–278]; (2) the design of an on-chip real–time closed-loop controller [24, 64, 279–282]. Addressing both challenges is important for the successful implementation of a bidirectional closed-loop neural interface. Thus the goal of this chapter is to review and analyze

X. Liu, J. Van der Spiegel, *Brain-Machine Interface*,
https://doi.org/10.1007/978-3-319-67940-2_5

the practical issues related to the design of a bidirectional neural interface and a closed-loop controller. Novel system-level and circuit-level designs are proposed to improve the state-of-the-art.

The chapter is organized as follows. Section 5.2 analyzes the origins of the stimulation artifacts in a bidirectional neural interface, reviews the prior works, and presents a study on the stimulation artifacts. The study focuses on different electrode configuration and circuitry topologies. Both in vitro and in vivo experimental results are presented. Section 5.3 summarizes the mechanisms of different closed-loop neural interface systems, and presents the design of a general–purpose PID controller for a closed-loop neural interface system-on-chip (SoC). All experimental procedures used in the studies presented in this chapter were approved by the institutional animal care and use committee (IACUC) of the University of Pennsylvania.

5.2 Stimulation Artifacts in the Bidirectional Neural Interface

5.2.1 Introduction

The stimulation artifact is a known issue in simultaneous neural stimulation and recording. A long-lasting stimulation artifact blanks the recording front-end and corrupts the evoked neural response. Thus, suppressing the stimulation artifact is critical in the design of bidirectional neural interface. Several techniques have been proposed in the literature to cancel or attenuate the stimulation artifacts, including using recording front-end blanking [57], symmetrical electrode placement [276], temporary frequency shifting [247, 277, 278], real–time signal processing in the computer [283] or on-chip [59]. However, most proposed techniques have certain constraints, which are usually not suitable for a general–purpose neural interface design.

The goal of this study is to find the stimulation artifacts in different cases. The combination of different stimulator and recorder configurations, namely monopolar and bipolar stimulation, single-ended and differential recording, with common and separate grounds were studied. In addition, different power supply configurations (dual-supply and single-supply), stimulator architectures (type-I and type-II) were taken into account. To the best of our knowledge, this work presents the first analysis of the stimulation artifacts considering both neural interface configuration and architectures of the electronic system. The main sources of the stimulation artifacts were first analyzed, and a testing board was designed to verify the analysis. Both in vitro and in vivo experimental results are presented. The conclusions and the design recommendations are given at the end of this section.

5.2.2 *Review of Prior Work*

Several studies of the stimulation artifacts in bidirectional interfaces have been reported in the literature. G.A. DeMichele et al. from Sigenics Inc. and Illinois Institute of Technology proposed a stimulus-resistant neural amplifier in 2003 [277]. The amplifier had an artifact suppression mode, which shifted the input frequency corner to 10Hz by a DC servo loop in the second stage. This work used a low gain (x4) pre-amplifier, which will not be easily saturated. Many custom neural front-end designs prefer to use a high gain in the first-stage amplifier to achieve a high overall noise efficiency. In these cases, this artifact suppression technique cannot resolve the long-lasting saturation from the first stage.

R.A. Blum and E.A. Brown et al. from Georgia Institute of Technology and University of Illinois at Urbana-Champaign proposed a stimulation artifacts model and a circuit module for removing the artifact in 2004 [247], with a follow-up work in 2008 [278]. The model only considered a voltage-mode stimulation without a charge balancing design, and assumed the recording and stimulation circuits are using the same electrode. A *pole-shifting* technique was used in the first stage, while the highpass pole was set at 200 Hz. A *soft-switching* technique was used to ensure a smooth transition between different switch phases, which required additional custom circuit design and optimization.

L. Rossi et al. from University of Milan proposed an artifact suppression device for recording the local field potential during a deep brain stimulation in 2007 [276]. The work used separate grounds for the recorder and the stimulator. The recorder's ground was placed on the scalp, while the stimulator's ground was placed on the shoulder. In addition, the stimulation electrodes were placed in the middle of the differential recording electrodes. Because the recording frequency was 2–40 Hz and the stimulation frequency was 130 Hz, a ten-pole lowpass filter was used to filter the stimulation artifacts. In many other applications, however, the recording and stimulation frequency ranges have overlaps, and the stimulation electrode location cannot be chosen in favor of the differential recording.

A.E. Mendrela et al. from University of Michigan, Ann Arbor and the University of Minnesota, Minneapolis proposed a bidirectional neural interface circuit with an active stimulation artifact cancellation circuit in 2016 [59]. An on-chip digital adaptive filter was used to remove the stimulus artifacts. The proposed design removed the stimulation artifact if the recording electrodes were not saturated or out of the linear range.

In summary, stimulation artifact is a complex issue, which depends on the types of the recording and stimulation electrodes, the electrical circuitry, the configuration of the ground, the characteristics of the recording front-end's input stage, the methods and parameters used for the stimulation, and so on. Most existing investigations on the stimulation artifacts are restricted for a certain application or in a certain configuration. It is important to understand the difference between the configurations and their effects on the stimulation artifacts, which is the goal of this study.

5.2.3 *Analysis of Stimulation Artifacts*

5.2.3.1 Origins of Stimulation Artifacts

The origins of the stimulation artifacts can be quite complicated, as reviewed in the previous section. The analysis and models in this section are not intended to give an accurate electrochemical description or a precise estimation of the stimulation artifacts, but to provide insights in developing techniques for reducing or canceling the stimulation artifacts.

- **Electrode Saturation**

During the stimulation phase, a portion of the charges will be stored on the double layer capacitors on the recording electrodes [244]. Ideally, after the reversal phase, the charges will be removed and a charge balance will be achieved. The tissue environment should return to the potential before the stimulation. However, this is not always achievable. If the recording electrode's potential is within the linear input range of the amplifier after the reversal phase, the recorder will usually return to the normal operating point reasonably fast. But if the electrode potential is pushed away from the recorder's linear input range, it might take a very long time for the recorder to recover, since the input stage of the circuits usually has a very large time constant needed for a high input impedance.

Due to the small amplitude of the neural signal, the neural recording amplifier is usually designed with a linear input range in tens of millivolts, and the common mode input range is usually limited to hundreds of millivolts, depending on the supply voltage and the circuit architecture. Capacitively coupled instrumentation amplifier usually has an extended input range, however, it is also limited by the ESD circuits and breakdown voltage of the input transistors. In addition, amplifiers employing chopping technique or DC servo loop suffer more from the voltage headroom constraint [78]. It should be noticed that even these limitations may not be a problem for the neural recording purpose alone, they may become a significant problem in a bidirectional neural interface.

- **Voltage Gradients**

Voltage gradients can be easily understood by considering the tissue is conductive, and all electrodes inserted in the tissue environment are interconnected. Figure 5.1 shows the circuit model proposed to analyze the stimulus artifacts in the bidirectional neural interface. A monopolar stimulation with a single-ended recording is used for illustration.

Firstly, consider the case where the recorder and stimulator share a common ground, as illustrated in Fig. 5.1a. Assume the electrode impedance for the stimulator, the recorder, and the common ground is Z_1, Z_2, and Z_3, respectively. The

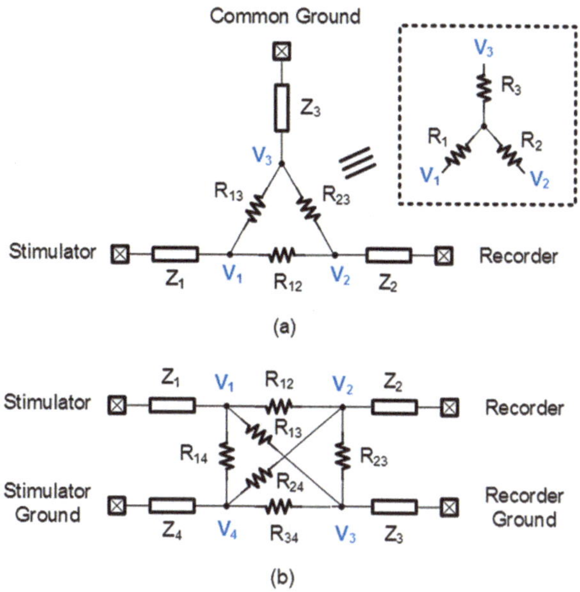

Fig. 5.1 The circuit model for the bidirectional neural interface with (**a**) a shared ground and (**b**) separated individual grounds

spreading resistances between the three electrodes are represented by R_{12}, R_{13}, and R_{23}. Assume that the stimulator and the ground electrodes have a low impedance, the recording electrode has a high impedance, and the instrumentation amplifier has a high input impedance. The stimulation artifact can be expressed as:

$$\begin{aligned} V_{\text{artifact}} &\approx \frac{R_3}{R_1 + R_3} V_{\text{stim}} \\ &= \frac{R_{23}}{R_{12} + R_{23}} V_{\text{stim}} \end{aligned} \tag{5.1}$$

If the R_{12} and R_{23} are in a comparable magnitude, the artifact can be as large as half of the stimulation's compliance voltage, which will easily saturate the recording electrodes or push the instrumentation amplifier out of the linear input range.

Secondly, Fig. 5.1b shows the case where the recorder and stimulator have separated grounds. In this case, the stimulation artifact can be expressed as:

$$V_{\text{artifact}} \approx \frac{(R_{13}R_{24} - R_{12}R_{34})R_{23}}{R_{\text{eq}}} V_{\text{stim}} \tag{5.2}$$

where

$$\begin{aligned} R_{\text{eq}} = {} & R_{12}R_{23}(R_{13} + R_{34}) + R_{13}R_{24}(R_{12} + R_{23}) \\ & + R_{24}R_{34}(R_{12} + R_{13}) \end{aligned} \tag{5.3}$$

If the recording and stimulating electrodes are well separated, $R_{12} \approx R_{13}$, $R_{23} \approx R_{24}$, then the V_{artifact} is approximately zero. Even if the electrodes are not separated far away, or the recording electrodes for signal and ground have significant different impedance, the term $(R_{13}R_{24} - R_{12}R_{34})R_{23}$ is still much smaller than R_{eq}, and V_{artifact} will be a very small portion of V_{stim}.

From the above analysis, the stimulation artifact due to the voltage gradient can be in the same order of the stimulating compliance voltage when the recording and stimulation share a common ground, and the artifacts can be minimized if the grounds are separated in the circuits and in the tissue environment.

- **Capacitive Coupling**

The capacitive coupling between the stimulating and recording leads also contributes to the stimulation artifact. Even though the capacitance between the leads is usually less than 1 pF, the amplitude of the simulation voltage can be six orders higher than the neural signal, so that the coupled stimulus may still be visible in the recorded signal. The coupling is worse if there is no shielding on the recording and stimulation electrodes, or the connecting wires are long.

5.2.3.2 Configuration of the Interface Circuits

The configuration of the stimulator and the recorder, and the placement of the electrodes have a high impact on the stimulation artifacts. In this section, different configurations of the neural recorder and the stimulator, and the related circuit implementations are discussed.

There are two typical configurations for a multi-channel neural signal recording front-end: *single-ended* recording and *differential* recording, as illustrated in Fig. 5.2a and b, respectively. In a single-ended recording, a reference electrode and a ground electrode are shared among all channels. The ground electrode can be used as the reference in some cases. In a differential recording, two electrodes collect the signal between of them, a third electrode provides the ground, and no reference electrode is required.

It should be noticed that *fully-differential* is a widely used term in circuit design. It refers to the circuit module which has both a differential input and a differential output. However, a differential recording doesn't necessarily require a fully differential circuit.

Similar to the recorder, there are two typical configurations for electrical neural stimulators: *monopolar* stimulation and *bipolar* stimulation. In a monopolar stimulation, the electrical charges are injected from one electrode for stimulation, and pulled from the same electrode to achieve the charge balancing (see Sect. 4.2.3). A low impedance counter electrode provides the return path. Usually, the ground electrode is used as the counter electrode. In a bipolar stimulation, the electrical charges are passed between two electrodes for stimulation. An additional low impedance ground electrode is usually connected to the tissue and provides the

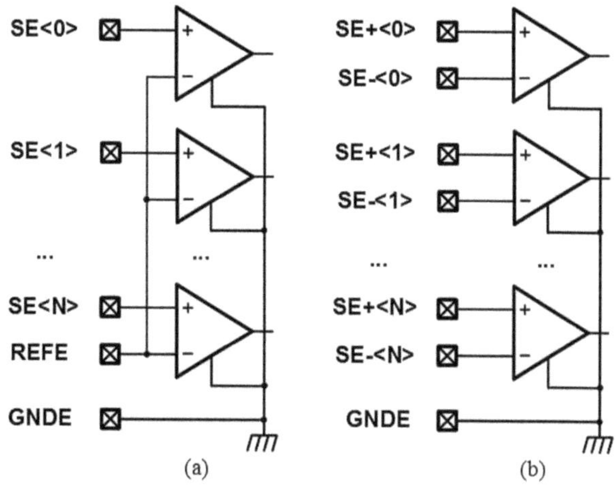

Fig. 5.2 Configuration of typical multiple channels neural recording front-end. (**a**) Single-ended recording configuration, and (**b**) differential recording configuration

electronic ground. However, the ground electrode is not necessarily required since only a potential difference is needed between the bipolar electrodes to generate the current. It should be noticed that there is a difference between a *biphasic* stimulation and a bipolar stimulation. In a biphasic stimulation, a reversal phase follows the simulation phase (see Sect. 4.2.2). Both monopolar and bipolar stimulator can perform a biphasic stimulation.

There are two typical methods to implement a stimulator. If both a current sink and a current source are connected to the working electrode and used to generate the stimulation and reverse phase, the stimulator circuit is referred to as *Type-I* stimulator in this work. If only a current sink or a current source is connected and used in the working electrode, and a voltage buffer is connected to the counter electrode, the stimulator circuit is referred to as *Type-II* stimulator in this work. It should be noticed that:

- A Type-I stimulator can perform both a monopolar stimulation (Fig. 5.3a) and a bipolar stimulation (Fig. 5.3b). In a same supply voltage, the bipolar configuration gives twice the equivalent compliance voltage range for the stimulation.
- A Type-II stimulator can only be used to perform a bipolar stimulation (Fig. 5.3c and d);
- Since only a current sink (or a current source) is implemented in a Type-II stimulator, it saves the voltage headroom for one current source (or a current sink), which is an advantage over the Type-I stimulator for a low-voltage and low-power design;
- A current sink is usually easier to implement than a current source since its circuit components are operating at a low voltage with respect to the ground. So most of the Type-II stimulators use current sinks (Fig. 5.3c) instead of current sources (Fig. 5.3d).

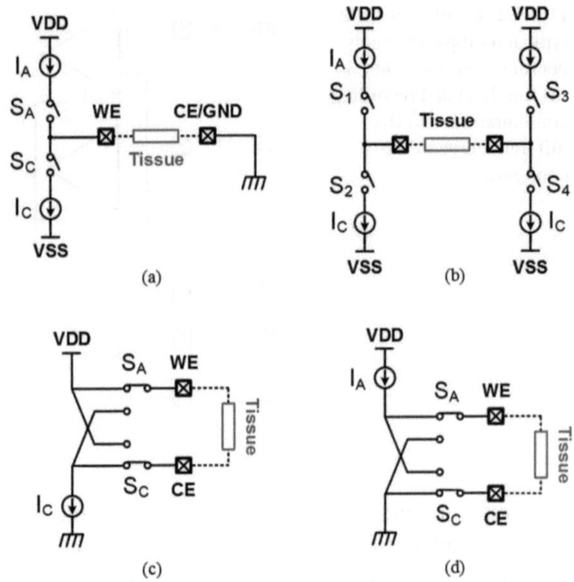

Fig. 5.3 Typical topologies of neural stimulators. (**a**) shows a monopolar stimulator, (**b**), (**c**), and (**d**) are bipolar stimulators. (**a**) and (**b**) are Type-I stimulators, (**c**) and (**d**) are Type-II stimulators

5.2.3.3 Practical Issues in Circuit Design

One of the practical design challenges is the different supply voltage requirements in the stimulator and the recorder modules. The stimulator module usually requires a high supply voltage for driving the high impedance electrodes. At the same time, the recorder and the digital modules need a low supply voltage to reduce the power consumption. Both dual-supply and single-supply are commonly used in bidirectional neural interface system, as shown in Fig. 5.4. In a dual-supply system, the grounds of all circuit modules are connected together. In a single-supply design, the lowest supply in each module is connected together. The design is also limited by the CMOS process. Dual-supply design sometimes require deep N-well or silicon on insulator (SoI) process. A level-shifter is required if low voltage digital signals are to be used to control high voltage stimulator switches.

The two configurations can be equivalent in the circuit design, however, when connecting the electronics ground to the tissue ground, there is a problem in the single-supply system. If the tissue ground is connected to the stimulator's ground, the DC common-mode neural signal will be much higher than the recorder's ground, and may be even higher than the recorder supply in many cases. So in a DC-coupled recorder design, or designs require a DC servo loop (see Chap. 2), this configuration is not feasible. If the tissue ground is connected to the recorder's ground, the tissue ground will not be centered between the highest and lowest supplies of the stimulator. Thus, the compliance voltage of the stimulator cannot be fully used, which will be a waste of power consumption. It should also be noticed that if the stimulator tries to discharge the local tissue to the stimulator ground GNDH, the potential difference between GNDH and GNDA will cause a DC current, which

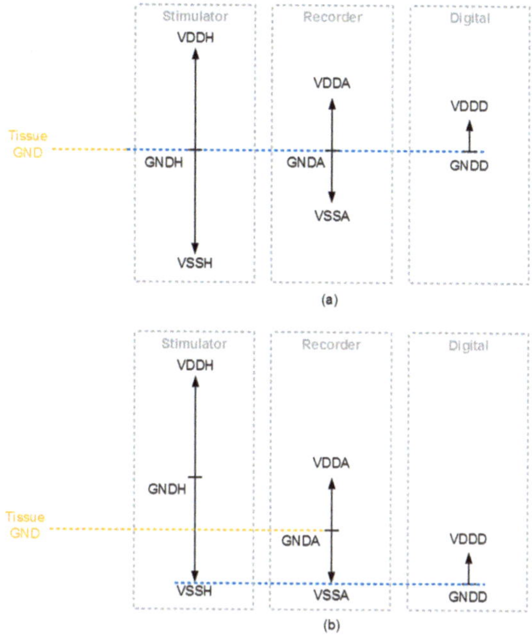

Fig. 5.4 Two typical supply configurations for the recorder and the stimulator: (**a**) dual-supply and (**b**) single-supply. In a dual-supply system, the grounds of all circuit modules are connected together. In a single-supply system, the lowest supply in each module is connected together. The design is also limited by the CMOS process

may damage the tissue. This must be avoided. Even though a single-supply design is simpler in certain cases, especially in bulk CMOS process, a dual-supply design is necessary for the bidirectional neural interface design.

5.2.4 Methods

A bench-test board was designed to study the stimulation artifacts in different configurations. The block diagram of the board is shown in Fig. 5.5, and the photograph of the assembled board is shown in Fig. 5.6.

The board is carefully designed with isolated ground for each module. The recorder, stimulator, and digital processor have individual ground and power management unit. The recorder and stimulator are powered by two batteries in series. Positive and negative regulators are used to provide regulated supply voltages of ±6 V. The digital module is powered from a 5 V USB cable, and regulated to 3.3 V to power the microcontroller. Optical isolators TLP292 from Toshiba [284] are used to provide a digital control signal to the recorder and simulator modules without connecting the ground. The isolation amplifier uses a duty cycle modulation-demodulation technique to convert the input signal to the output in separate grounds. A 50 kHz two-pole lowpass filter is integrated on board to remove the ripples. The data acquisition devices and the oscilloscope share the ground with the digital module. Using the isolation amplifier, the oscilloscope can monitor the analog output of the recorder without connecting the mains ground to the animals.

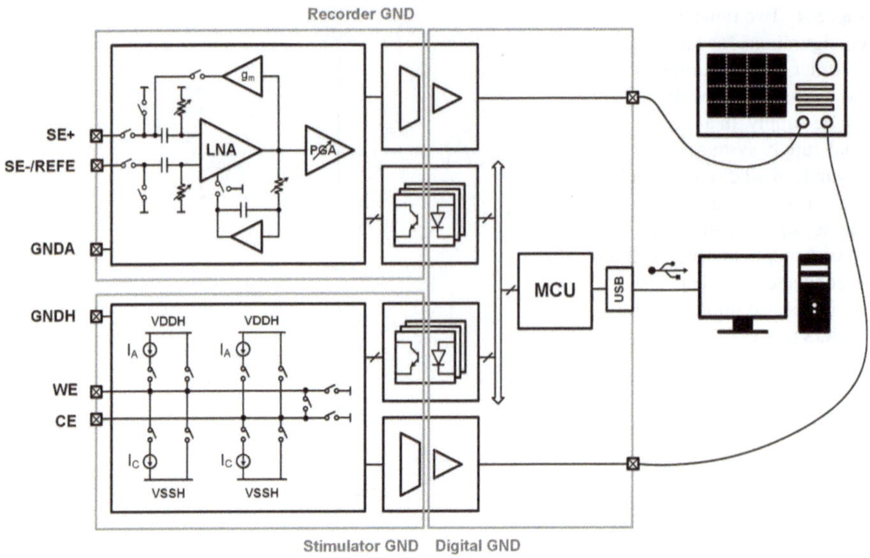

Fig. 5.5 The block diagram of the bench testing board. The supplies and grounds for the recorder, the stimulator, and the digital modules are intentionally separated on the board

Fig. 5.6 The photo of the assembled bench testing board. The dimension is 216 mm × 171 mm

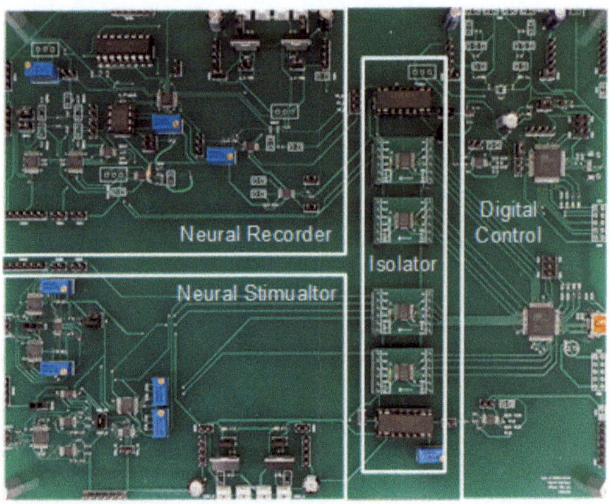

The recorder has two stages. The first stage uses a low-noise instrumentation amplifier INA111 from Burr-Brown [285]. The amplifier has a noise density of 13 nV/rtHz at 100 Hz. The gain is set at 11 by an external resistor. The input stage is biased by large resistors to provide a very low cut-off frequency. Switches are integrated to be able to disconnect or blank the recording electrodes during the stimulation. A DC servo loop is also implemented which can move the highpass corner frequency. The second stage provides an additional gain of 50. The recorder can perform either a single-ended recording or a differential recording depending on the connection of the electrodes.

The stimulator has two output stages, each has a current source, a current sink, and two switches for shorting the electrodes to the two power lines. Additional switches are also integrated to short the two electrodes, or short the electrode to the stimulator ground. By controlling the timing of the switches, the stimulator can perform a monopolar or a bipolar stimulation in either Type-I or Type-II configuration. The amplitude of the current source and sink are programmable. Tunable resistors are used to match the current source and sink. Another isolation amplifier can be connected to the electrode for monitoring the compliance voltage of the electrodes.

The digital module mainly consists of a microcontroller ATxmega128A4U from Atmel to generate the timing and control signal for the stimulator and the recorder. The microcontroller has an integrated USB 2.0 module. A computer user interface is developed in Matlab, and commands are sent to the bench test board via USB. In addition to the oscilloscope, a pre-developed data acquisition board is used to collect the data and send to the computer for online and offline analysis.

5.2.5 Experimental Results

5.2.5.1 In Vitro Experiment

A series of in vitro studies was conducted in 0.9 g/100 mil Sodium Chloride. The placement of the electrodes is illustrated in Fig. 5.7. A 75 μm tungsten low impedance electrode was used as the reference electrode. A stripped copper wire was used as the ground. A pair of electrodes were inserted in the saline for differential recording. Similarly, a pair of electrodes were used for bipolar stimulation. When performing a single-ended recording or a monopolar stimulation, only one of the electrodes was selected.

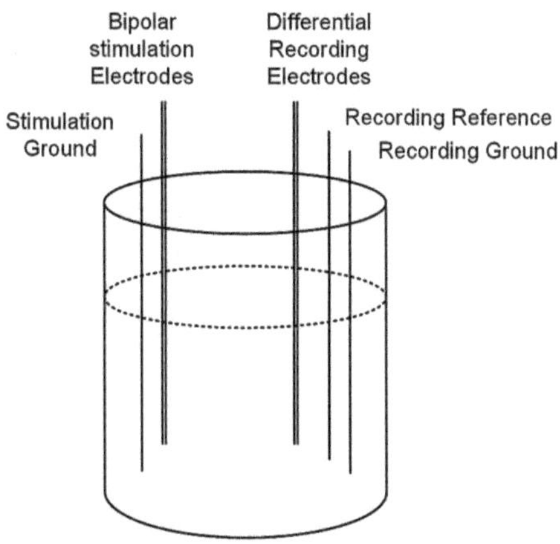

Fig. 5.7 Electrode setup for the stimulation artifacts experiments

Table 5.1 In vitro experiments for stimulation artifact study

				Recording	
				Single-ended	Differential
Stimulation	Type-I	Common GND	Monopolar	Fig. 5.8a	Fig. 5.8b
			Bipolar	Fig. 5.8c	Fig. 5.8d
		Separate GND	Monopolar	Fig. 5.9a	Fig. 5.9b
			Bipolar	Fig. 5.9c	Fig. 5.9d
	Type-II	Common GND	Bipolar (current sink)	Fig. 5.10a	Fig. 5.10b
			Bipolar (current source)	Fig. 5.10c	Fig. 5.10d
		Separate GND	Bipolar	Fig. 5.11a	Fig. 5.11b

Different configurations for recording and stimulation, different types of stimulators, with a common or separate grounds were explored in the in vitro study. Table 5.1 lists the experiment configurations and the corresponding results.

Firstly, the stimulator was tested in the Type-I configuration. The stimulator was configured to do a 10-pulse biphasic stimulation. The pulse amplitude was 100 μA, the pulse width was 200 μs, and the pulse interval was 3 ms. The discharge time was 1 ms, and the discharging resistor was 1 kΩ. The highpass frequency corner of the amplifier was set to be 0.159 Hz by a 10 nF coupling capacitor and a 100 MΩ biasing resistor. The lowpass frequency was 7.2 kHz, and the overall gain was 550. Figure 5.8 shows the recorded stimulation artifacts with the recorder and the stimulator share a common ground. The figures in the top row show the results from a monopolar stimulation, and the figures in the bottom row show the results from a bipolar stimulation. The figures in the left column show the single-ended recording, and the figures in the right column show the differential recording. In summary, a monopolar stimulation with a single-ended recording gives the worst result, and a bipolar stimulation with a differential recording results in the minimum artifact.

Figure 5.9 shows the recorded stimulation artifacts with the recorder and the stimulator having separated grounds. It should be noticed that the scale in this figure (one grid is 50 μV) is much smaller than Fig. 5.8 (one grid is 1 mV). The figures in the top row show the measurement results from a monopolar stimulation, and the figures in the bottom row show the measurement results from a bipolar stimulation. The figures in the left column show the single-ended recording, and the figures in the right column show the differential recording. In summary, a monopolar stimulation with a single-ended recording gives the worst result, similar to the conclusion in the common ground. However, with separated grounds, even the worst artifacts will not saturate the recording electrodes.

Secondly, the stimulator was tested in the Type-II configuration. The amplitude and timing of the stimulation pulses were the same as in the Type-I experiment. Figure 5.10 shows the recorded stimulation artifacts with the recorder and the stimulator sharing the ground. It should be noticed that the Type-II stimulator can only perform a bipolar stimulation. The figures in the top row show the measurement results using a current sink, and the figures in the bottom row show the measurement results using a current source. The figures in the left column show the single-ended

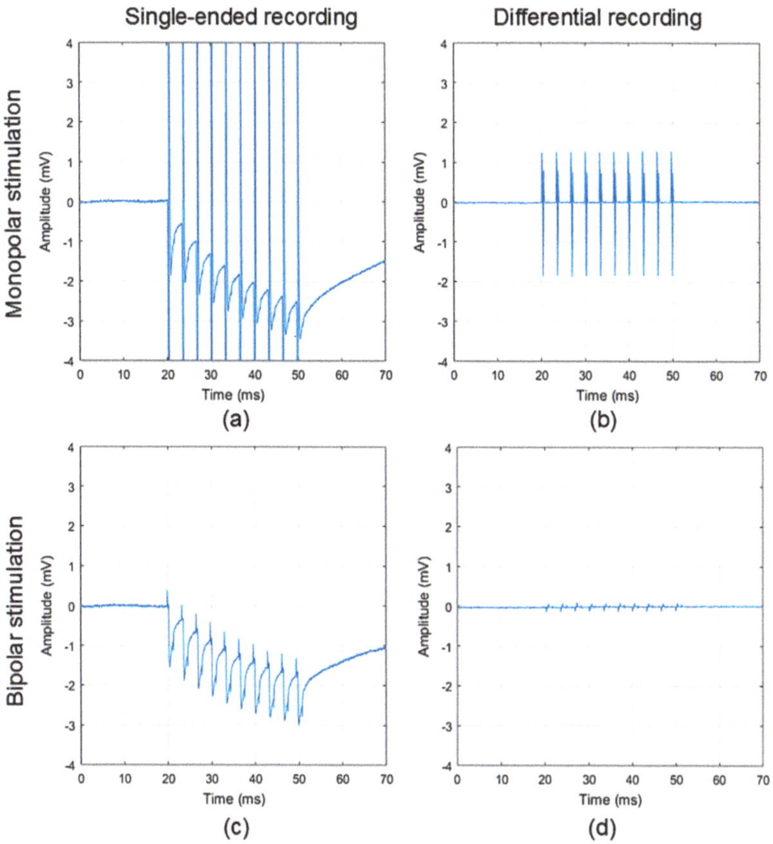

Fig. 5.8 The measured stimulation artifacts of a Type-I stimulator with a common ground between the recorder and the stimulator. (**a**) Monopolar stimulation and single-ended recording, (**b**) monopolar stimulation and differential recording, (**c**) bipolar stimulation and single-ended recording, and (**d**) bipolar stimulation and differential recording

recording, and the figures in the right column show the differential recording. In general, the Type-II stimulator using a current source gives better results than the one using a current sink. Differential recording also gives better results than single-ended recording.

Figure 5.11 shows the recorded stimulation artifacts with the recorder and the stimulator having separate grounds. One grid in this figure is 200 μV. The stimulators with current source and the current sink had a similar performance when their grounds were separated from the recorders. It should be noticed that if the recording amplifier is not saturated, signal processing techniques for removing the artifacts can be applied. In general, the stimulation artifacts are minimum when the grounds of the recorder and the stimulator are separated. When the grounds are connected, a differential recording helps reduce the artifacts. The worst case would be a common ground with a single-ended recording.

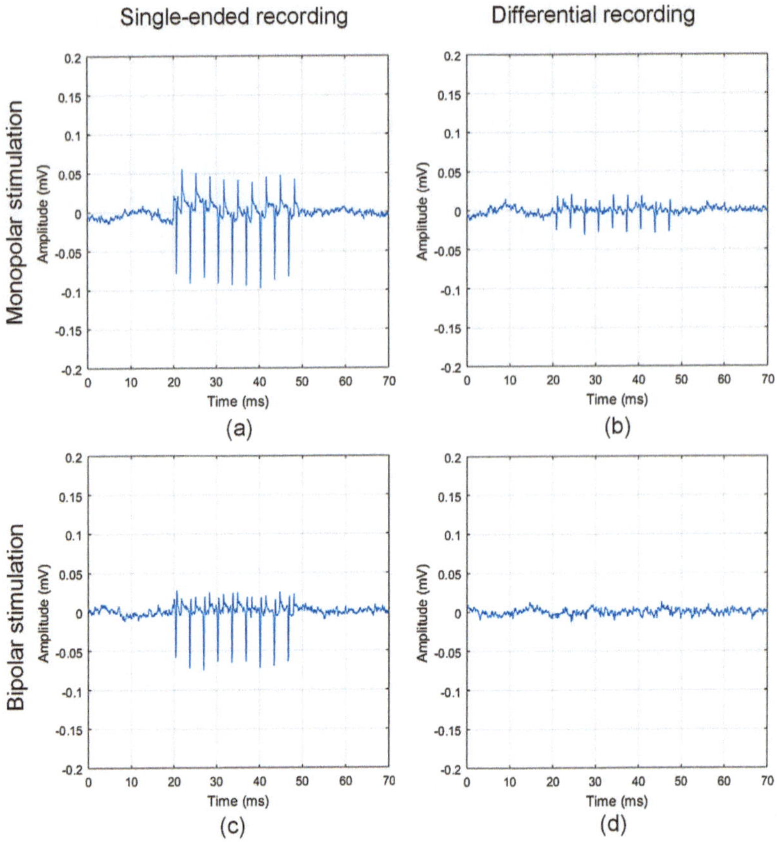

Fig. 5.9 The measured stimulation artifacts from a Type-I stimulator, which has separate grounds with the recorder. (**a**) Monopolar stimulation and single-ended recording, (**b**) monopolar stimulation and differential recording, (**c**) bipolar stimulation and single-ended recording, and (**d**) bipolar stimulation and differential recording

5.2.5.2 In Vivo Experiment

Two animal experiments were conducted to further verify the results from the in vitro study. A female Long-Evans rat received two implants, one in the motor cortex, the other in the sensory cortex. A ground stew was connected to the skull of the rat to provide the ground for recording. Figure 5.12 shows the experimental setup.

A bipolar stimulation was performed in the sensory cortex while a single-ended recording was conducted in the motor cortex. Figure 5.13a shows the recording when the grounds of the recorder and stimulator were connected. A large stimulation artifact appeared after the stimulation with a long recovery time. And Fig. 5.13b shows that when the grounds are separated, the stimulus artifact was minimized and an evoked potential was clearly visible, which is buried by the large stimulation artifact in the top figure.

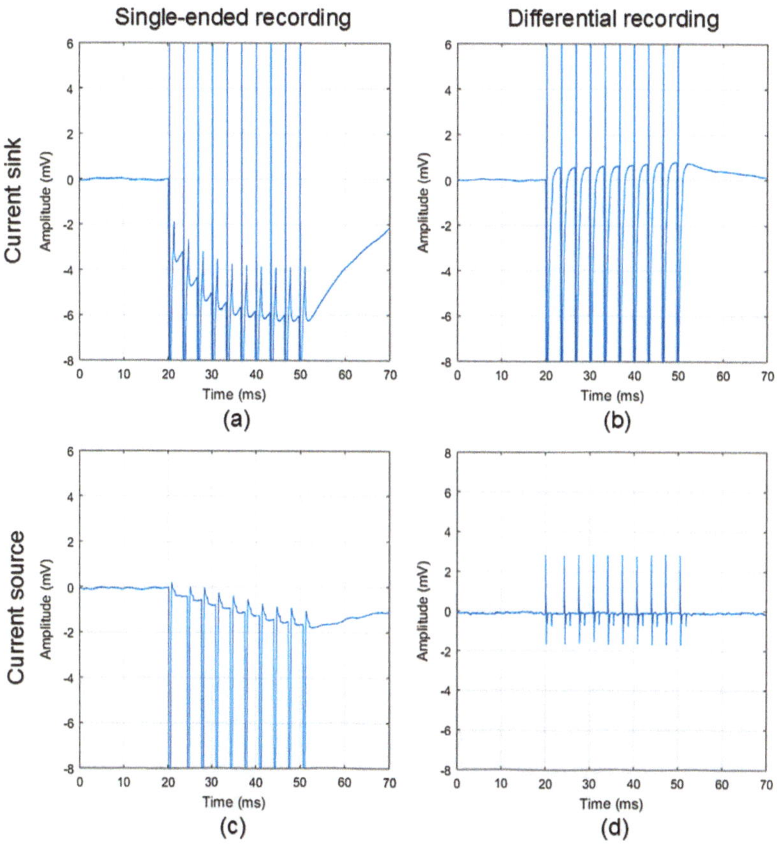

Fig. 5.10 Stimulation artifacts from a Type-II stimulator, which has a common ground with the recorder. The stimulator uses a current sink in (**a**) and (**b**), and a current source in (**c**) and (**d**). Recorder is configured to do single-ended recording in (**a**) and (**c**), and differential recording in (**b**) and (**d**)

Another bidirectional experiment was conducted in a macaque. The recording electrodes were chronically implanted in the left hippocampus while the stimulating electrodes in the upstream areas. The stimulation pulse train contains 10 pulses with an amplitude of 2 mA. Figure 5.14a shows the case when the grounds of the recorder and the stimulator were shorted together. The recording amplifier was saturated right after the stimulation and the recovery took hundreds of milliseconds. Figure 5.14b shows the case when the grounds were separated, and the artifacts were minimized. Figure 5.14c shows the output of the recording with an additional low-pass filter with a frequency corner of 200 Hz. The stimulation artifacts were completely removed by the filter since they were out of the signal band.

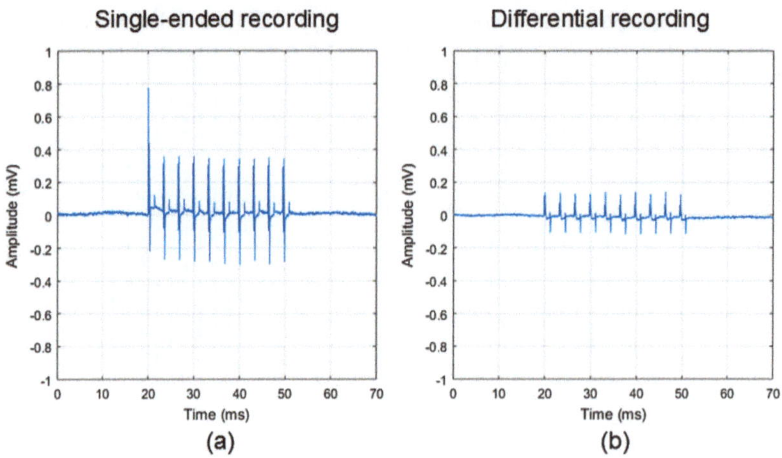

Fig. 5.11 Stimulation artifacts from a Type-II stimulator with separate grounds between the recorder and the stimulator. Recorder is configured to do single-ended recording in (**a**) and differential recording in (**b**)

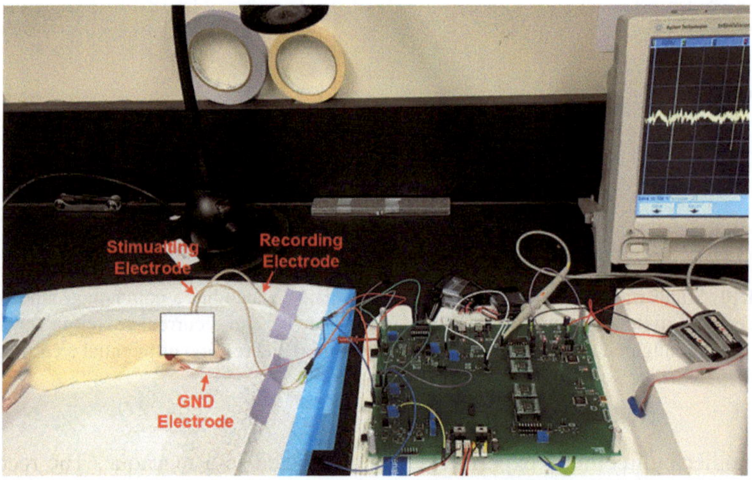

Fig. 5.12 The photo of the in vivo experimental setup. A female Long-Evans rat received two implants, one in the motor cortex, the other in the sensory cortex. The measurement was conducted using the same testing board as used in Sect. 5.2.4

5.2.6 Conclusion

In this work, the stimulation artifact in a bidirectional neural interface has been studied. Different electrode and circuit configurations were taken into account in this study. Both in vitro and in vivo experiments were conducted. Several conclusions are summarized as follows:

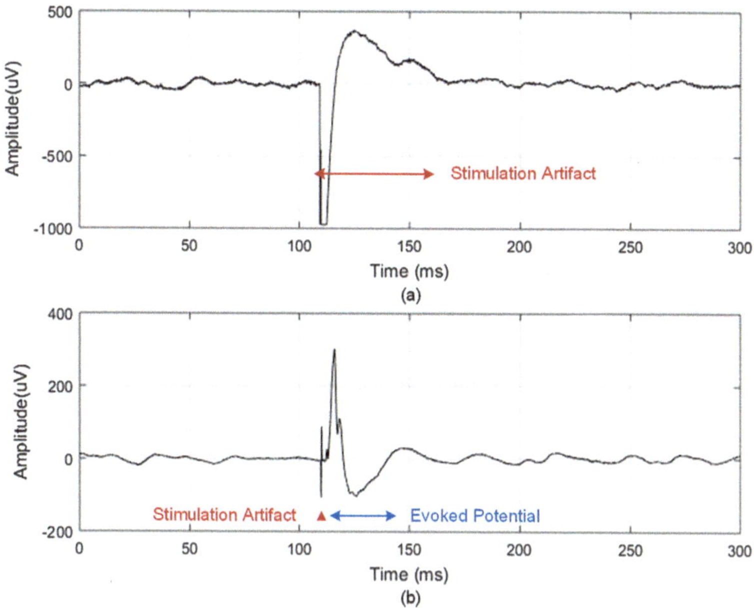

Fig. 5.13 A bidirectional neural interface experiment in a Long-Evans rat. Local field potential was recorded in the motor cortex while stimulating the sensory cortex. (**a**) When the grounds of the recorder and stimulator were connected, there was a large artifact and a long recovery time; (**b**) when the grounds were separated, the artifact is minimized. An evoked potential was clearly visible

- Stimulation artifacts can be minimized if the grounds of the recorder and stimulator are separated. Circuit techniques can be used to enable the isolation even if the bidirectional interface system-on-chip (SoC) is implemented on a single die.
- A charge balanced stimulation results in much smaller artifacts in recording than an unbalanced stimulation. Since charge balancing is not guaranteed in a voltage regulated stimulation, stimulation artifacts from voltage-regulated stimulator is usually much worse than from current-regulated stimulator.
- A bipolar stimulation usually results in smaller artifacts in the recording than a monopolar stimulation, since the stimulation is restrained in the area between the two electrodes. However, a bipolar stimulation cannot replace a monopolar stimulation in triggering certain physiological response.
- A differential recording usually suppresses the stimulation artifacts. However, attention must be paid to make sure the differential electrodes are within the linear input range of the recording amplifier. If the signal is distorted, it might lead to wrong analysis results. The input range of the recording amplifier is usually limited by the supply voltage and the ESD circuits. Extending the input range of the recording amplifier is very helpful in preventing the input stage from saturation. However, this is quite challenging in a low voltage front-end design.

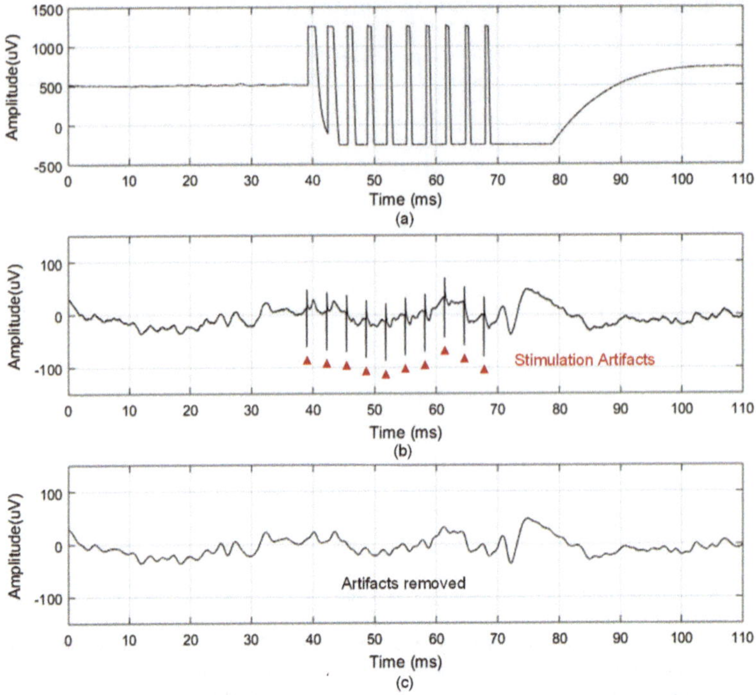

Fig. 5.14 A bidirectional experiment in a Rhesus macaque. (**a**) The grounds of the stimulator and the recorder were shorted together, the stimulation artifact saturates the amplifier, and it takes hundreds of milliseconds to recover. (**b**) The grounds of the stimulator and recorder were separated, and only minor artifact appeared in the recording. (**c**) A 200 Hz filter was applied to the recording in (**b**), which completely removed the artifacts

- There are several methods and circuit techniques which can speed up the recovery of the input stage of the neural amplifier. One solution is to temporarily shift the high-pass frequency corner to a higher frequency during or right after the stimulation. However, the signal will be corrupted and cannot be recovered without the knowledge of the exact timing, frequency and phase change.
- With a common ground, a Type-II stimulator with a current sink gives a larger artifact than a current source. This is because with a current sink in the working electrode, the counter the electrode will need to be connected to a high voltage with respect to the ground, which gives a large step input to the recording amplifier.
- Discharging the stimulation electrode might also give a step response to the recording amplifier.

In summary, separating recording and stimulation ground is highly recommended for a bidirectional neural interface design, especially if a monopolar stimulation and

Table 5.2 Stimulation artifacts with common ground

	Single-ended recording	Differential recording
Monopolar stimulation	– – –	+
Bipolar stimulation	–	++

a single-ended recording is necessary. If a common ground must be used, Table 5.2 gives an estimation of the stimulation artifacts (both amplitude and duration). In this table, "+" means good and "–" means bad.

5.3 Closed-Loop Neural Interface System

5.3.1 Introduction

There are two fundamental types of control systems, open-loop control system and closed-loop control system [286]. In an open-loop control system, the control action signal is independent of the output of the plant under control. In a closed-loop control system, the control action signal is dependent on the output of the plant through the feedback loop. Figure 5.15a shows the block diagram of a typical closed-loop control system. The system consists of a sensor, an actuator, a closed-loop controller, and the plant under control. Ideally, the closed-loop controller generates the control action signal for the actuator to set the output of the plant to be the same as the reference. Closed-loop control finds its applications in almost everywhere, not only in electrical or mechanical engineering, but also in biology, climate science, social science, economics and finance, and other applications [287]. Figure 5.15b shows the block diagram of a bidirectional closed-loop neural interface system, where the neural recorder works as the sensor, and the neural stimulator works as the actuator.

There are two types of feedback, the positive feedback and the negative feedback [287]. In a system with a positive feedback, the fed-back signal is in phase with the signal, while in a system with a negative feedback, the fed-back signal is out of phase with respect to the input signal. Both positive and negative feedback find applications in circuits and system design, but negative feedback is usually more applicable, because it improves the stability and accuracy of a system by correcting or reducing the unwanted changes. This is especially important for a neural interface since a positive feedback induced oscillation may cause permanent damage to the neural system.

There are many well-established control theory and stability compensation methods [115, 288]. The most commonly used closed-loop controller is a proportional-integral-derivative (PID) controller. More than 95% of the closed-loop industrial

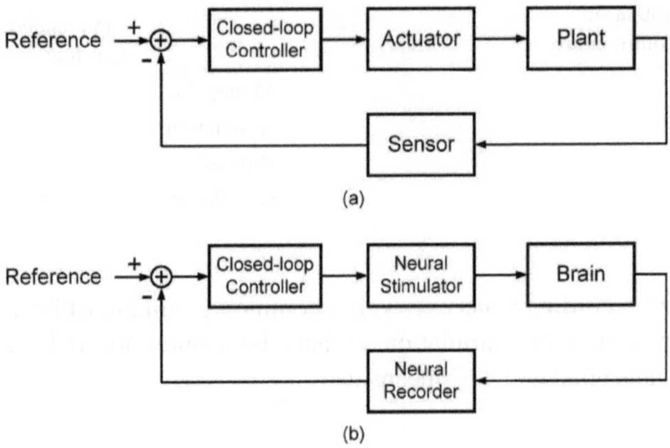

Fig. 5.15 (**a**) The block diagram of a typical closed-loop control system. (**b**) The typical block diagram of a bidirectional closed-loop neural interface system. The neural recorder works as the sensor, and the neural stimulator works as the actuator

processes use PID controllers [289]. The terms of a PID controller can be interpreted as: the proportional term depends on the present error, the integral term depends on the accumulation of past errors, and the derivative term is a prediction of future errors, based on the current rate of change [290].

This section presents the analysis and design of closed-loop neural interface systems. The mechanisms of different closed-loop neural interface systems are first reviewed. Then, the design of a closed-loop neural interface with a general–purpose PID controller is presented. Some of the figures and tables presented in this section were originally published in [57] ©IEEE. Reused, with permission.

5.3.2 *Mechanism of Closed-Loop Neural Interface System*

Figure 5.16 shows typical closed-loop brain–machine interface (BMI) systems with different control mechanisms. A BMI for prosthetic arm is used as an example for illustration, and the PID controller represents any closed-loop controller here. Figure 5.16a shows a basic bidirectional BMI. The electrical stimulator encodes the sensory information in sensory cortex, and the brain generates the motor intent. The motor intent is then decoded by the processor for controlling the actuators. In other words, the closed-loop control policy origins from the brain. J. Liu et al. from Michigan State University proposed an application aiming to improve the sensory encoding capacity of the BMI in 2011 [291]. Figure 5.16b illustrates the proposed method. The method involves an encoder mapping the sensory data acquired from the prosthetic to the desired patterns related to the somatosensory cortex activity. The errors between these desired patterns and those recorded in the somatosensory cortex are used in a PID controller to update the stimulation of the sub-cortical

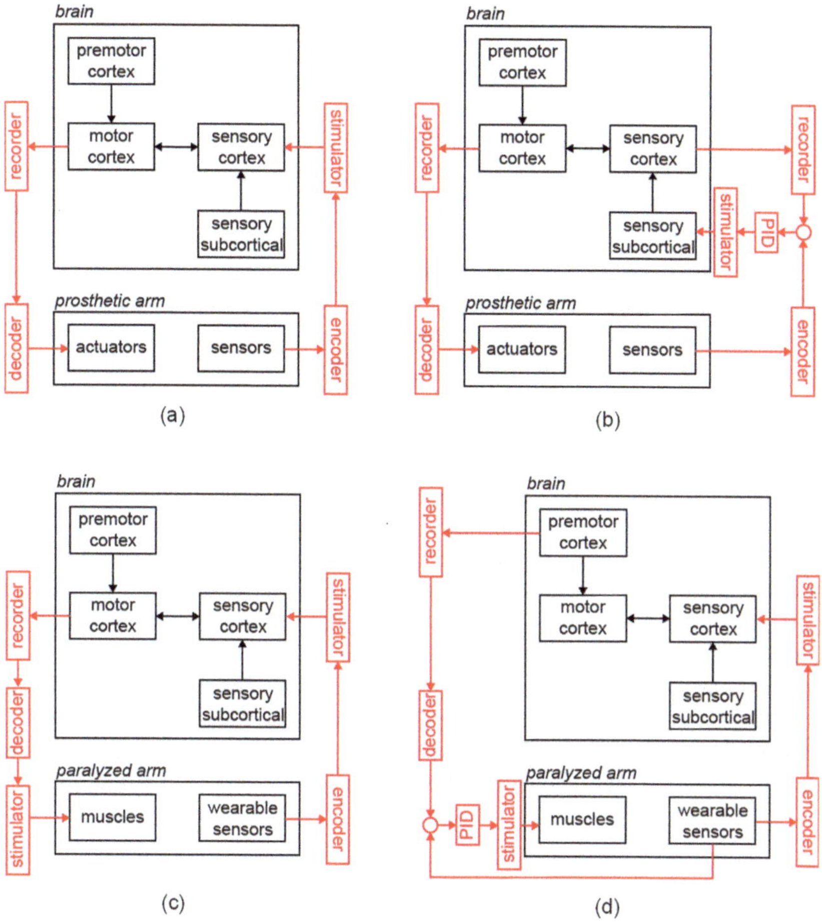

Fig. 5.16 The block diagrams of different closed-loop BMI applications. The functions of the proposed neural interface system are shown in *red*. (**a**) Standard bidirectional BMI for a prosthetic arm. (**b**) Same as (**a**) but with improved sensory encoding method using a PID controller. (**c**) Bidirectional BMI to re-animate paralyzed arm by decoding desired arm trajectory. (**d**) Same as (**c**) but decoding motor goal and implementing arm trajectory with a PID controller

somatosensory areas in the thalamus or brainstem. This approach could elicit more continuous, natural sensory percepts compared to those evoked by the limited set of pre-programmed typical stimulation patterns [292].

Another closed-loop control mechanism is illustrated in Fig. 5.16c. In this case, the BMI system uses electrical stimulation to control a paralyzed arm rather than a prosthetic arm. Brain-controlled muscle stimulation has been proved to be a viable method of re-animating paralyzed arms in monkeys and humans [24, 279, 280]. In these studies, the muscle stimulation, and thus the arm movement trajectory, was entirely driven by motor cortex activity. However, prior work has shown that recording from pre-motor cortical areas to decode motor goals, not entire intended trajectories, can improve performance and lower cognitive demand [64, 281]. Thus, another potential BMI application for a closed-loop controller could be to update the muscle stimulation based on the error between a decoded goal and the recorded state of the re-animated arm [293], as illustrated in Fig. 5.16d.

Besides prosthetics, other examples of closed-loop bidirectional BMI applications include the deep brain stimulation (DBS) for Parkinson Disease and epilepsy. H. Rhew et al. from the University of Michigan, Ann Arbor proposed a closed-loop DBS system in 2014 [48]. The system detects the abnormal energy in the LFP and adjusts the stimulation current using a PI controller. W. Chen et al. from the National Chiao Tung University proposed a closed-loop neural prosthetic in 2014 [44]. The proposed design detects the seizure event and delivers a deep brain stimulation with parameters modulated from the internal brain state.

In addition, closed-loop neural interfaces are also important for electrophysiological studies [294]. A. Wallach et al. from Technion proposed a neuronal response clamp in 2011 [274]. In this work, a closed-loop technique enabling control over the instantaneous response probability of a neuron was proposed using a PID controller. It has been used to characterize the input–output neuronal relationship. Sense-stimulate devices with closed-loop controllers have also been proposed for neuromodulatory applications [40]. Moreover, closed-loop stimulation of sleep slow oscillation has been proposed to enhance memory [282].

In summary, there are many different configurations and mechanisms for using a closed-loop neural interface, and it is critical for a wide range of prosthetics and neuroscience research. Thus, the design of an energy efficient real–time bidirectional neural interface with a closed-loop controller is highly desirable.

5.3.3 Design of a Closed-Loop Neural Interface with a PID Controller

5.3.3.1 Introduction

As reviewed in Sect. 5.3.1, the PID controller is the most commonly used control loop feedback mechanism [290]. Brain is a highly non-linear, dynamic time-variant system, which can hardly be accurately modeled. A PID controller needs only the process variables and the target value, not requiring the knowledge of a system model or the underlying process. Thus, the PID controller is suitable for a wide range of applications in neuroscience research and neuroprosthetic development. In this work, a programmable PID controller in the analog domain has been designed to support a variety of closed-loop experiments.

The basic working principle of the PID controller is briefly reviewed here. A PID controller calculates the difference between the desired reference and the measured output of the plant under test as the error value $e(t)$. The output of a PID controller $u(t)$ in the time domain is:

$$u(t) = K_p e(t) + K_i \int e(t)dt + K_d \frac{de(t)}{dt} \quad (5.4)$$

where K_p, k_i, and K_d are coefficients for the proportional, integral, and derivative terms, respectively. By tuning the three parameters of the model, a PID controller can meet different control requirements.

There are four main characteristics of a closed-loop step response, including: (1) rise time, which is the time it takes for the system's output to rise to 90% of the desired level; (2) overshoot, which is the peak level higher than the steady state, usually normalized against the steady state; (3) settling time, which is the time it takes for the system to converge to the steady state; and (4) steady-state (S-S) error, which is the difference between the steady-state output and the desired output. The effect of each controller parameter K_p, K_i, and K_d are summarized in Table 5.3 [290]. It should be noticed that the tuning process is usually more complicated in practice.

5.3.3.2 System and Circuit Implementation

The PID controller has been reported for implementation in both analog [295, 296] and digital domains [297, 298]. The analog implementation has the advantages of low-power consumption and compact layout, especially when the input and output are both analog signal. In this work, the PID controller is implemented using Gm-C blocks. The block diagram of the overall closed-loop system is shown in Fig. 5.17. The system consists of the closed-loop controller, a neural recorder, a neural feature extraction unit, a neural stimulator, and buffers for connecting them. In the PID controller, the error signal is the difference between the extracted neural feature and a pre-set reference value. The output of the PID controller is a weighted sum of the

Table 5.3 General effects of PID parameters

Parameter	Rise time	Overshoot	Settling time	S-S error	Stability
K_P	Decrease	Increase	Small change	Decrease	Degrade
K_I	Decrease	Increase	Increase	Eliminate	Degrade
K_D	Minor change	Decrease	Decrease	No effect in theory	Improve if K_d is small

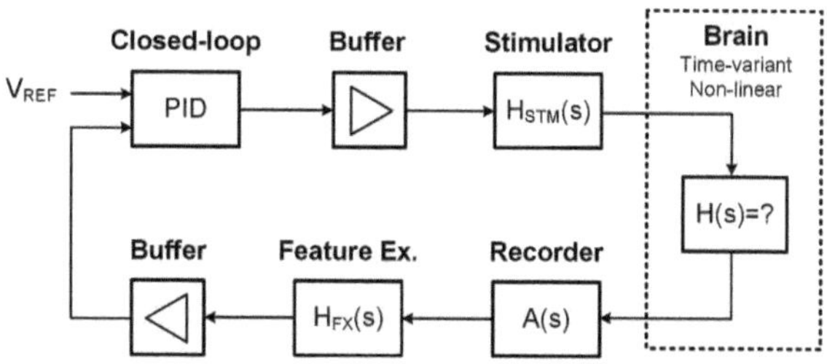

Fig. 5.17 The block diagram of the overall system using the designed PID controller

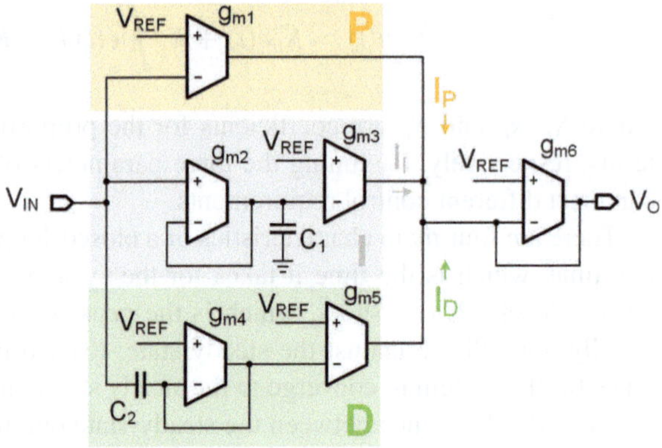

Fig. 5.18 The circuit schematic of the designed PID controller

proportional, the derivative, and the integral terms of the error signal. The neural stimulator works as the actuator in the system. The output of the PID controller can be used to modulate the stimulating current amplitude, the stimulating frequency, or other parameters. The sensor in the system is the neural recorder and the neural feature extraction unit. Neural features, for example, spectral energy, entropy and action potential fire-rate, can be used as the input of the PID controller.

The circuit schematic of the PID controller is shown in Fig. 5.18. The PID controller consists of six programmable Gm blocks and two capacitors. The circuit schematic of the Gm block and the tunable biasing current generation module can be found in Figs. 3.4 and 3.6 in Chap. 3, respectively. The parameters for each of the *P*, *I*, and *D* components are independently programmable. The transfer function of the PID controller is given by:

$$\frac{V_{\text{out}}(s)}{V_{\text{err}}(s)} = \frac{g_{m1}}{g_{m6}} + \frac{g_{m3}}{g_{m6}\left(1 + \dfrac{sC_1}{g_{m2}}\right)} + \frac{g_{m5}}{g_{m6}\left(1 + \dfrac{g_{m4}}{sC_2}\right)} \tag{5.5}$$

where $V_{\text{err}}=V_{\text{ref}} - V_{\text{in}}$. The gain of the *P*, *I*, and *D* components are $K_P=g_{m1}/g_{m6}$, $K_I=g_{m3}/g_{m6}$, and $K_D=g_{m5}/g_{m6}$.

The basic parameter choosing and tuning of the PID controller have been reviewed in Sect. 5.3.1. For a complex neural system where an accurate model can hardly be achieved, the plant exploration based method can be used. The initial estimation of the optimal operating points can be learned from the Zeigler-Nichols tuning method [299]. The final controller parameters can be determined by using an iterative procedure, based on the least-root-mean-square error. Considering the requirement of a BMI system, sufficient phase margin for stability must be guaranteed. Since an in-depth study of the control theory is not the focus of this work, the well-established PID tuning theory will not be discussed here. Readers can find information in references [300, 301].

5.3.3.3 Experimental Results

The proposed design has been fabricated in IBM 180 nm CMOS technology. The micrograph of the fabricated chip and the layout of the PID module are shown in Fig. 5.19, with major blocks highlighted. The PID module occupies a silicon area of 100 μm × 75μm, including digital registers. Bench testing was conducted to evaluate the designed PID controller. In vivo bidirectional closed-loop experiments will be discussed in the next chapter.

The basic function and tuning of the PID controller were tested with a second-order RC ladder network, as shown in Fig. 5.20. The transfer function of the RC network is given by:

$$H(s) = \frac{\frac{1}{R_1 R_2 C_1 C_2}}{s^2 + s\left(\frac{1}{R_1 C_1} + \frac{1}{R_2 C_1} + \frac{1}{R_2 C_2}\right) + \frac{1}{R_1 R_2 C_1 C_2}} \tag{5.6}$$

The output of the RC ladder network is fed back to the PID controller and compared with a preset reference signal to find the error signal. A step change from 0.8 to 0.9 V was given at the reference. The measured transient response of the system in different configurations is shown in Fig. 5.21. The design is programmable over a large range, and useful in versatile closed-loop applications.

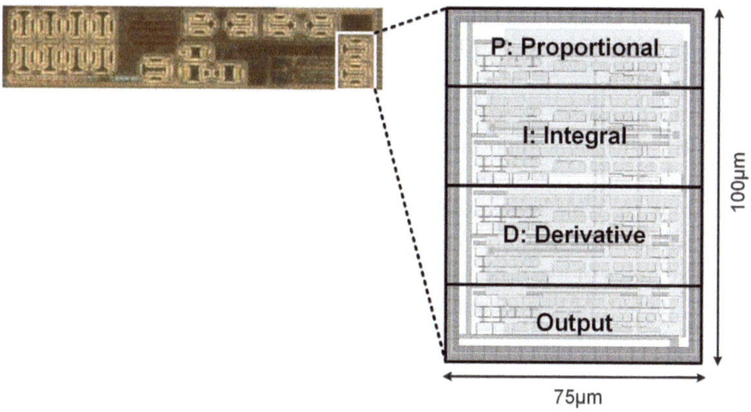

Fig. 5.19 The microphotograph and the layout of the PID controller module

Fig. 5.20 The second-order RC ladder network used for testing the PID controller

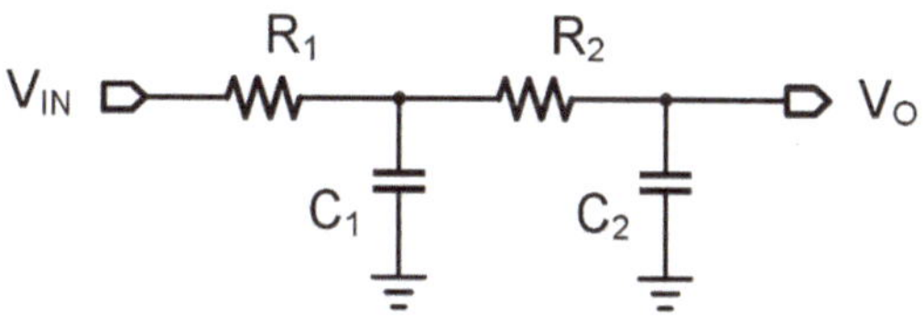

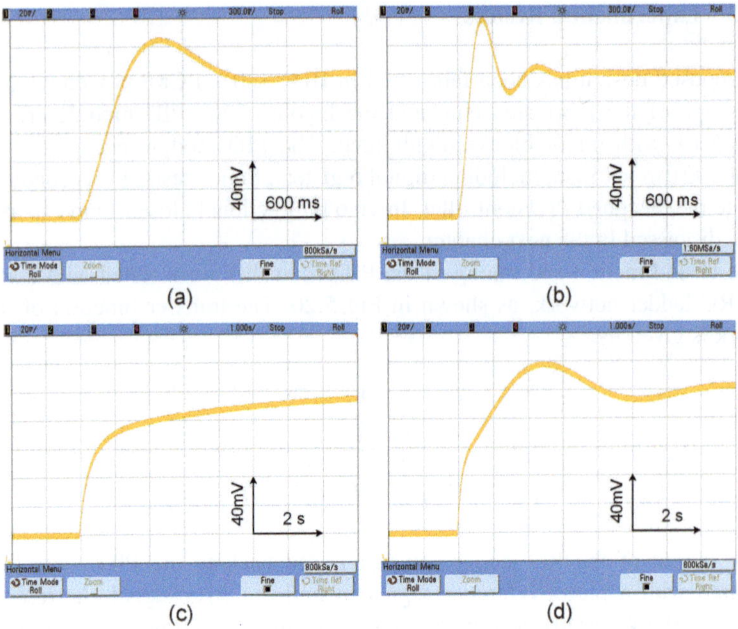

Fig. 5.21 The measured transient response of the PID controller in different configurations. The ratio of the K_p, K_i, and K_d are (**a**) 1-2-0, (**b**) 2-1-0, (**c**) 1-0-1, and (**d**) 1-1-1

In addition, a closed-loop neuronal response clamp experiment [274, 294] was set up to test the proposed PID controller. Nervous systems' response to the rapidly changing sensory information is highly variable with complex dynamics. The dynamic response is reflected from a single neuron level to a neuronal network level. Thus, it is important to study the behavior in a closed-loop approach with an appropriate context of realistic input–output dependency. Voltage-, and current-clamps are well-known techniques [294] in closed-loop electrophysiology. Recently, a dynamic neuronal response clamp technique was proposed to study the threshold dynamics of a neuron using extracellular stimulation and recording [274]. A modified version of this technique is employed to test the proposed closed-loop system.

The diagram of the designed testing system is illustrated in Fig. 5.22. The major blocks used are the PID controller, a neural stimulator, an action potential detector, a neuron model, and a lossy integrator for finding spike-rate. The integrate-and-fire model [208] for the single neuron employed in this experiment can be expressed as:

$$\tau_m \frac{dV}{dt} = V_m - V(t) + R_m I_s(t) \tag{5.7}$$

where $\tau_m \approx 10\,\text{ms}$ is the membrane time constant, V_m is the resting membrane potential, $V(t)$ is the actual membrane potential as a function of time, $R_m \approx 10^7\Omega$

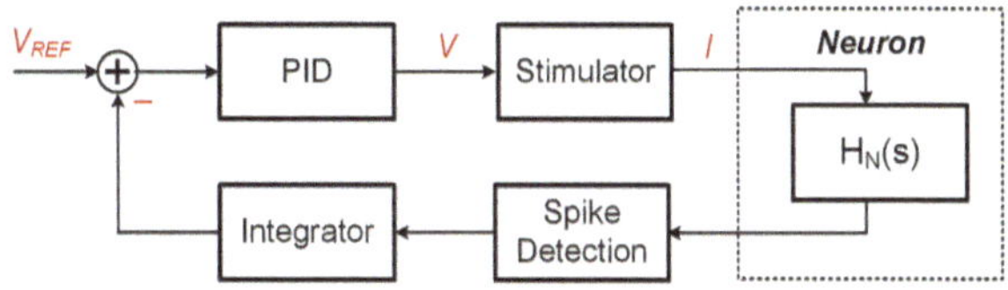

Fig. 5.22 The block diagram of the dynamic neural clamp experiment using the PID controller. The major blocks used are the PID controller, a stimulator, an action potential detector, a neuron model, and a lossy integrator for finding spike rate. The on-chip stimulator is configured in the test mode to output a continuous current

and $I_s(t)$ is the stimulation current. Once the membrane potential reaches a certain threshold V_{TH}, an action potential occurs and reset the potential back to its resting membrane potential. In this test, an off-chip microcontroller (Atmel XMEGA 128A4U [302]) with integrated ADC and DAC was used to model the neuron. The MCU is running at a sampling rate of 100 kHz, corresponding to a time resolution of $dt = 10\,\mu\mathrm{s}$. The continuous time differential equation is simplified by a discrete difference equation for the implementation in the MCU. The MCU's ADC measures the $R_m I_s[t]$, and the DAC generates $V[t]$ based on the following equations:

$$V[t] = \begin{cases} \dfrac{V_m + R_m I_s[t-1] + \tau_m V[t-1]}{1+\tau_m} & V[t-1] < V_{\mathrm{TH}} \\ V_m & V[t-1] > V_{\mathrm{TH}} \end{cases} \tag{5.8}$$

The stimulator was reconfigured in a testing mode to output a continuous stimulation current to meet the requirement of intracellular stimulation. The stimulation current amplitude was modulated by the output voltage of the PID controller. The neural model responded to the stimulation current, generating the membrane potential. The action potential detector module evaluated the membrane potential voltage with a predefined threshold voltage. The output of the detector was a PWM wave, which was sent to the integrator and converted to a voltage proportional to the spike rate. In this work, the spike rate was converted to a voltage in the embedded MCU. The difference between the integrator's output voltage and the reference voltage was sent to the PID controller.

Figure 5.23 shows 12 test trials with different proportional-integral-derivative parameters. Each dot indicates an action potential. The same reference was set for all trails at the start time. The neuron responded to the stimulation current until it settled at a constant firing rate, in a manner based on the choice of the P, I, and D terms. At the same time, Fig. 5.24 shows 12 testing trials with different references. The neuron settled at a relative constant firing rate proportional to the reference. The test results showed that by programming the parameters, one can control the behavior of the neuron without the knowledge of the exact model [294].

In summary, in this work, the importance of closed-loop control in neuroscience study and neuroprosthetic development is analyzed, and the mechanisms of different

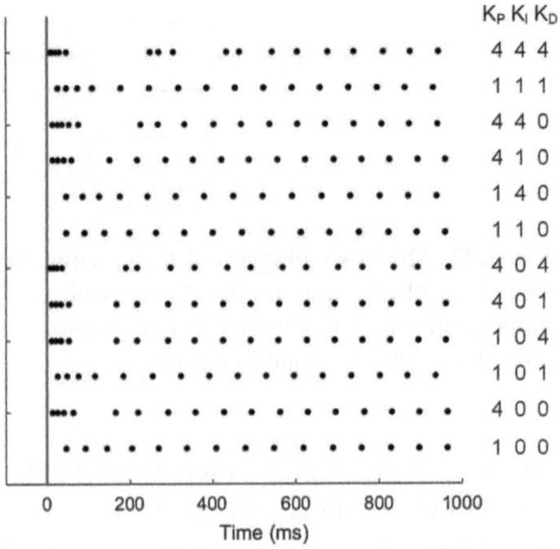

Fig. 5.23 The measured transient responses of the dynamic neuronal clamp with different PID parameters. Relative values of *P*, *I*, and *D* components are shown on the *right*

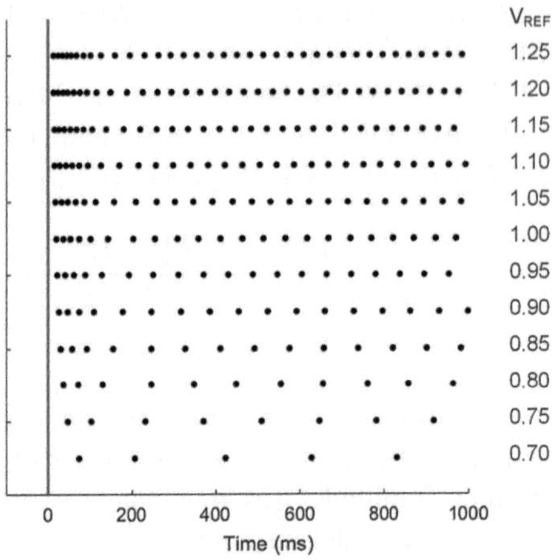

Fig. 5.24 The measured transient responses of the dynamic neuronal clamp with different references

closed-loop BMI systems are reviewed. The most commonly used closed-loop controller, a general–purpose PID controller, is selected to be implemented in analog integrated circuits, and integrated in a bidirectional closed-loop BMI system on chip (SoC). The system and circuit implementation of the PID controller are presented, and the function of the designed chip has been evaluated in bench testing. The proposed design provides a promising solution for a wide range of neuroscience and neural engineering investigations.

Chapter 6
System Integration and Experiments

6.1 Introduction

Behavioral and in vivo animal experiments have been used through the history of biomedical research. Non-human animal experiments have become one of the most important methods in modern neuroscience research, and are highly valuable for the development of the brain–machine interface (BMI). However, most of the available medical instruments are designed for human medical treatments, which may not work well on animals. It is especially challenging if the designed experiment requires monitoring of the animal's brain activities or giving real–time stimulation feedback while the animal is freely behaving. In addition, the study of neural modulation and closed-loop control also requires a custom designed wearable or implantable BMI device to perform on-chip signal processing, feature extraction, classification, machine learning, and neuromodulation. In summary, the design of a wireless BMI system for freely behaving animals is of great practical value and provides a very powerful tool for neuroscience research.

Previous chapters have discussed the neural interface circuit design from several perspectives. However, a complete system is more than a simple summing of individual building blocks. More importantly, many practical design issues are overlooked and underemphasized in the literature. A system that has been perfectly characterized in bench testing may not work well in an actual animal experiment. In this chapter, custom system integration for animal experiments, especially for free behavior animal experiments is discussed. The methodologies and experimental results are presented in detail.

The chapter is organized as follows. Section 6.2 presents a general–purpose experimental platform, namely the PennBMBI. A custom designed command and communication protocol is presented, with a user-friendly interface. Wireless neural recording, stimulating and sensing functions have been verified in both anesthetized and awake rats. Section 6.3 presents a custom designed watermaze experiment for the study of augmenting perception through modulated electrical

X. Liu, J. Van der Spiegel, *Brain-Machine Interface*,
https://doi.org/10.1007/978-3-319-67940-2_6

stimulation of somatosensory cortex. A waterproofed wireless neural stimulator and a complete animal tracking and neuromodulation experimental system are presented. Section 6.4 describes a bidirectional neural interface system for freely behaving monkeys. Long-term neural stimulation and recording during monkey's awake, sedated, and sleeping states are presented. A study on the hippocampal gamma-slow oscillation coupling using the developed system was also described. All experimental procedures used in the studies presented in this chapter were approved by the institutional animal care and use committee (IACUC) of the University of Pennsylvania.

6.2 The PennBMBI: A General–Purpose Experimental Platform

6.2.1 Introduction

In this section, the design of a general–purpose wireless brain-machine-brain interface (BMBI) system is presented [43, 51]. The system integrates four battery-powered wireless devices designed for a closed-loop sensorimotor neural interface, including a neural signal analyzer, a neural stimulator, a body-area sensor node, and a user-friendly graphic interface implemented on a PC. The neural signal analyzer features a four-channel analog front-end with configurable passband, gain stage, digitization resolution, and sampling rate. The target frequency band is configurable from EEG band to action potential band. A noise floor of 4.69 μVrms is achieved over a bandwidth from 0.5 Hz to 6 kHz. Digital filtering, neural feature extraction, spike detection, sensing-stimulating modulation, and compressed sensing measurement are realized in a central processing unit integrated into the analyzer. A flash memory card is also integrated into the analyzer. A two-channel neural stimulator with a compliance voltage up to ± 12 V is included. The stimulator is capable of delivering unipolar or bipolar, charge-balanced current pulses with programmable pulse shape, amplitude, width, pulse train frequency and latency. The system also includes a multi-functional sensor node, consisting of an accelerometer, a temperature sensor, a force sensor, and a general sensor interface port. A computer interface is designed to monitor, control, and configure all aforementioned devices via a wireless link, using the custom designed communication protocol. Wireless closed-loop operation between the sensory devices, neural stimulator, and neural signal analyzer can be configured. The proposed system was designed to link two sites in the brain, bridging the brain and external hardware, as well as creating new sensory and motor pathways for clinical practice. Some of the figures and tables presented in this section were originally published in [51] ©IEEE. Reused, with permission.

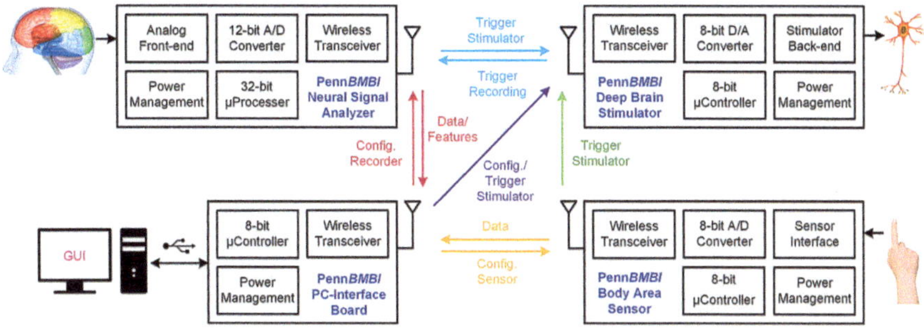

Fig. 6.1 The block diagram of the PennBMBI system. The system mainly consists of a neural signal analyzer, a neural stimulator, a multi-functional sensor node, and a graphic user interface. Closed-loop operation paths between devices are shown

6.2.2 System Overview

Figure 6.1 illustrates the block diagram of the PennBMBI system. In general, there are four types of devices, including (1) neural signal analyzer, (2) neural stimulator, (3) multi-functional sensor node, and (4) PC interface with the graphic user interface (GUI).

The first block is a four-channel neural signal analyzer (NSA). The proposed NSA consists of an analog front-end that is capable of recording neural signal from the EEG band to the action potential band, with an input signal amplitude varying from $10\,\mu V$ to $1\,mV$. Configurable analog bandpass filters are used to suppress the electrode offset, and to bandpass the signal in the frequency of interest. An additional programmable gain stage and an analog-to-digital converter (ADC) with programmable sampling rate and resolution are also integrated. Digital filtering, neural feature extraction, action potential detection, sensing-stimulating modulation, and compressed sensing [153] are realized in a central processing unit integrated on board.

The second block is a dual-channel stimulator with a high driving capability that enables a charge-balanced current stimulation up to $400\,\mu A$ with a compliance voltage of $\pm 12\,V$. The device can be wirelessly controlled to deliver current-regulated stimulation pulses with programmable pulse shape, width, pulse train frequency and latency.

The third block is a multi-functional body-area sensor node. The sensor enables the communication of sensory information to the brain with sensor-controlled wireless neural stimulation. The sensor node integrates a 3-axis accelerometer, a temperature sensor, a force sensor, and a general–purpose analog interface port. The port can be used with different commercial sensors, such as pressure sensor, motion sensor [303, 304], etc.

In addition, a graphic user interface (GUI) has also been implemented for all device configurations, data acquisition, and online and offline signal analysis. A wireless link is implemented between all the devices for device configuration and data transmission.

Table 6.1 Organization of the memory bank

Bank Addr	Word Addr	Description	Value
BANK 00	word 00	Password	
BANK 01	word 00 bit[1:0]	ID—Property	00—GUI 01—Neural analyzer 10—Stimulator 11—Sensor node
	word 00 bit[7:2]	ID—Serial #	0x00–0x3F
BANK 10	word 00	Bank 10 Length	
	word 01	Bank 11 Length	
	word 10-END	Configuration	
BANK 11	word 00-END	User memory	

Wireless links have been built between all devices for configuration, data transfer, and closed-loop operation. The transceiver nRF24L01+ from Nordic Semiconductor is used in all devices. The transceiver features a maximum on-the-air data rate of 2 Mbps in 2.4 GHz ISM band using GFSK modulation. Embedded Enhanced ShockBurst^{TM} baseband protocol engine and automatic packet transaction handling are integrated in the transceiver. Commands and data were sent with two-byte Cyclic Redundancy Check (CRC) scheme with an acknowledgement and auto-retransmit ability.

A customized protocol has been designed for the communication with all devices in the system. The four devices are divided into two classes: (1) central unit, which is realized in the PC through a wireless dongle; and (2) satellite devices, including the NSA, the stimulators, and the sensor nodes. The memory in each satellite device is organized in four banks, consisting of 8-bit words as illustrated in Table 6.1. A password is saved in Bank 00 for kill and/or lock functions. Bank 01 is reserved for device ID number and class identity. Bank 10 records all the configuration information for online processing. Bank 11 is a user memory which can be freely organized in any fashion depending on the application.

All of the satellite devices can be configured by the central unit via the wireless communication channel. The commands are all organized in a "header + argument + data" format, including (priority from high to low)

1. CMD RST. Reset commands include global reset command and local reset command. Global reset command terminates all undergoing procedure in all devices, while a local reset command only works on selected devices with a matched ID in the header. Global reset command can only be sent by the central unit, while local reset command can be sent from any host device. Acknowledgement from the slave will report the state before the reset operation. The slave device will enter IDLE state after the reset operation.

2. CMD STD. Standby command is similar to the reset command. It pauses the undergoing procedure in a target device without reset it. Acknowledgement from the slave will report the state it stops at.
3. CMD WKP. Wake-up command is used to return a device from the IDLE state or to continue a current procedure which was previously paused by a standby command.
4. CMD CFG. Configure command is used for online configuration of a target device. The specifications will be embedded in the argument section. The translation of the argument value varies while a different target device is applied. Table 6.2 only lists the argument section for the configuration of the NSA and the stimulator.
5. CMD ACC. Access command is a request to get the communication channel access. Usually data transfer is followed by this command after a proper acknowledgement is received. Any device with an approved flag received from an acknowledgement and the host of the acknowledgement are denoted as a "matched pair." The transmitting device is defined as the client device, while the receiving device is defined as the host device.
6. CMD DTX. Data transmission command carries the repackaged memorized data from a matched client device to a matched host device.

6.2.3 *Hardware Implementation*

6.2.3.1 Neural Signal Analyzer

A Neural Signal Analyzer (NSA) is designed to perform neurological signal recording and analysis. The NSA integrates a four-channel analog front-end, a central processing unit (CPU), a 2.4 GHz wireless transceiver, a removable Micro SD card, a power management unit, and peripheral modules. Figure 6.2 shows the assembled device. The overall dimension including the battery is 56 mm × 36 mm × 13 mm.

The analog front-end integrates four independent amplifier channels and a tissue ground driving circuit. The circuit block diagram of the front-end circuit is shown in Fig. 6.3. A supply voltage of 3.3 V is used. Configurable gain stages and filters are designed to meet the requirements of recording neurological signals with different bandwidth and amplitude levels [78]. The input signal is capacitively coupled to the instrumentation amplifier. The input impedance is over 100 MΩ, which is important when using high impedance electrodes. The highpass frequency corner is lower than 0.5 Hz, which rejects the DC offset from the electrode polarization, and also preserves the very low frequency signal component [90].

The gain of the first stage instrumentation amplifier is set to be 200 [305]. Resistors with a low-temperature coefficient (TC) are used to minimize the gain drift. An integrator implemented by amplifier A4 with a configurable capacitor is used as a DC servo loop. Amplifier A5 is used to provide an additional

Table 6.2 Organization of the customized communication command

CMD RST [a]	Header	CMD ID, 8 bit
		Host ID, 8 bit
		Client ID, 8 bit (0x00 for global reset)
ACK RST	Header	ACK ID, 8 bit
		Device ID, 8 bit
	Argument	STATE ID
CMD STD	Header	CMD ID, 8 bit
		Host ID, 8 bit
		Client ID, 8 bit (0x00 for global pause)
ACK STD	Header	ACK ID, 8 bit
		Device ID, 8 bit
	Argument	STATE ID
CMD WKP	Header	CMD ID, 8 bit
		Host ID, 8 bit
		Client ID, 8 bit (0x00 for global wake-up)
ACK WKP	Header	ACK ID, 8 bit
		Device ID, 8 bit
	Argument	STATE ID
CMD CFG	Header	CMD ID, 8 bit
		Host ID, 8 bit
		Client ID, 8 bit
	Argument (Analyzer)	Channel #, gain, stop frequency 1, stop frequency 2, filter gain, and CRC, all 8 bit format
	Argument (Stimulator)	Channel #, pulses #, pulse width, stimulation/reversal amplitude, stimulation/interphasic/interval time, pulse shape, and CRC, all 8 bit format
ACK CFG	Header	ACK ID, 8 bit
		Device ID, 8 bit
	Argument	STATE ID, 8 bit
		Flag, 0x01 for configuration success
CMD ACC	Header	CMD ID, 8 bit
		Host ID, 8 bit
		Client ID, 8 bit
ACK ACC	Header	ACK ID, 8 bit
		Device ID, 8 bit
	Argument	Flag, 0x01 for approval
CMD DTX	Header	CMD ID, 8 bit
		Host ID, 8 bit
		Client ID, 8 bit
	Data	Repackaged memorized data
ACK DTX	Header	ACK ID, 8 bit
		Device ID, 8 bit
	Argument	Flag, 0x01 for successful transmission client device will resend package if failed

[a] Argument section and data section are not required in CMD RST

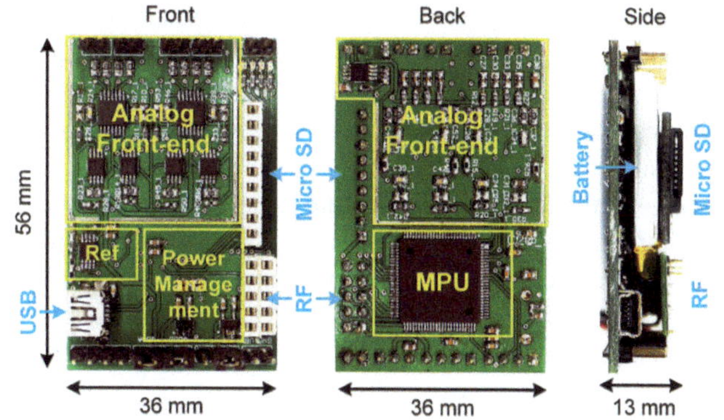

Fig. 6.2 Photograph of the PennBMBI neural signal analyzer (NSA) in front, rear, and side view. The wireless module and Micro SD card are not shown in the front view

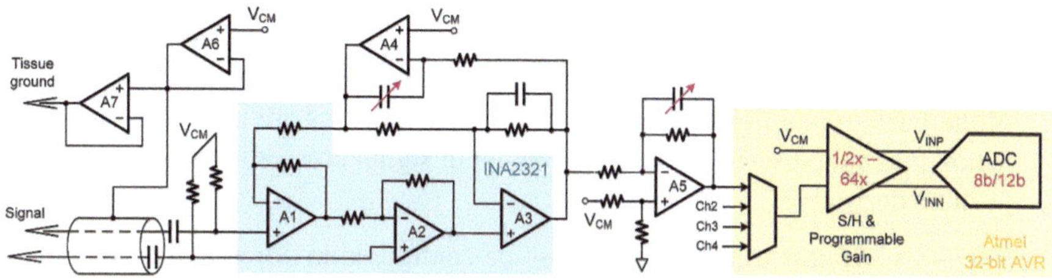

Fig. 6.3 The circuit block diagram of the analog front-end. The analog front-end consists of four-channel instrumentation amplifiers, a programmable gain amplifier, and a programmable ADC

programmable gain and a configurable lowpass frequency. An integrated 12-bit pipeline ADC digitizes the amplified neural signal at a configurable sampling rate. The ADC can also be configured to work in 12 bit or 8 bit. In addition, A6 and A7 are shared by the four channels to drive the tissue ground and the metal shield.

An Atmel 32-bit AVR Microcontroller AT32UC3C1512C [306] is integrated in the NSA device. In the recording mode, a peripheral direct memory access (DMA) controller is used for digital data acquisition, data buffering, and serial peripheral interface (SPI) accessing. The DMA controller handshakes with peripheral interfaces directly, while the central processor core is in the sleep mode to save power. The recorded data can be sent out via the wireless module or to the on-board Micro-SD card through SPI.

Online neural signal processing is performed in the 32-bit floating point DSP core in the microcontroller (MCU). Various functions are implemented, including:

- **Digital bandpass filter**: A Type-I real Finite Impulse Response (FIR) bandpass filter implemented. Six frequency bands are predefined for different applications, and the filter coefficients with 24 taps and 10 taps are stored in the flash memory.

- **Time-domain feature extractor**: Common time-domain features, such as line-length, area, energy, maximum/minimum, and zero-crossing, can be extracted in real–time by a proper configuration of the sliding window length and overlay.
- **Spectral energy feature extractor**: 16/128-point FFT is implemented for spectral analysis.
- **Compressed sensing**: Neural signal features sparsity in certain basis/dictionaries [148], enabling a near lossless reconstruction under sub-Nyquist sampling. A signal agnostic compressed sensing measurement is implemented in the CPU. The input signal vector length N is set to be 512, and the measurement number M (M<N) can be programmable to 256, 128, 64, or 32. The compressed sensing measurement is realized as $y = \Phi x$, where $x \in \mathbb{R}^{N \times 1}$ is the input neural signal, $y \in \mathbb{R}^{M \times 1}$ is the measurements, and $\Phi \in \mathbb{R}^{M \times N}$ is the Pseudo-random measurement matrixes. The Φ is stored in the flash memory. The reconstruction is performed on the receiver end using a convex optimization algorithm.
- **Action potential detection**: For action potential detection, the filters are configured to first bandpass the signal in a frequency range of 300–6 kHz. An amplitude threshold S_{th} is firstly used for a coarse spike detection. The value of S_{th} is $4\times\ \sigma$ of the background noise. Two time-amplitude windows are used to perform the action potential discrimination after the input signal crosses the threshold with a positive derivative.

The NSA wireless transceiver can also be configured in different data streaming modes, including: raw data, extracted neural features, action potential time stamps, and compressed sensing measurements. It can also send stimulation commands wirelessly to the stimulator, or receiving triggers from other devices.

The NSA is powered by a rechargeable 3.7 V lithium-ion battery (Ultralife UBP002). A supply voltage of 3.3 V is used for the analog front-end, digital microcontroller, and wireless transceiver. The quiescent current of the analog front-end is 380 μA per channel. The CPU consumes 490 μA per MHz. A 950 mAh battery supports the device for an overnight continuous recording.

6.2.3.2 Neural Stimulators

A dual-channel neural stimulator is designed to deliver bipolar or unipolar, charge-balanced current pulses with programmable pulse shape, amplitude, width, pulse train frequency and latency. The overall device has a dimension of 43 mm × 27 mm × 8 mm, as illustrated in Fig. 6.4. The stimulator integrates a current driving output stage, an MCU with integrated DAC and ADC, a wireless transceiver, a power management unit, and peripheral circuits.

A dual DC–DC converter is used for boosting the voltage from the 3.7 V lithium-ion battery to ±12 V in order to drive the output stage with high impedance electrodes. The converter will be switched to idle mode when no stimulation is to be delivered to reduce power consumption. A modified Howland current source is used as a bidirectional current driving stage, as shown in Fig. 6.5. Amplifiers A1–A4 are

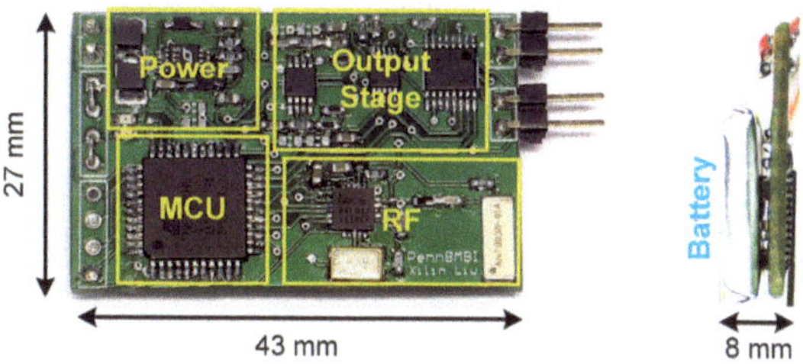

Fig. 6.4 Photograph of the neural stimulator. The stimulator mainly consists of a current driving output stage, an MCU with integrated DAC and ADC, a wireless transceiver, a power management unit

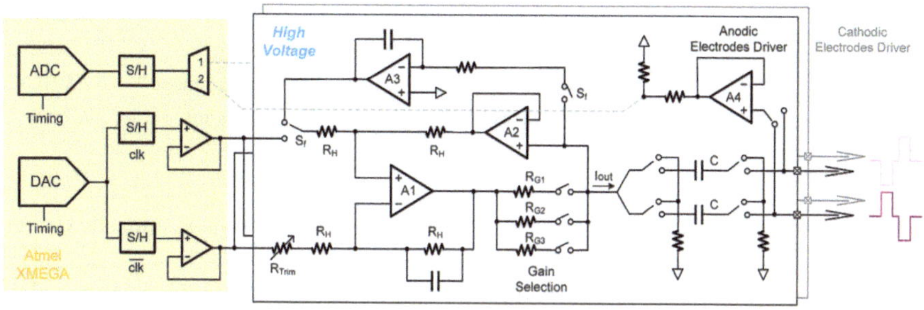

Fig. 6.5 The circuit schematic of the high compliance voltage output stage. Arbitrary stimuli waveform can be generated from a DAC. V-to-I gain is programmable

implemented using high-voltage dual supply op-amps with JFET inputs. A resistor trimmer is used to tune the resistor network for a good common mode rejection ratio (CMRR) and a high output impedance. A feedback capacitor is added for stability. Different transconductances can be selected by setting the gain resistors. Amplifier A2 is a unity-gain buffer which isolates the gain resistor from the resistor network.

The two channel DACs are used to generate differential input to the Howland current source to minimize the offset. The DAC is shut down and both inputs are grounded in idle mode to minimize the power consumption. A feedback integrator A3 is enabled in the idle mode to improve the stability as well as to reduce the current leakage [36]. The amplifier A4 is implemented for buffering the electrode's compliance voltage. An integrated ADC is triggered twice during the stimulation phase to read the compliance voltage after a resistor divider. The impedance of the electrode is calculated in the MCU. If the calculated electrode impedance is lower than a predefined threshold, the MCU generates an alert to the user, and stops the stimulation in order to prevent potential tissue damage from electrode shorting. A blocking capacitor is also used in each channel to prevent direct current injection and limits the maximum net charges.

6.2.3.3 Body Area Sensors

The designed multi-functional body area sensor node has a dimension of 31 mm × 13 mm × 8 mm, as shown in Fig. 6.6. The sensor node integrates a

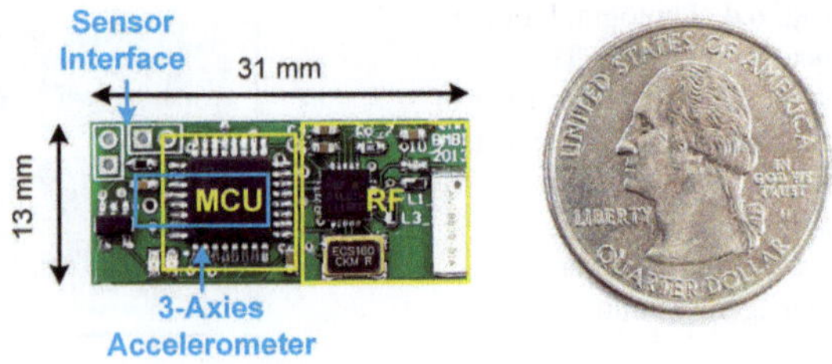

Fig. 6.6 The photograph of the sensor node side by side with a quarter dollar

Table 6.3 Power consumption of the sensor node

Microcontroller	240 μA	Accelerometer [307]	23 μA
RF sleep [308]	0.9 μA	Thermistor	<7 μA
RF transmit	7 mA	Flexiforce sensor	<50 μA
Total working	321 μA	Total transmit	7.3 mA

microcontroller, a 3-axis digital accelerometer, a temperature sensor, and a flexiforce sensor.

The accelerometer interfaces with the MCU through the I2C protocol. The outputs of the thermistor and the force sensor are analog signals, which are digitized by an 8-bit SAR ADC integrated in the MCU. General–purpose ports are reserved for extensions of the sensor node. The sensor node is powered by a 2.65 g, 110 mAh rechargeable lithium battery. The power consumption of all the modules used in the sensor node is listed in Table 6.3.

6.2.3.4 Computer Interface

The computer interface includes a custom designed wireless dongle and a graphic user interface (GUI). The dongle integrates a microcontroller, a wireless transceiver, and a USB 2.0 port. The microcontroller receives commands from the computer and sends to all other devices accordingly. A Matlab-based GUI has been built for wirelessly monitoring, controlling, and configuring all devices. The GUI is shown in Fig. 6.7. The GUI includes seven panels, including: (1) PC configuration panel, where the communication port can be configured. All of the configurations (including other panels) can be exported or loaded; (2) NSA configuration panel, where the gain, sampling rate/resolution, filter passband can be configured for each individual channel. For the signal processing modes performed in hardware, the time window size and threshold for spike detection can also be configured; (3) stimulator configuration panel, where the amplitude, pulse width, pulse train number, and time interval of the stimulation can be configured; (4) sensor configuration, where parameters for sensor nodes can be configured; (5) signal processing configuration, where the parameters for real–time signal processing can be configured; (6) closed-loop configuration, where closed-loop operation between different devices can be configured; (7) display windows, where the output from analyzers and sensor nodes can be displayed in real-time.

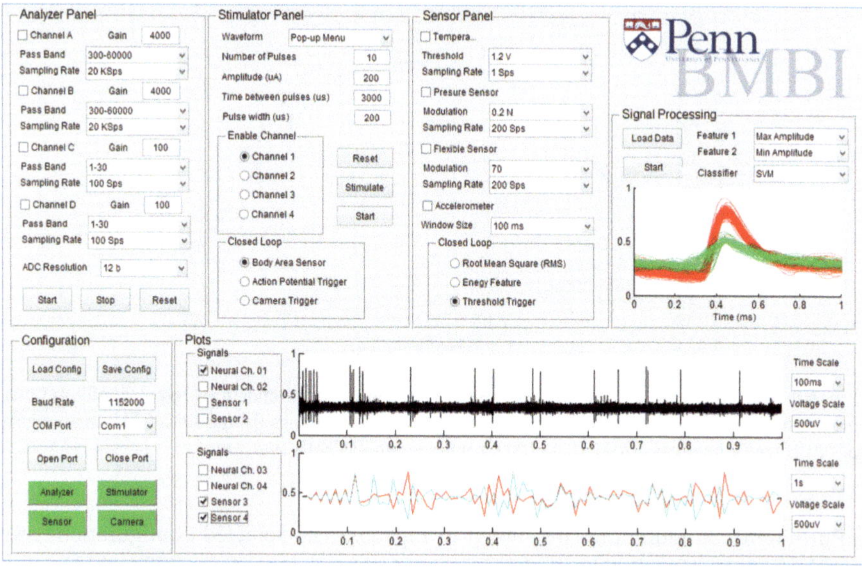

Fig. 6.7 The designed Matlab-based Graphic User Interface (GUI). Seven panels are included in the GUI: (1) PC configuration; (2) recording device configuration; (3) stimulator configuration; (4) body-area sensors configuration; (5) signal processing configuration; (6) closed-loop configuration; (7) display windows

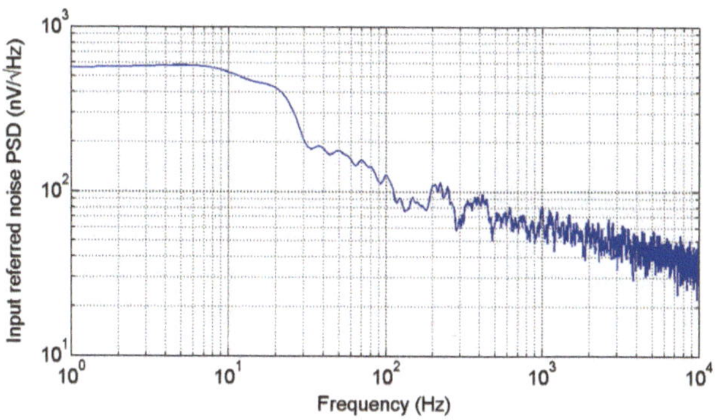

Fig. 6.8 Input referred noise spectrum of the analog front-end

6.2.4 *Experimental Results*

6.2.4.1 Bench Testing

Figure 6.8 shows the measured input referred noise spectrum of the analog front-end in the NSA. The integral noise from 1 Hz to 10 kHz is 4.69 μVrms. The calculated noise efficiency factor [93] is 14.6. The mid-band gain error is 0.87% and the measured CMRR at 1 kHz is 67.4 dB. The measured frequency responses

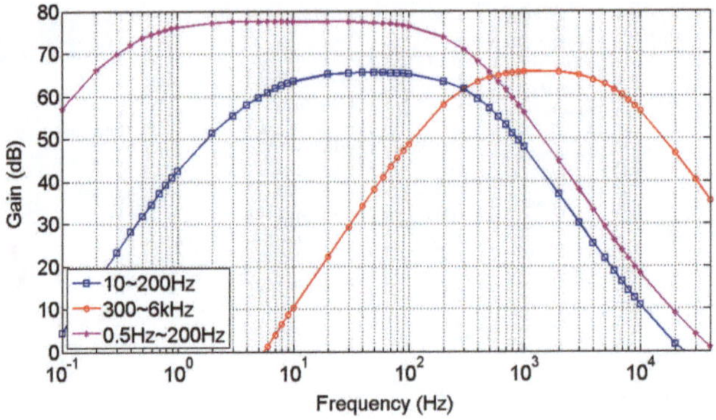

Fig. 6.9 The measured frequency response of the analog front-end in different configurations: bandpass from 10 to 200 Hz with a gain of 66 dB (*blue*), bandpass from 300 to 6 kHz with a gain of 66 dB (*red*), bandpass from 10 to 200 Hz with a gain of 78 dB (*magenta*)

in different configurations are shown in Fig. 6.9. The selected configurations are: bandpass from 10 to 200 Hz with a gain of 66 dB, bandpass from 300 to 6 kHz with a gain of 66dB, and bandpass from 10 to 200 Hz with a gain of 78 dB.

The output currents of the neural stimulator were measured in different loads. Figure 6.10a shows the standard deviation σ of the output current from the anodic and cathodic drivers with different loads. The σ is calculated for each output current in different loads. Figure 6.10b shows the average current mismatch between the anodic and cathodic electrodes in different loads. The average mismatch with respect to the corresponding output current is 0.75%. The stimulator was also tested in 0.9 g/100 mil Sodium Chloride using a 75 μm tungsten electrode. Figure 6.11 shows the measured voltage across the bipolar electrodes in different stimulation current levels. A blocking capacitor of 1 μF was used. The characteristics of the neural signal analyzer and the neural stimulator are summarized in Table 6.4. A bit error rate (BER) lower than 10^{-3} was measured using the wireless module at a distance of 3m in a normal animal experiment environment.

Two open-loop experiments were conducted to verify the system-level operation. In the first experiment, the NSA to stimulator pathway was tested. As shown in Fig. 6.12a, the neural signal was firstly recorded by the NSA. Real–time action potential detection was performed using the time-amplitude window discrimination method, as illustrated in Fig. 6.12b. Once an action potential was detected, a CMD CFG was wirelessly sent from the NSA to the stimulator, triggering a group of pulse stimulation.

In the second experiment, the sensor to stimulator pathway was tested. As shown in Fig. 6.13, the amplitude of the sensing result was mapped into the stimulation pulse frequency. A CMD CFG command was generated from the sensor node and wirelessly sent to the stimulator. The command argument is encoded according to the digitized output of the sensor node.

Fig. 6.10 The measured output currents in different loads (1 kΩ, 2 kΩ, . . . , 8 kΩ). (**a**) shows the standard deviation of the output currents. (**b**) shows the current mismatch between anodic and cathodic electrodes

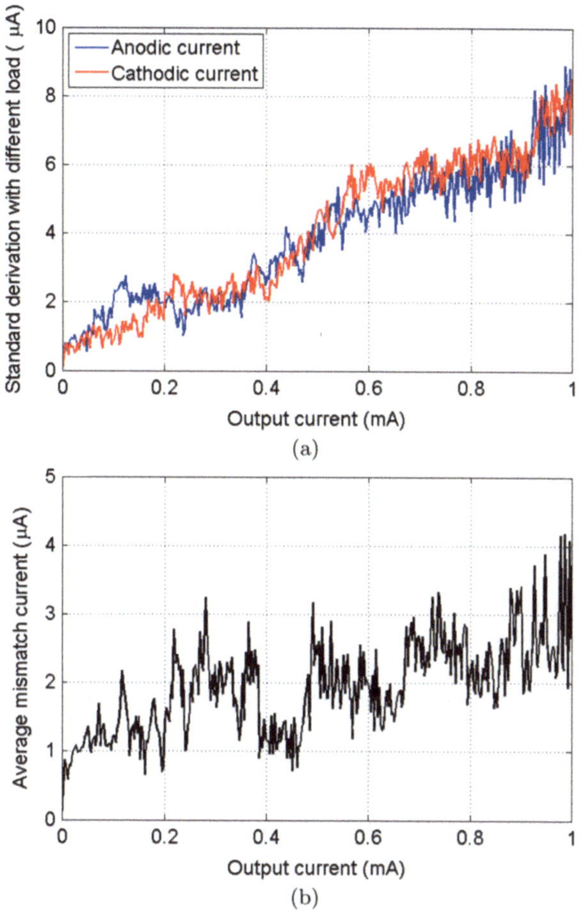

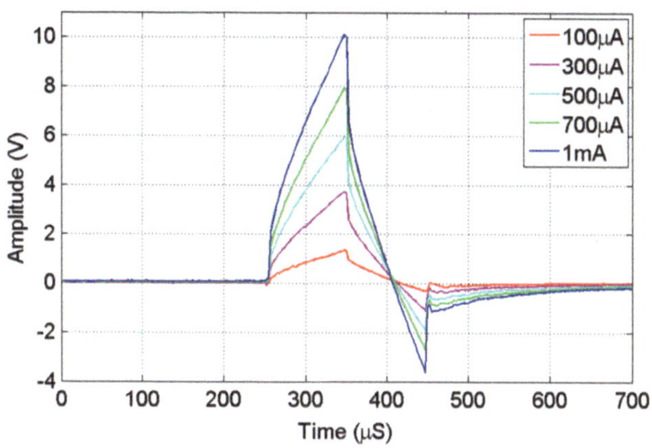

Fig. 6.11 The measured stimulation pulses with different current amplitudes in Sodium Chloride solution

Table 6.4 Specifications of the PennBMBI system

Neural signal analyzer			
Supply voltage	3.3 V	Supply current	380 μA/ch
Input impedance	>200 MΩ	Offset tolerance	1 V
I-Amp noise floor	4.69 μVrms	I-Amp CMRR	>61 dB
ADC resolution	12 bit		
Neural stimulator			
Output current	0–1 mA	DAC resolution	6 bit
Compliance voltage	±12 V	Output impedance	>100 MΩ
Standard deviation	1.71 μA	Driver mismatch	0.75%

6.2.4.2 In Vivo Experiments

To further evaluate the PennBMBI system, several experiments of wireless neural recording, stimulating and sensing were conducted in both anesthetized and awake rats. The neural recording was performed in an anesthetized rat with a tungsten microelectrode placed in the whisker motor cortex. The NSA was configured to have a passband of 300–6 kHz, a sampling rate of 21 kSps, and a gain of 72 dB. The recorded action potentials are shown in Fig. 6.14. The recording shows two different neurons firing action potentials in a close succession.

In order to evaluate the quality of the recording, we did a recording session using the PennBMBI NSA and a rack-mounted commercial medical instrumentation (RZ2 Workstation, Tucker-Davis Technologies) simultaneously. Figure 6.15 shows the two recordings together. A close match between the two recordings can be observed, which proves that the NSA can faithfully record the action potentials with a signal-to-noise ratio (SNR) comparable to a commercial medical instrumentation.

To demonstrate the sensor and stimulator nodes, an awake rat with a chronically implanted stimulating microelectrode in the lateral hypothalamus was placed in an operant conditioning chamber with a lever press. The sensor node detected the lever press and wirelessly sent a trigger to the stimulator worn on the rat's back to deliver a stimulus train (30 of 100 μA, 200 μs constant current pulses). This setup allowed the rat to associate the lever press with the rewarding sensation of hypothalamic stimulation. The result provides one example of how the various nodes of the PennBMBI, in this case, the sensor and stimulator, can be flexibly combined to enable a wide range of neuroscience and neural engineering experiments in freely behaving animals.

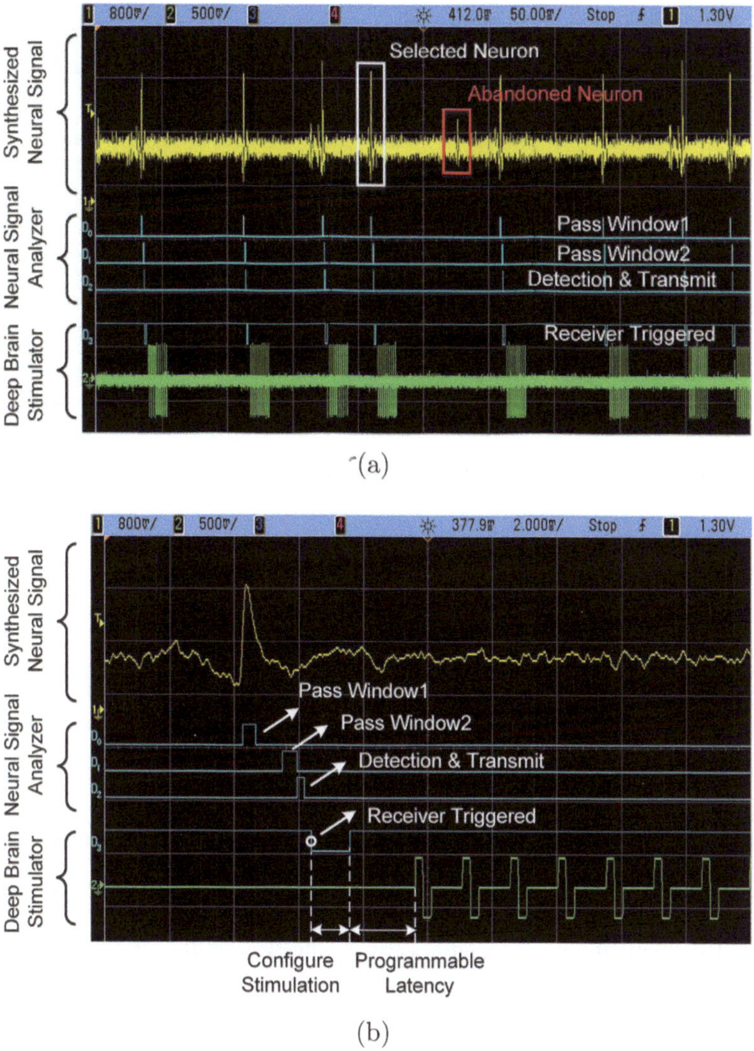

Fig. 6.12 The measured wireless closed-loop operation from the neural signal analyzer and the stimulator. (**b**) is a zoomed-in view of one action potential detection process in (**a**)

6.3 The Watermaze

6.3.1 Introduction and Background

Sensation and perception are essential abilities for human and animals. Loss of sensation due to nerve damage prevents even basic activities of daily living. Even though many recently developed neutrally controlled prosthetics successfully

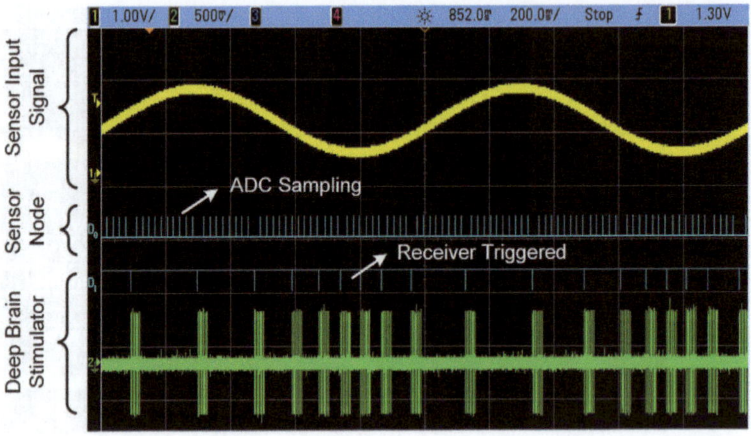

Fig. 6.13 The measured wireless operation between the sensor node and the neural stimulator

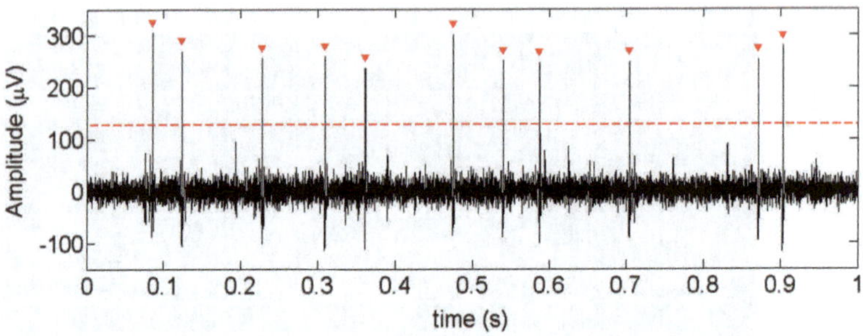

Fig. 6.14 Neural signal recorded by the PennBMBI NSA. The action potentials are marked by *red triangles*

replaced motor pathways, somatosensory feedback is critical for paralyzed individuals to adequately use them, which has often been underscored. Recently, there has been an increased interest in conveying lost information through direct brain stimulation using a neuroprosthetic device [16]. These strategies rely on the brain learning to use the remapped or artificial stimuli to inform actions. A common paradigm to study this learning process involves using brain stimulation to guide rats through a maze [309]. These so-called "rat-robot" studies have mapped a number of different navigation signals to brain stimulation [310]. All reported land-based mazes are designed with a discrete number of actions and goal locations. A concern with these studies is that the rats could simply memorize a few stimulus-response contingencies rather than learn a more generalized stimulus-dependent navigation strategy.

In this work, a new rat-robot paradigm is developed using a classic test of rodent navigation: the Morris water maze (MWM). In the MWM, the rat swims in a large circular tank looking for a hidden, submerged platform on which to stand

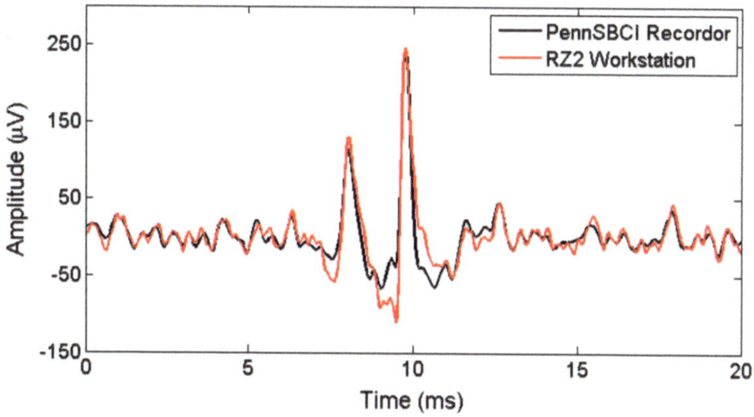

Fig. 6.15 Comparison between recordings from the PennBMBI NSA (*black*) and the RZ2 Neurophysiology Workstation (*red*)

[311]. In our task, the submerged platform was positioned randomly on each trial to dissociate visual cues from the platform location, and the rats navigated to the platform using only the sensation encoded from the brain stimulation [63, 312]. The experiment setup is illustrated in Fig. 6.16. For simplicity, the proposed experiment system is referred to as *watermaze system*, and the wireless neuroprosthetic device is referred to as *watermaze stimulator*. Custom hardware and software were developed to support the watermaze experiment. The findings suggest that rats can quickly interpret artificial percepts to guide behavior. Some of the figures and tables presented in this section were originally published in [63] ©IEEE. Reused, with permission.

6.3.2 System Overview

The block diagram of the watermaze system is shown in Fig. 6.17. The developed watermaze system includes both hardware and software. The hardware system consists of: (1) a wireless neuroprosthetic device (watermaze stimulator), and (2) a computer interface with a tracking camera. The software system mainly includes (1) the animal tracking and modulation algorithm, and (2) the communication and stimulation protocol. It should be noticed that, even though the system is developed for this experiment, it can be generalized to perform various neuroscience experiments.

The developed watermaze experiment can also be understood as a typical closed-loop system, as illustrated in Fig. 6.18. The wireless neuroprosthetic device works as the actuator, the tracking image sensor finds the error signal, and the PC determines the closed-loop algorithm.

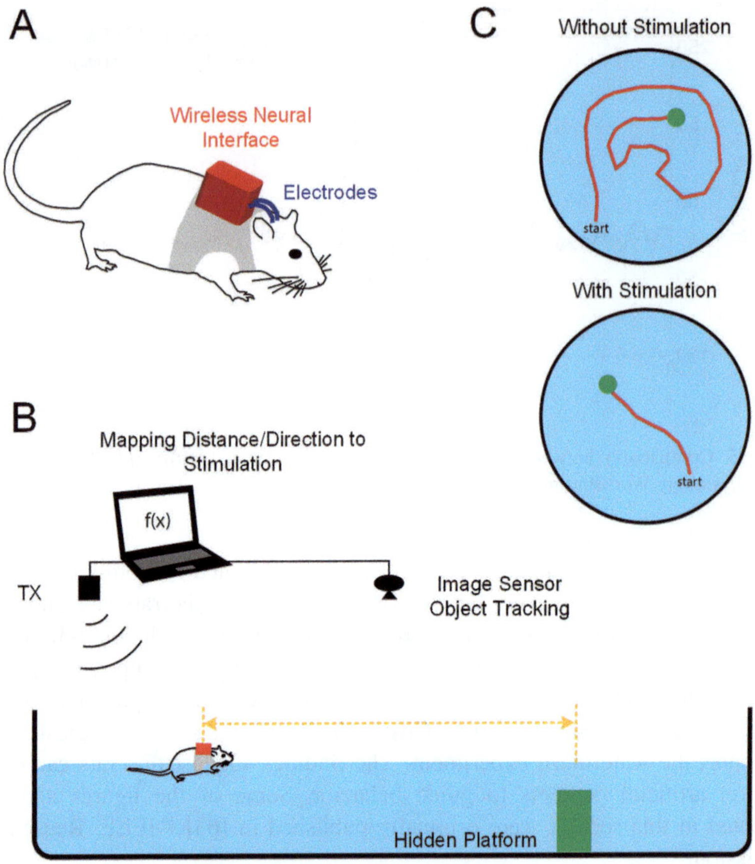

Fig. 6.16 Illustration of the developed perception augmentation experiment. (**a**) shows a rat wearing the developed wireless waterproof neuroprosthetic. The electrodes are chronically implanted in the somatosensory cortex. (**b**) shows the experimental setup. A rat navigates to a hidden platform using only the perception established from the stimulation. (**c**) illustrates the rat's swimming traces with/without the simulation guidance

6.3.3 Hardware Implementation

6.3.3.1 Design of the Watermaze Stimulator

The block diagram of the watermaze stimulator is shown in Fig. 6.17. The watermaze prosthetic consists of (1) a microcontroller for overall control and processing, (2) a fully programmable neural stimulator with an impedance monitoring module, (3) a wireless transceiver, and (4) a power management unit.

In this work, a microcontroller ATxmega128A4U [302] from Atmel is used as the central processor. It communicates with a 2.4 GHz wireless transceiver from Nordic Semiconductor nRF24L01 [308] for wireless communication, including retrieving

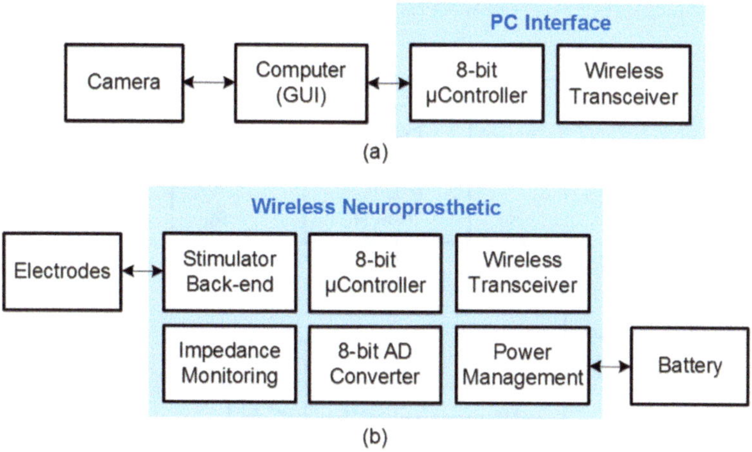

Fig. 6.17 The block diagram for the watermaze system, including: (**a**) the PC interface with a tracking camera, and (**b**) the wireless neuroprosthetic

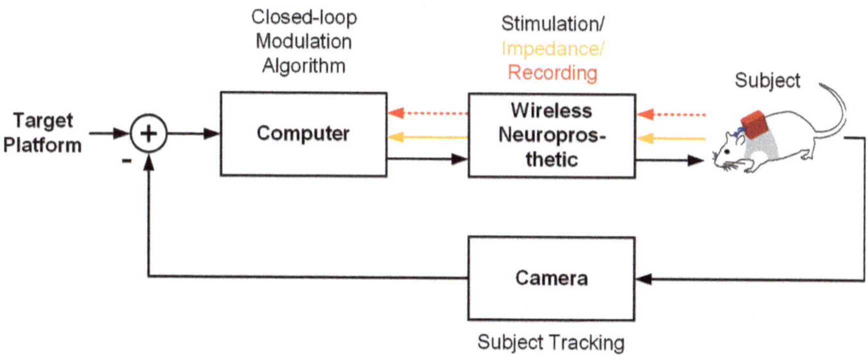

Fig. 6.18 The typical closed-loop diagram for the developed watermaze experiment for perception augmentation

device configuration and stimulation commands, and sending back the compliance voltage for estimating the electrode impedance during the stimulation.

The power management unit includes a single channel LDO TPS791 [313] from Texas Instruments for powering the microcontroller and some peripheral circuits. The chip has a full-scale output current up to 100 mA, with a very low dropout voltage of 38 mV. The root mean square noise is 15 μV. A dual channel DC/DC converter LT1945 [314] from Linear Technology is used to generating a high voltage for powering the stimulating output stages. The chip takes an input voltage as low as 1.2 V, so potentially it can be powered by coin batteries. Regulated positive and negative outputs can be generated up to ±34 V, setting by feedback resistors' ratio. The converter consumes 12 μA in the active mode and less than 1 μA in the shutdown mode. In the first two generations of the watermaze stimulators, an

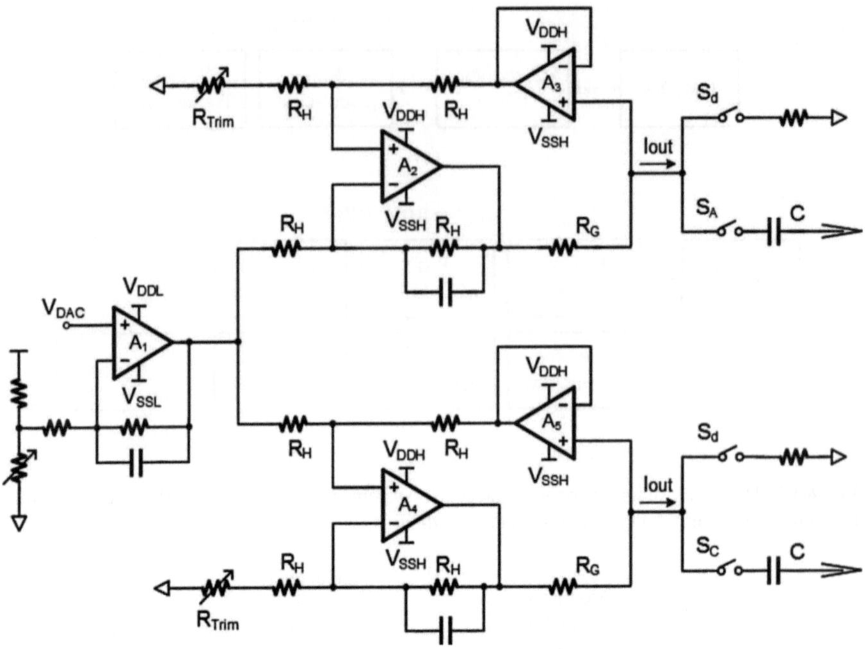

Fig. 6.19 The circuit schematic of the first generation watermaze stimulator

inverting charge pump LTC1983 [315] from Linear Technology is used to generate the negative supply. The chip gives a fixed −3 V with ±4% accuracy with an input voltage from 2.3 to 5.5 V. The full-scale output current is up to 100 mA, with a flyback capacitor of 1 μF. The battery used in the first generation watermaze stimulator is LP-402025 from Sounddon. A 150 mAh Lithium Ion Polymer battery from Pkcell, which gives higher peak current, is used in the following generations.

The first generation watermaze stimulator uses three stack PCB boards, including the transceiver with a PCB antenna. The circuit schematic for the output stage is shown in Fig. 6.19. The core circuits are two Holland current sources (A_1–A_5). The output current is determined by the input voltage across the gain resistor, which is independent of the load impedance. A blocking capacitor C is used to block the DC current path to the brain. The DAC integrated in the microcontroller is used to generate the input voltage. The output range of the DAC is from ground to a reference voltage, which is set to be VDD. The amplifier A_1 works as a level shifter. A trimming resistor is used to tune the output voltage. The DAC generates the stimulation waveform in voltage, and the driving sites convert the voltage waveform into a current with a programmable transimpedance determined by the gain resistor R_G. The amplifiers A_2–A_5 are designed using high voltage op-amps OPA2140 from Texas Instruments [316]. These op-amps operate with dual supplies up to ±18 V. In this version, the supply voltages are designed to be ±15 V in order to drive up to 300 μA stimulation current with a load impedance of 50 kΩ. The switches S_C, S_A,

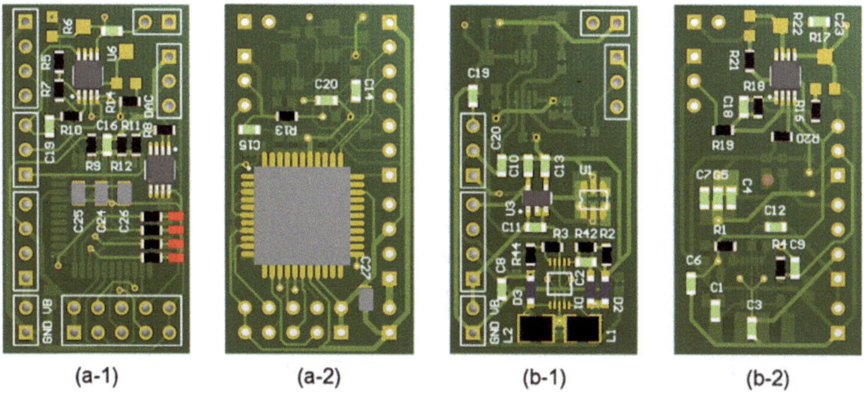

Fig. 6.20 The 3D construction of the first generation watermaze stimulator board. (**a-1**) and (**a-2**) are the top boards, and (**b-1**) and (**b-2**) are the bottom board

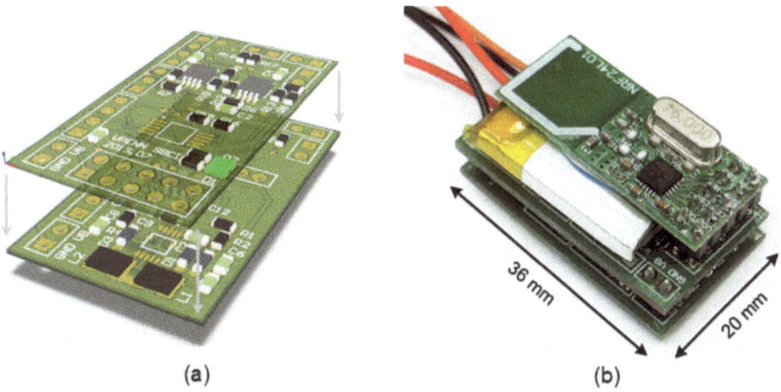

Fig. 6.21 (**a**) The 3D construction of the first generation of the watermaze stimulator board. (**b**) The photo of the assembled stimulator board. The wires are for electrodes and battery

and S_D connect the output stages to the driving electrodes. Multiple stimulation channels can be supported by adding a multiplexer without too much area and power penalty. The only drawback is the lack of ability in driving two stimulating sites simultaneously. However, a near simultaneous stimulation by switching the electrodes is more than sufficient in most cases. When no stimulus is to be delivered, the opamps are disabled to save power, and the switches S_D will short the electrodes to the ground in order to prevent current leakage. It should be noticed that since the blocking capacitors still isolate the circuits from the tissue, there is no DC current path even in this case.

The dimension of the final assembled devices is 36 mm × 20 mm × 19 mm. The 3D construction of the first generation watermaze stimulator is shown in Fig. 6.20. The photo of the assembled device is shown in Fig. 6.21.

The second generation watermaze stimulator is designed on a single PCB board. The circuit schematic is shown in Fig. 6.22. The second generation still uses two

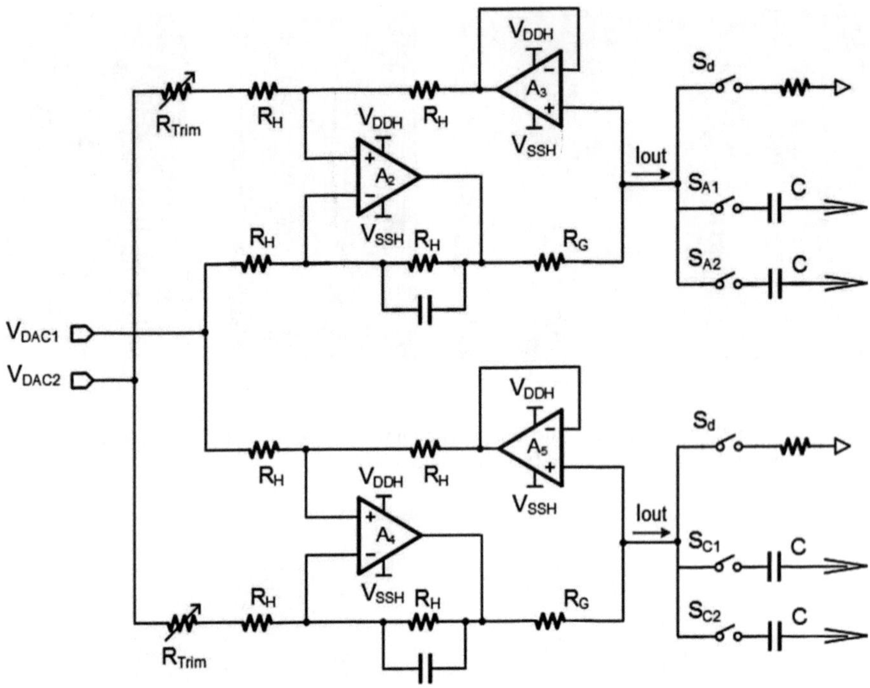

Fig. 6.22 The circuit schematic of the second generation watermaze stimulator

sets of dual supplies (VDDH, VSSH, VDDL, VSSL) and a Holland current source. Two DAC channels from the microcontroller are used to set a differential input to the Holland current source, which removes one trimming resistor, and reduces the risk from resistor drifting caused DC stimulation current. Two stimulation channels are designed in this version, using multiplexer ADG409 from Analog Devices. The wireless transceiver nRF24L01, antenna and related matching circuits are soldered directly on the PCB.

The PCB of the second generation watermaze stimulator is shown in Fig. 6.23. The photo of the assembled device is shown in Fig. 6.24. The whole device is coated with PDMS for waterproofing.

The third generation watermaze stimulator is designed with a goal to simplify the design and to improve the robustness. A single high supply voltage is used instead of dual high supplies, and the stimulation current is passed between the bipolar electrodes alternatively to generate stimulation and reversal phases. The circuit schematic is shown in Fig. 6.25. A_1 is a regulating op-amp which is used to produce a high output impedance. The output current is set by V_{DAC}/R_G, where the V_{DAC} is programmable. The opamp is powered by the low supply voltage, and the high supply voltage is only used to drive the stimulating electrode. Switches S_3 and S_4 are used to purge the blocking capacitors and discharge the residue charges.

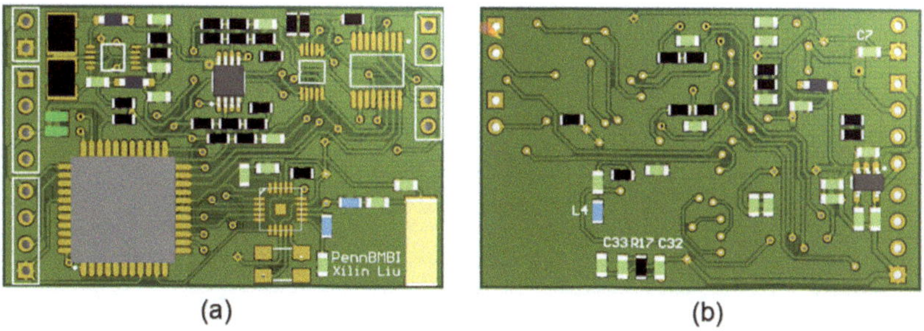

Fig. 6.23 The PCB of the second generation watermaze stimulator board. (**a**) is the *top view* and (**b**) is the *bottom view*

Fig. 6.24 The photo of the assembled second generation watermaze stimulator board. The whole device is coated with PDMS for waterproofing

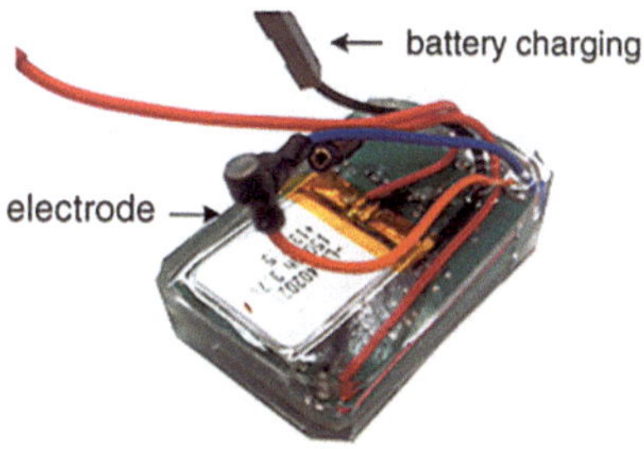

Fig. 6.25 The circuit schematic of the third generation watermaze stimulator. The tuning voltage V_{DAC} is generated from the microcontroller

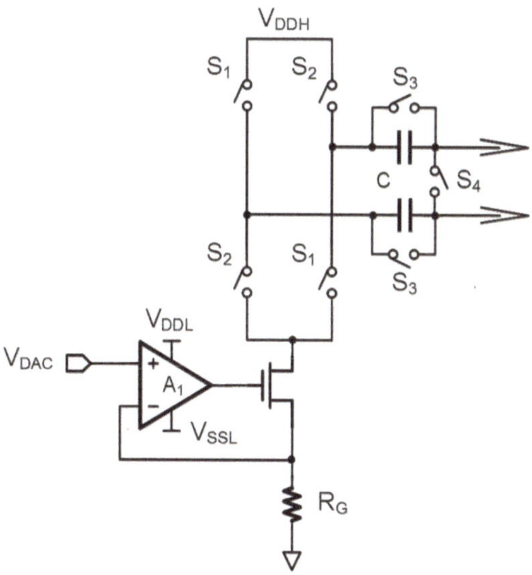

The PCB of the third generation watermaze stimulator is shown in Fig. 6.26. The photo of the assembled device is shown in Fig. 6.27. The whole device is coated with PDMS for waterproofing.

The electrodes of the stimulator are multiplexed to an ADC. The compliance voltages of the electrodes are measured at the beginning and the end of the

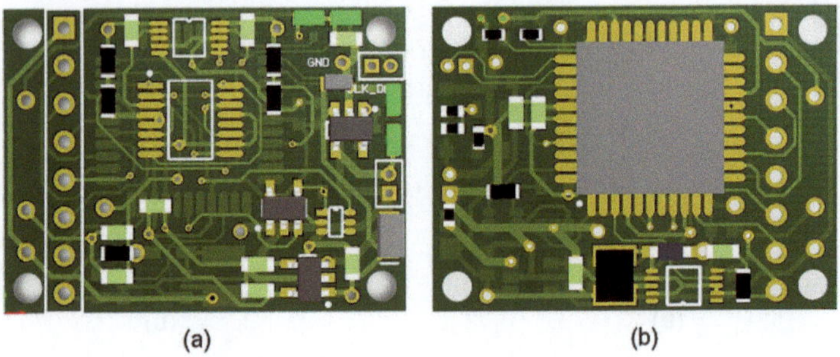

Fig. 6.26 The PCB of the third generation watermaze stimulator board. (**a**) is the *top view* and (**b**) is the *bottom view*

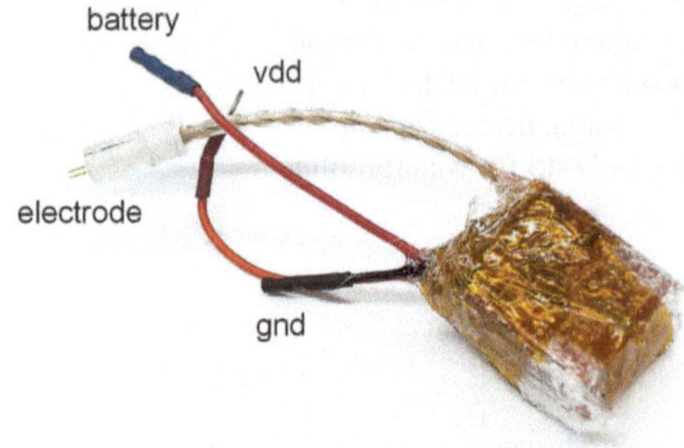

Fig. 6.27 The photo of the assembled third generation watermaze stimulator board. The whole device is coated with silicon for waterproofing

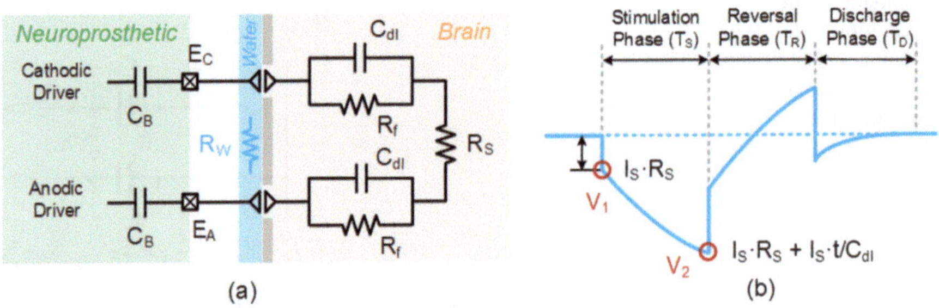

Fig. 6.28 (**a**) The equivalent circuit model for the electrode–electrolyte interface. (**b**) A typical stimulation waveform between the electrodes E_A and E_C. The compliance voltages at the beginning and end of the stimulation phase are measured for estimating the impedance

stimulation phase, as shown in Fig. 6.28. The spreading resistance can be estimated by $R_s = V_1/I_S$, where V_1 is the voltage between the two electrodes at the beginning of the stimulating phase, and I_S is the stimulation current. An impedance baseline is measured every time before an experiment. During the experiment, if V_1 is much less than the baseline compliance voltage, it indicates that the equivalent resistance between the electrodes drops significantly, possibly because the electrodes are shorted by water. The experiment should stop since little current is actually passing between the electrodes. On the other hand, if V_1 is much larger than the baseline, the electrodes may lose connection with the tissue, or a much larger current than the designed value is passing the tissue, possibly because of an electronic failure. The experiment must be halted in both cases to keep the animal safe from tissue damage.

6.3.3.2 Electrode and Electrode Connector

Electrodes were chronically implanted in the sensory cortex. Different electrodes have been tested in this project, including commercial and custom-made tungsten and stainless steel electrodes. The final selected electrode is a two-channel commercial electrode with a relatively low impedance.

In practice, it takes a considerable amount of practice and time to put the watermaze stimulator device on an awake rat every time before an experiment. The process can be easier after the rats got used to the jacket, but it can still be time-consuming. So a magnetic connector was used in the early stage of this experiment, which makes the docking process much easier. However, we later found that the magnets cannot provide a secure connection, especially when the rat is swimming. Besides, the water may short the electrodes through the magnets. The finally selected connector has a screw thread to secure the connection and also prevent water from shorting the electrodes. The impedance of the electrodes should still be checked before and during the experiment to make sure the animal receives the stimulation without potential damage to the brain tissue.

6.3.3.3 Image Sensor and Computer Interface

The developed computer program is compatible with most USB webcams on the market. However, there are two issues worth attention: the viewing angle and the autofocus and exposure feature. Because the camera faces the water all the time, many webcams have trouble in auto focusing. Moreover, webcams often give a wrong exposure due to the light reflection on the water. Different webcams have been tested, including a Microsoft LifeCam VX-5000, a Logitech HD Webcam C310, a Logitech HD Webcam C615, and Microsoft Q2F-00013 USB 2.0 LifeCam. The last one was eventually chosen for this project because it gives the programmability in focusing and exposure.

The camera was securely mounted on the ceiling above the water tank. Even though a resolution up to 1080p is supported by the camera, a resolution of 640×480 is used, which is more than sufficient for the tracking purpose in this task. A higher solution will potentially cause a processing delay.

The PC interface from the PennBMBI system was used as the wireless dongle in this project (Sect. 6.2.3). The interface mainly consists of a 16/8-bit XMEGA microcontroller, a 2.4 GHz wireless transceiver, and a USB 2.0 port. With the full-speed USB, the communication delay can be minimized.

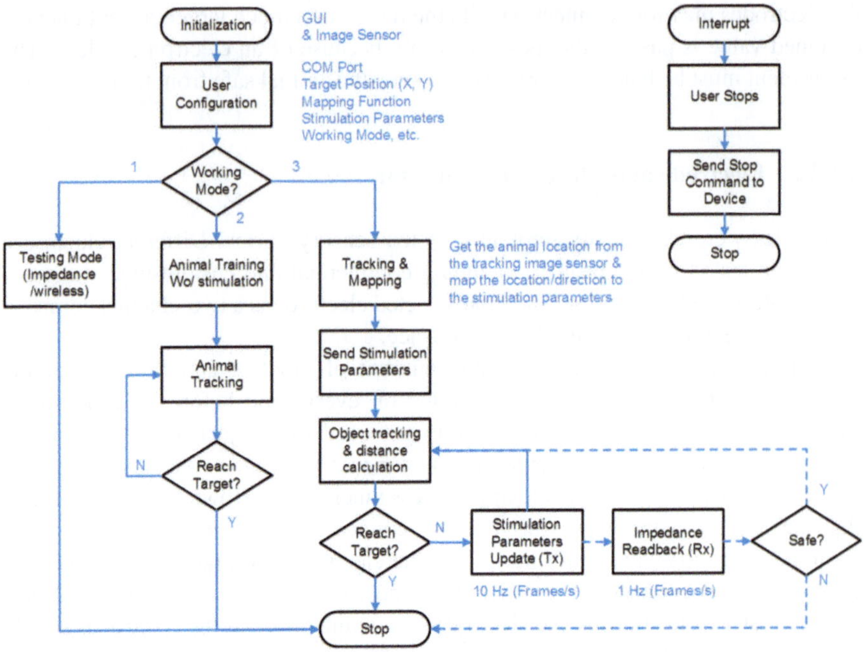

Fig. 6.29 The flowchart of the computer program developed for the watermaze experiment

6.3.4 *Software Implementation*

6.3.4.1 Communication Protocol

The software on the computer side was implemented using Matlab. C language was used in the microcontroller programming. The flowchart of the computer program is shown in Fig. 6.29. The program mainly has three operation modes: (1) testing mode, (2) animal training mode, and (3) experiment mode (mission mode).

The testing mode includes both wireless communication test and electrode impedance test. Both tests need to be run every time before the animal is set into the water for the experiment. In the animal training mode, the image sensor tracks the rat's swimming trace but no stimulation is delivered. The training mode helps the animal get used to swimming in the water tank, and also learn the existence of a hidden platform in the tank. After the rat reaches the hidden platform, it will be rewarded to be motivated. Also, the control data is collected in this mode for analysis and comparison purposes. In the experiment mode, the image sensor tracks the rat's swimming trace, and the computer maps the relative location and/or direction of the rat into a stimulation sequence. The established mapping algorithms include: (1) binary mapping, (2) linear mapping, and (3) Gaussian mapping. In the binary mapping, the rat receives a simulation train only if it's heading towards the hidden platform. In the linear mapping, the stimulation frequency is modulated by the distance between the rat's location and the hidden platform in a linear fashion, as:

$$f_{\text{stim}} = \alpha \cdot \sqrt{(x - x_0)^2 + (y - y_0)^2} + \beta \tag{6.1}$$

where x, y indicate the location of the rat, x_0, y_0 indicate the location of the platform. α is the gain factor, and β is the offset parameter. If α is positive, the rat receives a higher frequency stimulation when it swims away from the target; if α is negative, it receives higher stimulation frequency when it swims towards the platform. The offset β should be set so that the stimulation frequency is a positive parameter in the range from 0.5 to 300 Hz. In the Gaussian mapping, the animal's distance to the platform maps to the stimulation frequency according to a Gaussian distribution, as:

$$f_{\text{stim}} = f_{\max} \cdot e^{-\frac{(x - x_0)^2 + (y - y_0)^2}{2\sigma^2}} \tag{6.2}$$

where $f_{\max}$ is the designed maximum frequency, and σ is the standard deviation. Moreover, various mapping functions can be easily implemented in this program, which is important for research and investigation purpose.

The computer program updates 10 frames per second, and sends the updated stimulation parameters to the wireless watermaze stimulator. If the animal reaches the hidden platform, the program stops. The user can set the radius of the target. The program checks the load impedance every second to guarantee the safety of the animal. The experiment stops immediately if the measured impedance or compliance voltage is out of the safe range. In addition, an interrupt service allows the user to halt the experiment at any time.

The flowchart of the watermaze stimulator is shown in Fig. 6.30. The program has a main routine and an interrupt service routine. After powering on, the device performs the initializations. The wireless module will be configured in the receiving mode and then the CPU will be put in the sleep mode to lower the system's power consumption. Once an RF package is received, the device first checks if this is a stop command. If a stop command is received, the device disconnects the output driver from the electrodes to prevent any potential damage to the animal. Next, the device sends a signal back to the computer indicating the stop command has been executed. The stop command is also used for testing the wireless communication. A wireless communication is established if the computer can successfully read back the response from the device. The computer program retries to establish the wireless handshake ten times before timeout.

If the received package is not a stop command, the device checks the working mode, and proceeds accordingly. In the impedance testing mode, the device delivers one pulse train according to the received parameters and read back the compliance voltage. Since the impedance testing mode is a manually triggered stimulation mode, it can also be used to test the animal's reaction to the stimulation out of the water. In the experiment mode, a watchdog timer first starts. The timer counts for

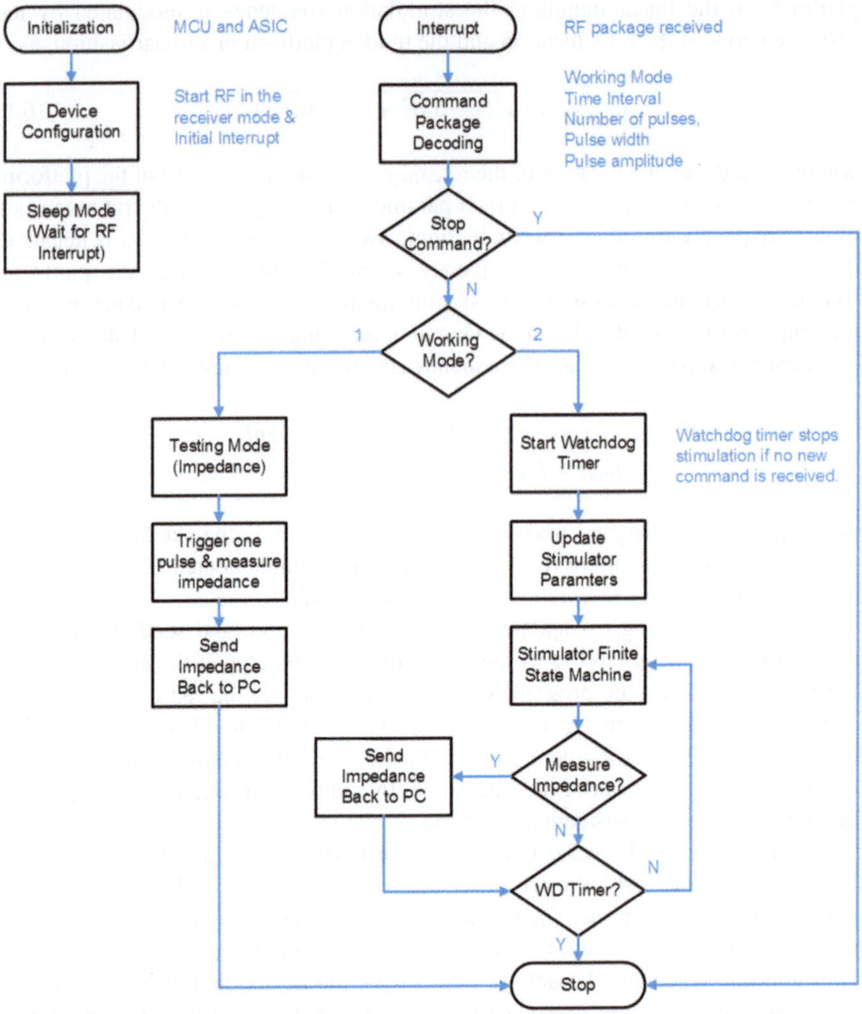

Fig. 6.30 The flowchart of the program implemented in the watermaze stimulator

6 s, during which time if no new RF package is received, the stimulation stops. The time out feature prevents an endless stimulation in case of wireless communication failure.

The DAC's output is set according to the stimulation amplitude, and the local finite state machine is set according to the timing parameters. The stimulator remains the stimulation based on the received time interval until the next command is received. Figure 6.31 illustrates the timing of the watermaze stimulator in two scenarios: the stimulation time interval is shorter or longer than the time per frame.

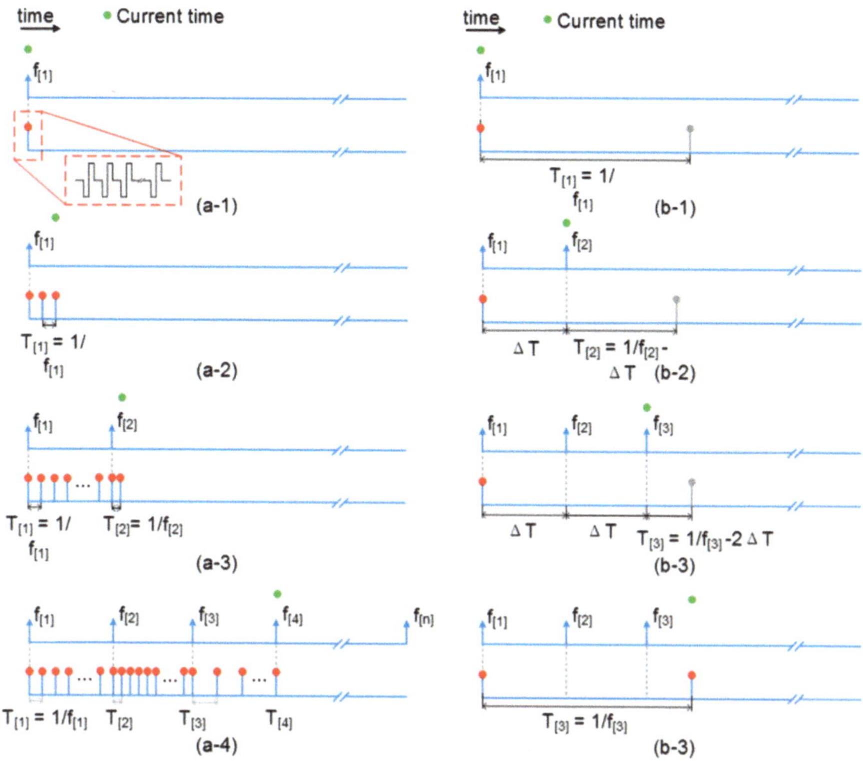

Fig. 6.31 Illustration of the watermaze stimulator's timing. Each *red dot* represents a stimulation pulse train, and the *green dot* represents the current time in each subplot. (**a**) shows the delivered pulses when the stimulation interval is shorter than the time per frame, and (**b**) shows the delivered pulses when the stimulation interval is longer than the time per frame. The finite state machine will correct the stimulation time interval according to the most recent command

6.3.4.2 Animal Tracking

A color based animal tracking algorithm was implemented in the Matlab program. After acquiring the image frame from the camera, the program first extracts the red components of the image. Since each pixel consists of R (red), G (green), and B (blue) components, the extracted image has the same dimension as the original image. The extracted image is then filtered by a median filter to suppress the noise [304]. Next, the image is converted to binary using a predefined threshold. The threshold can be used to tune how sensitive the algorithm is, and should be adjusted according to the environmental light condition. The center of the detected area is used as the location of the object. A drawback of this algorithm is that it won't be able to discriminate the animal if multiple red objects exist in the scene. This error can be avoided by not placing other red objects in the scene. Since most part of the scene is the water tank, the error can be avoided in practice.

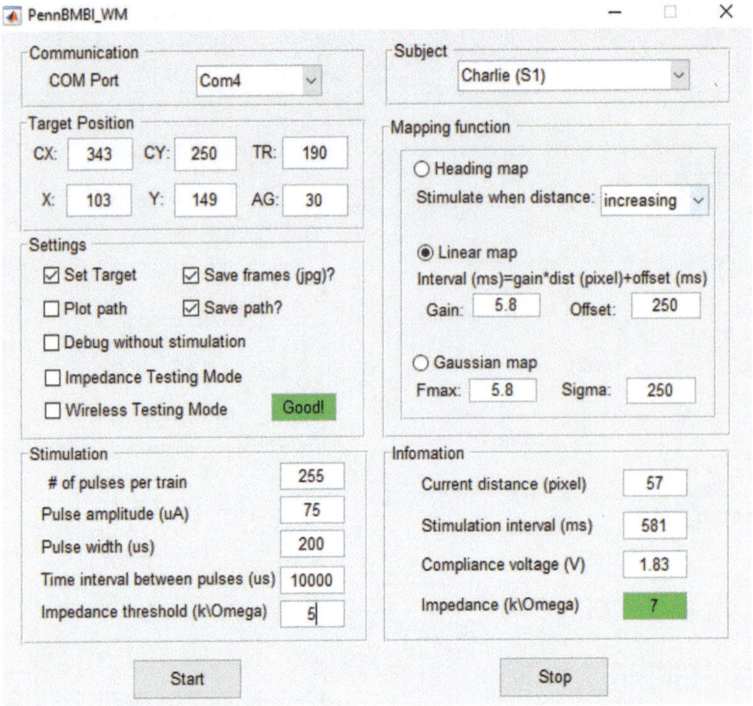

Fig. 6.32 Matlab based Graphic User Interface (GUI) for the watermaze project

6.3.4.3 User Interface

A graphic user-friendly interface has been designed for this project, as shown in Fig. 6.32. There are mainly six panels in the GUI. The communication panel sets the COM port used to communicate with the PC interface dongle. The target position panel sets the location and radius of the tank, the location and radius of the hidden platform. Initially, these positions will display NA. In the target setting mode, a set of random locations will be generated. But the user can manually set these parameters.

In the settings panel, there are several options of the program. Firstly, there are options for saving the frames, saving the path, and plotting the path. Figure 6.33 shows a frame of the captured video during the experiment. The large green circle shows the submerged platform, the yellow dot represents the start location. The red curve shows the swimming trace. The left top corner shows the current distance to the platform and the total path length the rat has traveled in the current trial.

In addition, there are several working modes the user needs to choose before running the experiment. Check the "Wireless Testing Mode" box, then press the "Start" button, the program will try to communicate with the device. If it successfully reads back from the device, the edit window will turn green and show

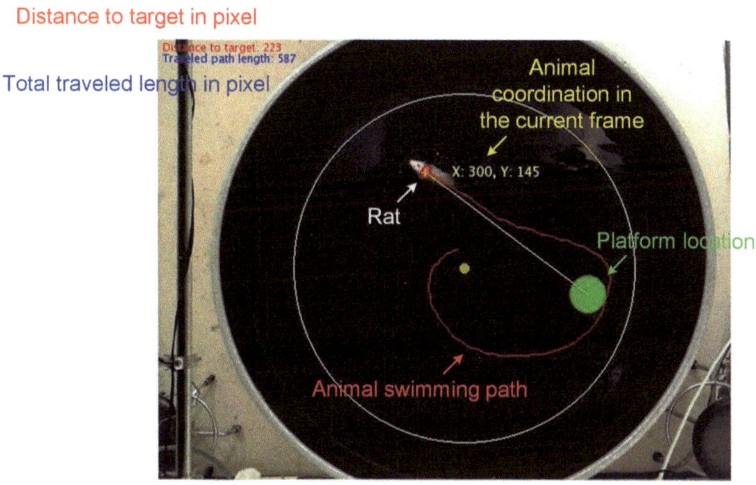

Fig. 6.33 One frame of the captured video during the experiment. The *large green circle* shows the submerged platform, the *yellow dot* represents the start location. The *red curve* shows the swimming trace. The *left top corner* shows the current distance to the platform and the total path length the rat has traveled in the current trial

"Good!". If the communication cannot be established after ten attempts, the window will turn red and displays "Bad!". Check the Impedance "Testing Mode box," then press the "Start" button. The program will send stimulation commands and read back the compliance voltage for estimating the electrode and load impedance. Check the "Debug without stimulation" mode box, then press the "Start" button, the program will load the camera and start tracking the rat, but no stimulation will be delivered. This mode can be used for training the animal and getting the control data.

In the simulation panel, several parameters for the stimulation can be set. The parameters include: (1) number of pulse per train (8 bit), (2) pulse amplitude in μA (8 bit), (3) pulse width in μs (8 bit), (4) time interval between pulses in μs (16 bit), and a threshold for the electrode impedance in kΩ. If the detected impedance is lower than the threshold, the impedance window panel will turn red for a warning. All of the input numbers will be truncated to the maximum number of bits allowable for the assigned registers in the microcontroller.

In the mapping function panel, the user can select which mapping function to use, as explained in Sect. 6.2. The panel also includes windows for entering parameters of the mapping functions. The established mapping functions include a binary mapping, a linear mapping, and a Gaussian mapping. More mapping functions can be added in the Matlab program easily.

The information panel is for displaying the tracking results and compliance voltage readouts in real–time. If the read back impedance is lower than the user-defined threshold, the window will turn red. If the compliance voltage cannot be read back, the window will turn yellow and display "999," which is the error code. Otherwise, the window is green during normal operation.

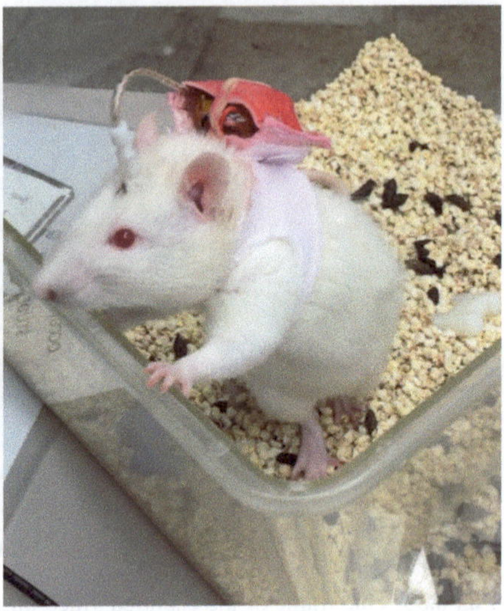

Fig. 6.34 A rat received an implant in the somatosensory cortex wearing the wireless waterproofed stimulator device

6.3.5 Experimental Results

The experimental results using the developed watermaze system will be presented in this section. A pair of electrodes was implanted in the somatosensory cortex of a Long-Evans rat. Figure 6.34 shows a rat wearing the wireless waterproofed stimulator device. The jackets used for housing the device are dyed red for the color-based object tracking.

A couple paradigms, modulation algorithms, and parameter combinations were studied in this work. Initially, the rats swam in random directions until the platform was found or the trial timed out (60 s). A typical example of the swimming trace before the simulation is shown in Fig. 6.35a. The performance significantly improved over the course of about 50 trials, as shown in Fig. 6.35b. The poor performance on the catch trials, in which no stimulation was delivered, confirmed that the learned behaviors were guided by the stimulation.

Figure 6.36 shows the stimulation pulses versus time, together with the rat's distance to the target platform. Each red vertical line represents a stimulus pulse. During this experiment, the rat receives a stimulation when it swims away from the platform. There are in total of 710 stimulus pulses delivered in this 12 s trial. The result clearly shows that the rat learned to turn around when it received the stimulation.

In order to give a better quantitative analysis, the platform's locations are restricted to one in each quadrant in the following trials. The four locations are separated by 90° with equal distance to the center of the tank, as illustrated in Fig. 6.37. In each trial, the platform is randomly placed in one of the four locations. The rat is initially set free at the center of the water tank, and it had no visual clue of the platform's location. A total of 139 trials were conducted in this setup, including 124 trials with stimulation and 15 catch trials without stimulation. Naturally, the

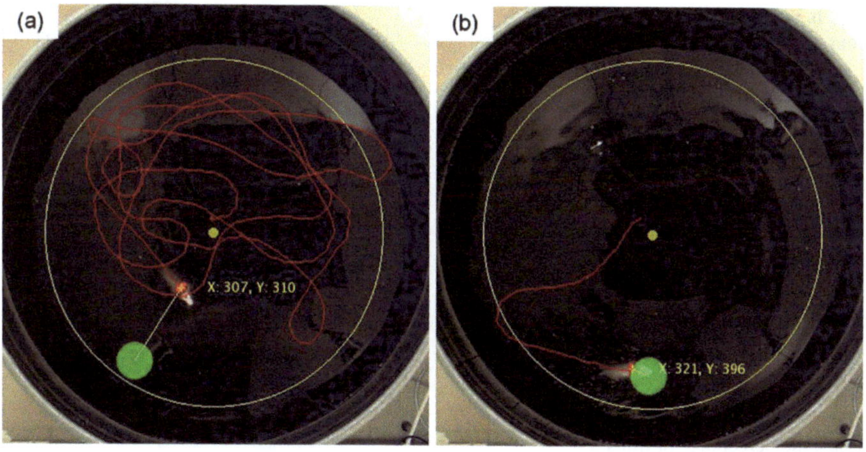

Fig. 6.35 The in vivo experimental results. (**a**) and (**b**) are webcam captured frames during the experiments. The *small yellow* and *large green circles* represent the start and platform locations, respectively. These were superimposed on the video frame and not visible to the rat. (**a**) shows the rat's swimming trace without the simulation, and (**b**) shows the rat's swimming trace with the simulation guidance

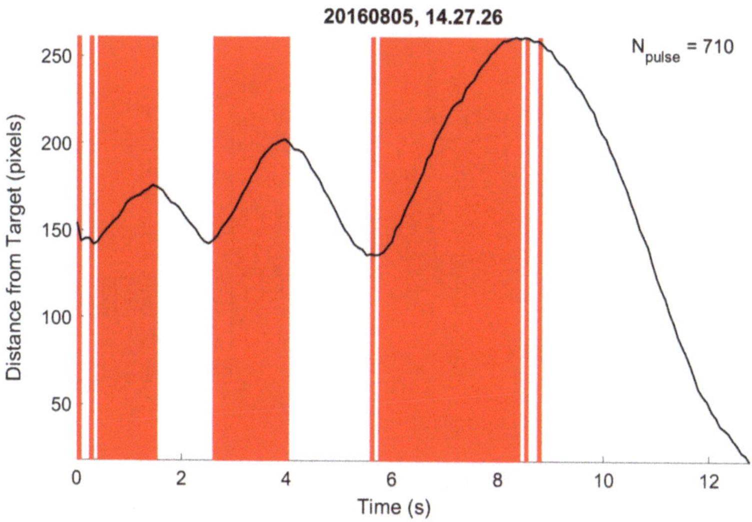

Fig. 6.36 A typical trial with stimulation. The animal receives a stimulation when it swims away from the platform. The stimulation pulses are marked by *red vertical lines* in this figure. It clearly shows the animal turned the direction when it received the stimulation

chance for the rat to visit each of the four locations should be equal, since the platform is randomly placed. Figure 6.38 compares the percentage of the trials when the rat finds the correct location on the first visit to one of the four locations with and without the stimulation. Without the stimulation, the percentage for the correct

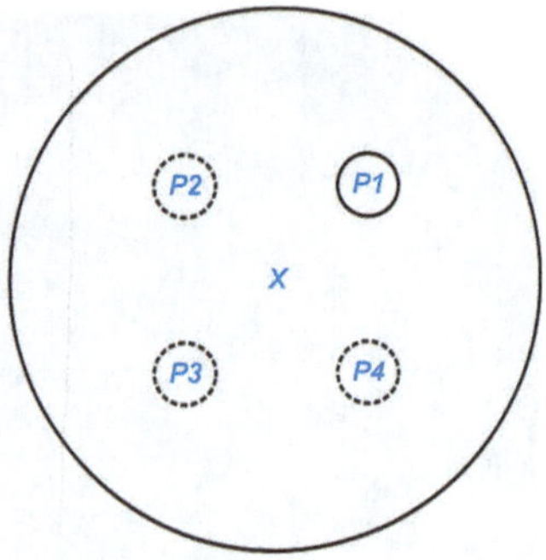

Fig. 6.37 Illustration of the four possible platform locations *P*1–4. In this experiment setup, the platform was randomly placed in one of them. The rat was initially set free at the center of the water tank in each trial

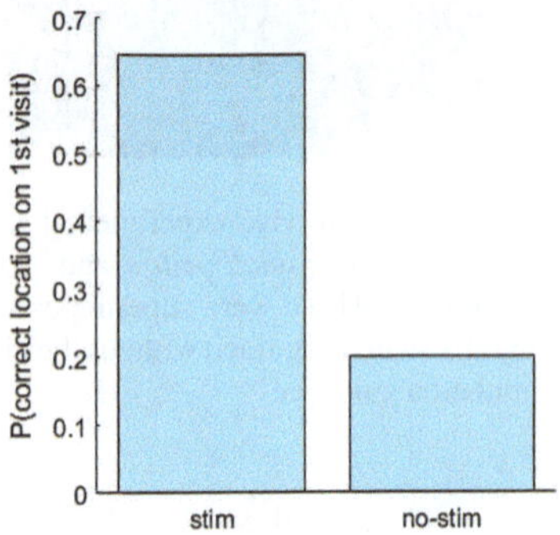

Fig. 6.38 In a total of 139 trials consisting of 124 stimulation trials and 15 catch trials, the percentage of trials in which the rat reaches the correct platform in its first visit is 65% with stimulation, and 20% without stimulation

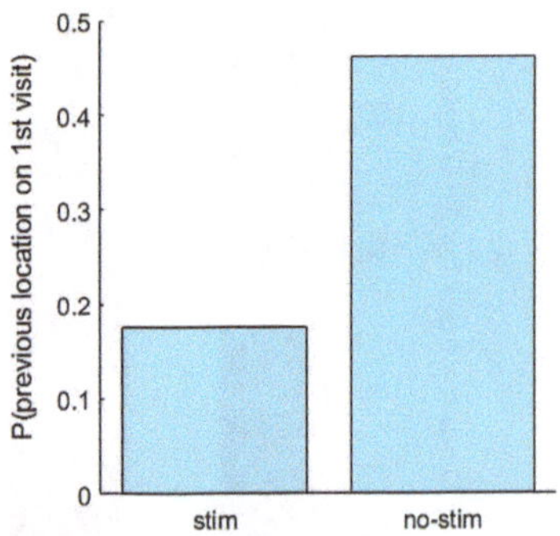

Fig. 6.39 The rat first attempted the platform location in the previous trail in 17.5% trails with stimulation, and 46.2% trails without stimulation

visit is around 20%, while with the stimulation, the percentage is about 65%. This result clearly indicates that the rat has learned to use the stimulation.

Figure 6.39 compares the percentages of trials when the rat's first visited location is the platform's location in the previous trial with and without the stimulation. In 17.5% of the trials with stimulation, the rat's first visited location is the platform's location in the previous trial, while the same scenario happens in 46.2% of the trials without stimulation. The result clearly indicates that the rat relies mainly on its memory to find the platform if no stimulation is presented.

Finally, Fig. 6.40 compares the total number of locations the rat visited until it found the actual platform. The result is again the average of the 139 total trials conducted in this setup, including 124 trials with stimulation and 15 catch trials without stimulation. The average visit times in trials with stimulation is 1.8, while

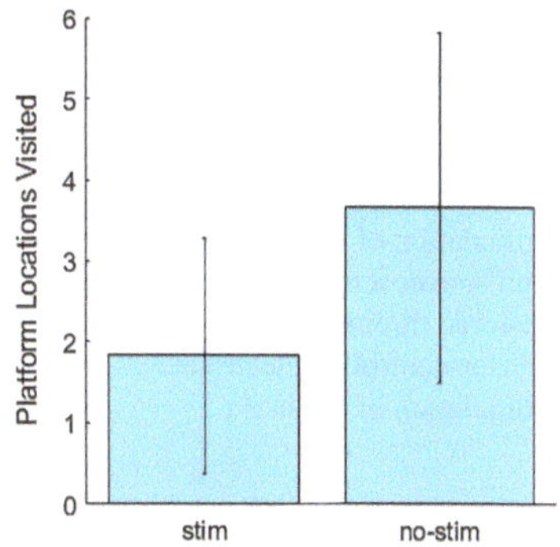

Fig. 6.40 The average times of platform visits are 1.8 times with simulation, and 3.7 times without stimulation guidance. The error bars show the standard deviation in the data

in trials without stimulation is 3.7. The error bars show the standard deviation of the data. The result indicates that for the trials with the stimulation, the rat finds the platform much faster than those trials without the stimulation.

In summary, this section has presented a custom designed wireless BMI platform consisting of a wireless waterproof neuroprosthetic, an animal tracking system and a user interface. The design features a failure prevention mechanism for animal safety. A custom software framework has also been developed to support the experiments. The experiment is the first reported wireless sensory encoding experiment conducted in a freely swimming animal. The experimental results indicate that animals can quickly interpret artificial percepts to guide behavior. The result is important for the development of sensorimotor neuroprosthetics. More importantly, with the fully programmable wireless interface to the neuroprosthetic, the developed system can be used as a general–purpose platform for investigating different sensory encoding experiments in freely behaving animals.

6.4 Bidirectional Neural Interface for Freely Behaving Macaque

6.4.1 Introduction and Background

Sensations and actions are inextricably linked. Behavioral goals are achieved by sampling the environment with the available sensory modalities and modifying actions accordingly [57]. Somatosensory feedback is especially important to the dexterous hand movement control. Hand prosthetics with motor pathway replacement alone are not adequate enough for the use of a paralyzed hand [15]. Artificial sensation restoration is needed for this technology to meet the performance required for clinical adoption. The sensation may be restored by a direct electrical microstimulation of the brain [16].

Figure 6.41 illustrates the envisioned bidirectional clinical hand prosthesis with motor function restored through brain-controlled stimulation of hand muscles, and somatosensation restored through sensor-controlled electrical stimulation of

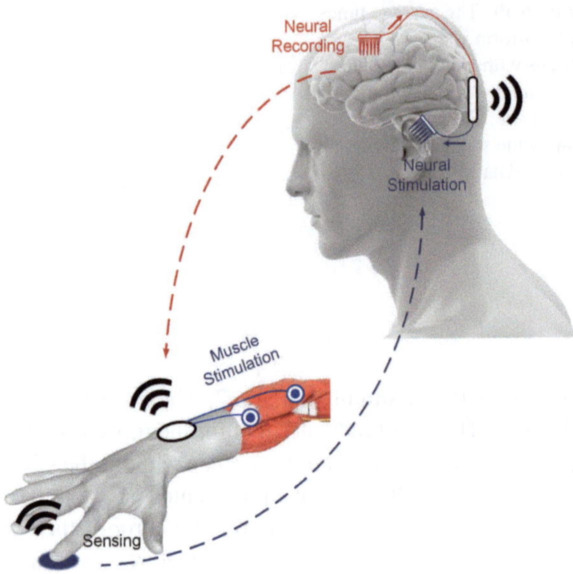

Fig. 6.41 Envisioned bidirectional clinical hand neuroprosthesis. Motor function is restored through brain-controlled electrical stimulation of hand muscles, and somatosensation is restored through sensor-controlled electrical stimulation of the brain

the brainstem [63]. The cuneate nucleus (CN) in the dorsal brainstem carries fine touch and proprioceptive information from the upper body, and is a suitable sensory encoding site. Besides, its compact representations may be reliably activated artificially. Recently, our collaborators from the Translational Neuromodulation Laboratory (TNL) at the University of Pennsylvania demonstrated the first successful chronic interface to the CN of macaques [317], which allows us to investigate the sensation encoding with CN microstimulation in monkeys [63, 318].

In this section, the design of a bidirectional BMI system for the operation in freely behaving monkeys is presented. It has been used to explore the aforementioned sensory mapping, hippocampal oscillation during sleep and awake, and related research. The detailed system architecture, circuit design, and the animal experimental methods are presented. Some of the figures and tables presented in this section were originally published in [57] ©IEEE. Reused, with permission.

6.4.2 Circuit and System Design

Figure 6.42 shows an illustration of the custom designed BMI device for experiments in freely behaving monkey. The overall BMI system includes multiple custom ICs and discrete electronic components. The detailed system configuration varies for different experiments. The electrodes are chronically implanted, with connectors cemented on the skull to mate with the BMI device. The custom IC performs the noise sensitive neural signal recording, the energy-efficient neural feature extraction, and the high safety electrical stimulation.

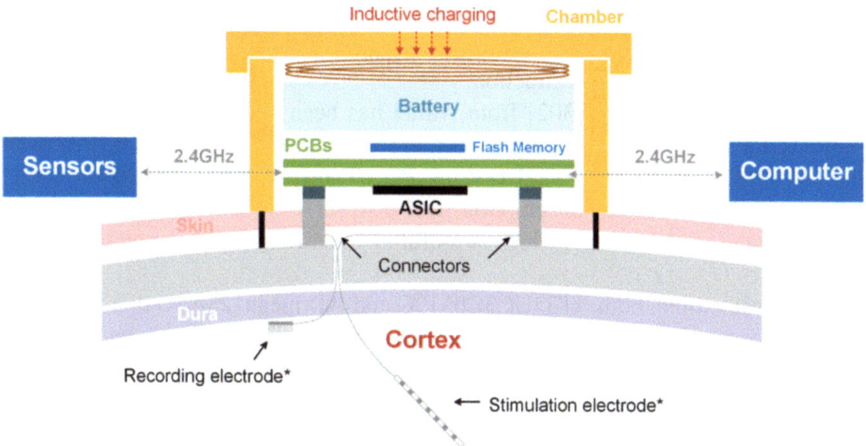

Fig. 6.42 Illustration of the BMI device designed for experiments in freely behaving monkeys (not to scale). The whole device is house in a chamber. The electrodes are chronically implanted, and the nano-connecters are secured by dental cement (*electrode type varies with different applications)

Table 6.5 Comparison of MCUs used in this work

Features	TM4C123GH6PM	ATxmega128A4U	nRF51822
CPU	32-bit	8/16-bit	32-bit
Voltage	3.3 V	1.6–3.6 V	1.8–3.6 V
Clock	80 MHz	32 MHz	16 MHz
Flash	256 kB	128 kB	128 kB
RAM	32 kB	8 kB	16 kB
EEPROM	2 kB	2 kB	NA
UART	8×	5×	1×
SPI	4×	2× 12-bit	2×
USB	USB 2.0	USB 2.0	NA
ADC	2× 12-bit	12× 12-bit	8× 10-bit
DAC	NA	2× 12-bit	NA
Wireless	NA	NA	2.4 GHz
Package	LQFP 64pin	TQFP 44pin	QFN 48pin

A general–purpose low-power MCU is integrated into the system for: (1) the configuration and control of the ICs, (2) handling of the data packets, and (3) performing closed-loop algorithms. Several MCUs with different performance, interface and features have been used in this project. Table 6.5 compares the key features of the three selected MCUs. The 32-bit Tiva TM4C123GH6PM [319] from Texas Instruments is the mainly used for this project. The TM4C123GH6PM is a high-performance ARM Cortex M4 based MCU. Compared with the 32-bit AT32UC3C0512C from Atmel, which was previously used to upgrade the PennBMBI system, the TM4C123GH6PM is more power efficient and has more open source libraries. However, the AT32UC3C0512C and several DSP from Analog Devices have better signal processing ability, which is important for heavy duty on-chip neural feature extraction.

The ATxmega128A4U [302] from Atmel has been introduced in the previous sections for the PennBMBI system and the watermaze project. Compared with the TM4C123GH6PM and AT32UC3C0512C, ATxmega128A4U features a smaller package, a lower power consumption, and a lower cost. In addition to ADC, it also integrates a 12-bit DAC, which is useful in generating a programmable voltage reference.

At the same time, both AT32UC3C0512C and ATxmega128A4U don't have on-chip wireless module, so an additional wireless module nRF24L01+ is integrated on board for many applications in this project, including the PennBMBI system and the watermaze project. A fully integrated wireless MCU, nRF51822 is used in a couple of applications with severe space restraint. The wireless protocol in the nRF51822 is air compatible with the nRF24L01+. The nRF51822 also features a Bluetooth 4.0 protocol for communication with commercial workstations and mobile devices. The nRF51822 has a 32-bit ARM Cortex M0 based core with rich hardware interfaces including UART, SPI, and ADC. However, the processing ability is much weaker than the AT32UC3C0512C and ATxmega128A4U. Very limited on-chip signal processing can be performed in real–time, especially when streaming data near the full wireless data rate.

A Micro-SD card module is integrated in the system for a low-power wireless recording when there is no need for real–time display or processing. An FAT32 file system is implemented for the Micro-SD card. An open source Generic FAT file system module FatFs [320] is modified for use in this work. The FatFs is a generic FAT/exFAT file system written in ANSI C and is independent of hardware platforms. Additional IO interface layers are written so that all MCU can use the file system. It should be noticed that wiring into the SD card doesn't necessarily require a file system. However, writing data without a file system will result in a limited sector address range. The maximum data sector range can be reached without a filesystem is 2 GB. Moreover, no file system means no direct access from a computer system. The reading and writing of the SD card would need a dedicated hardware or software. In this work, the FAT32 file is implemented for easy access, organization, and future extension. A configuration file can be easily edited and saved on the card. The configuration and parameters of the bidirectional BMI device can be set in the configuration file, for example, file name, file length, data buffer size, sampling frequency, front-end gain, filter corner frequency, stimulation pulse width, pulse interval, number of pulses per group, number of groups with different pulse configurations, and so on. The user can easily change these settings without programming the MCU.

The whole BMI device is powered by 3.7 V Lithium batteries. A reliable and high-capacity battery is a key component in a wearable device. A lot of batteries are available on the market. The Polymer Lithium batteries from Adafruit are used in this work. These batteries are lightweight and are among the highest energy density Lithium batteries on the market. Coin batteries and non-rechargeable batteries

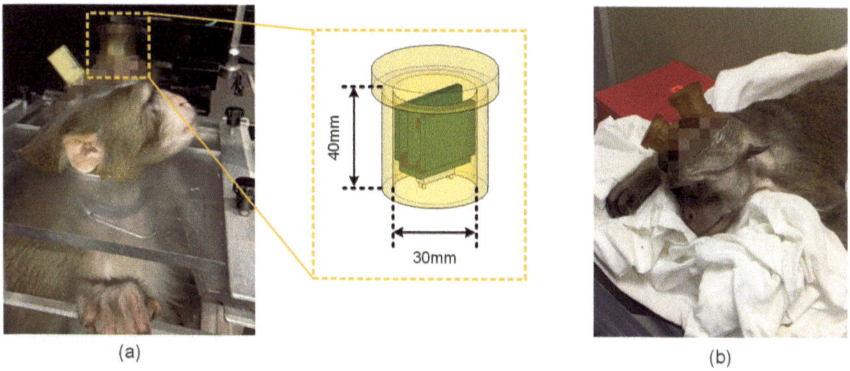

Fig. 6.43 The photograph of (**a**) monkey D with one of the custom designed chambers, and (**b**) monkey M with two of the custom designed chambers. Each chamber has a diameter of 30 mm and a height of 40 mm

are used in several low-power and small footprint sensor nodes in this project. The integrated power management module consists of: (1) battery protection, charging and management circuit, and (2) power management modules including switching converter and LDOs. The TM4C123GH6PM, ATxmega128A4U and AT32UC3C0512C are powered at 3.3 V, and the nRF51822 can be powered at 1.8 V. So nRF51822 has a power advantage over the other three MCUs. But it should be noticed that the most power-hungry components are the wireless transceiver module (36 mW during transmitting) and the Micro-SD card module (40 mW during writing).

Figure 6.43a shows a photograph of monkey D with one of the custom designed chambers. In preparation of the monkey for this experiment, magnetic resonance (MR) images of the brain were acquired with fiducial markers. A sterile surgery was performed to implant the electrode arrays using an MR-guided neuronavigation system. The chamber was attached to the skull with screws and acrylic. The chamber has a diameter of 30 mm and a height of 40 mm. The cap of the chamber is removable and secured by screws. Monkey D has three six-channel electrodes implanted, with three connectors in the chamber. Figure 6.43b shows a photograph of the monkey M with two of the custom designed chambers. Monkey M has four 32-channel electrode arrays implanted, with two connectors in each chamber. The chamber has the same dimension as the one used in Monkey D.

A couple of devices have been developed in this project for several different animal and experiments. The photographs of several developed devices are shown in Fig. 6.44. Figure 6.44a shows a bidirectional BMI with one recording channel and one stimulation channel. The device has an on-board Micro-SD card module for data storage. The recording and stimulation modules have separated grounds for stimulation artifacts suppression. Figure 6.44b shows a 32-channel wireless neural recording device. The device features a continuous 12-h recording with real–time data streaming. The MCU integrated is ATxmega128A4U. Figure 6.44c shows a 16-channel wireless bidirectional BMI device. The MCU integrated is nRF51822.

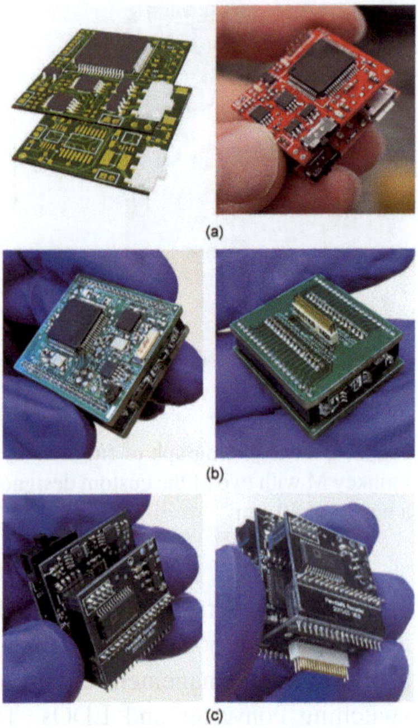

Fig. 6.44 Photographs of several assembled devices for the chamber. (**a**) A bidirectional BMI device with a Micro-SD card module, (**b**) a 32-channel wireless neural recorder, and (**c**) a 16-channel wireless bidirectional BMI

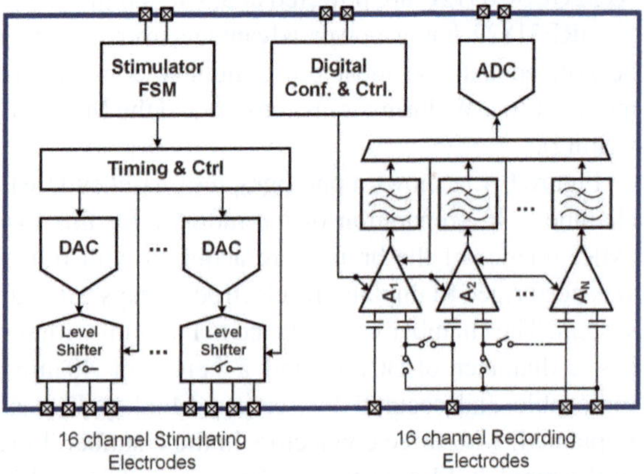

Fig. 6.45 The block diagram of the basic version of the bidirectional BMI ASIC

Custom designed ASIC have been integrated in these BMI devices. Figure 6.45 shows the block diagram of the basic version of the bidirectional custom IC. The basic version ASIC only integrates the analog interface to the brain. Signal processing can be performed in the general–purpose processor. The main building blocks include: (1) analog front-end, (2) stimulator back-end, (3) data converters, and (4) peripheral modules.

Figure 6.46 shows the block diagram of the advanced bidirectional neural interface SoC with the proposed energy-efficient neural feature extraction module and on-chip PID closed-loop controller. The SoC mainly consists of: (1) 16-channel

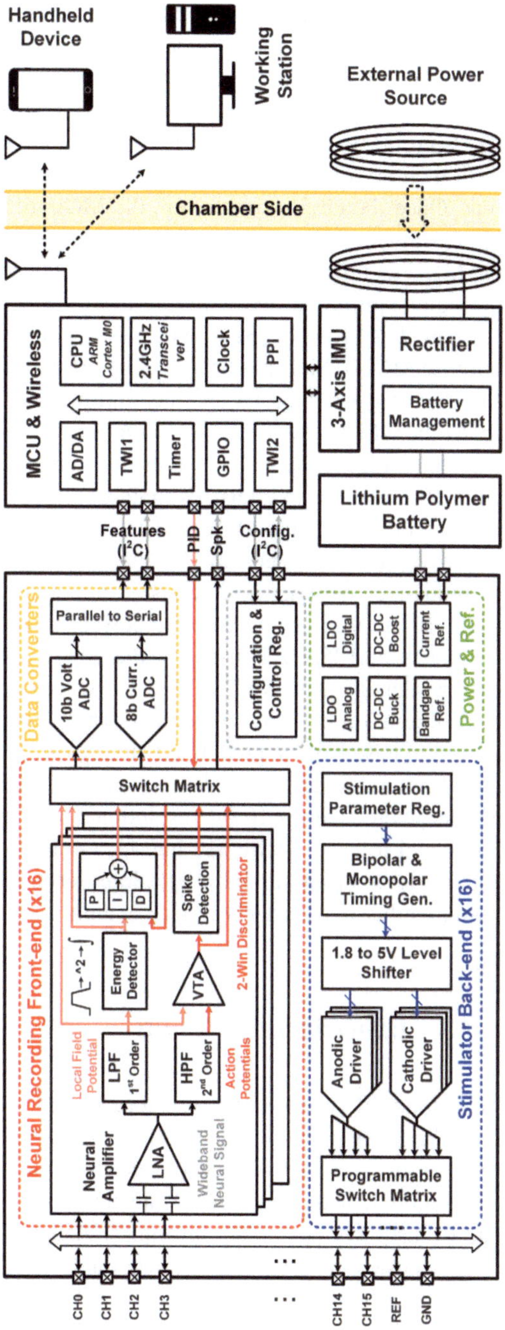

Fig. 6.46 Architecture of the bi-directional, closed-loop brain–machine interface system. The system includes a custom SoC and supporting electronics

neural front-end with in-channel neural feature extraction unit and closed-loop controller, (2) 16-channel independently programmable neural stimulators, (3) voltage and current mode data converters, (4) power management, analog voltage and current references, and peripheral circuits. The detailed circuit implementation has been presented in Chaps. 2–5. The configuration of the SoC is stored in the flash memory of the MCU, and it can be programmed wirelessly via the Bluetooth link. Once the device powers up, the MCU first reads the default configuration in the flash memory, and then configures the SoC accordingly. The interface between the MCU and the SoC is shown in Fig. 6.47. The configuration and data readout are through a simplified two-wire interface (TWI) module. The TWI module supports standard I^2C protocol [321], which is compatible with most off-the-shelf general–purpose MCUs. Multiple SoCs can be used together to support more channels. The MCU works as the master and the multiple SoCs work as the slaves. The MCU first sends the address, and the chip with the specified address responses. Two IOs are used to configure the address, thus, the current design setup can support up to four chips (64 channels in total). This can be easily expanded in the future, for example, 8-bit address can support up to 128 chips. The timing for the *START*, *STOP*, and *ANSWER* commands is shown in Fig. 6.47.

Inductive charging of batteries is supported in several BMI devices. T3168 and XKT510 are used as the wireless power transmitter and receiver ICs. A switching frequency of 125kHz is used for the power transmitter. MC73831 is used for the battery management. MC73831 is a linear charge management controller designed for Lithium ion batteries, which performs a constant current and a constant voltage charging based on a predefined policy. A 3-axis accelerometer ADXL345 has also been integrated into the system. ADXL345 is an ultra-low power digital accelerometer with a 10-bit resolution for ±16 g. It can be powered from 2.0 to 3.6 V, and it has a 3-wire SPI interface to the MCU. It can be used to monitor the activity level of an animal over a long term. Multiple accelerometers can also be used together to accurately model an animal's gesture.

6.4.3 Experimental Results

The proposed system has been fabricated in standard printed circuit board (PCB) and CMOS technology. The PCB process used in this project features FR-4 material, two or four layers, a thickness of 0.8 mm, a minimum trace and spacing of 0.15 mm, a minimum hole diameter of 0.2 mm, a minimum via diameter of 0.15 mm. The PCB surface uses a hot air solder leveling (HASL) lead-free finishing.

The basic version of the bidirectional neural interface IC have been fabricated in both On-Semi 0.5 μm CMOS technology and IBM 180 nm CMOS technology. Figure 6.48a shows the micrograph of the fabricated chip in 0.5 μm CMOS technology. The major building blocks are low-noise amplifiers and neural stimulator back-ends. The chip occupies a silicon area of 3 mm × 3 mm, including the IO pads. The supply voltage is from 3.3 to 5 V. Figure 6.48b shows the micrograph of the

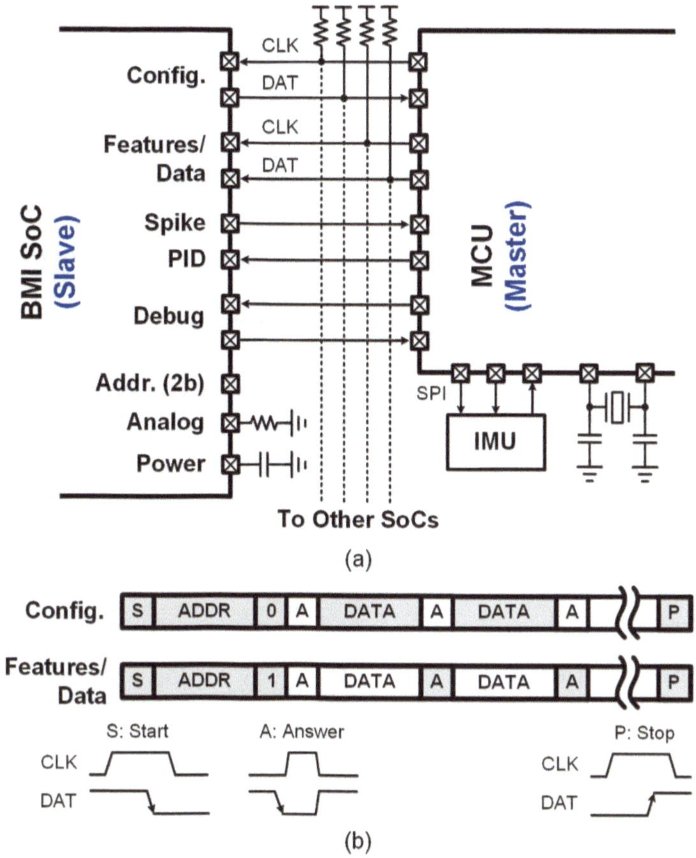

Fig. 6.47 (**a**) The communication interface between the SoC and the general–purpose MCU (not all pads are shown), (**b**) the communication data format. The MCU (master) writes the *gray* sectors

fabricated chip in 180 nm CMOS technology. The chip occupies a silicon area of 4.5 mm × 1.5 mm, including the IO pads. The major building blocks are: (1) analog front-end, (2) stimulator back-end, (3) data converters, and (4) peripheral modules.

Figure 6.49 shows the micrographs of the advanced bidirectional neural interface SoC with energy-efficient neural feature extraction module and PID closed-loop controller. Figure 6.49a is the first version including 12 channels with debugging and testing structures, and Figure 6.49b shows the second version including 16 channels. Both chips occupy a silicon area of 4.5 mm × 1.5 mm, including the IO pads. The major building blocks are highlighted in the figure, including: (1) neural front-end with neural feature extraction units and closed-loop controller, (2) programmable neural stimulators, (3) data converters, (4) power management, analog references, and peripheral circuits. The bench testing results of each module in the bidirectional neural interface SoC have been presented in Chaps. 2–5. Table 6.6 summarizes the measured key specifications of the SoC.

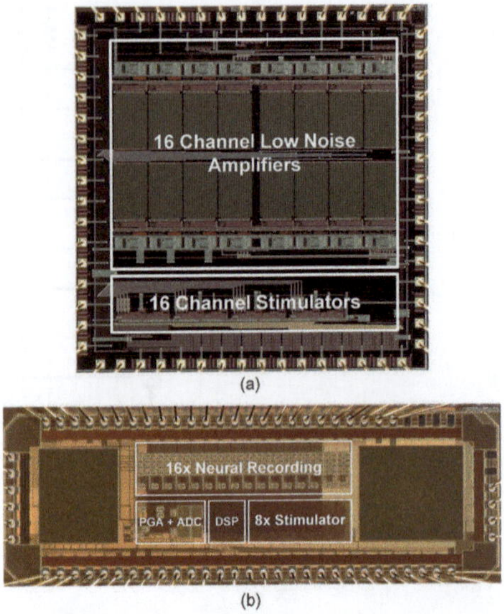

Fig. 6.48 The micrographs of the basic version of the bidirectional neural interface SoCs in (**a**) On-Semi 0.5 µm CMOS technology, and (**b**) IBM 180 nm CMOS technology. Major building blocks are highlighted in the figures

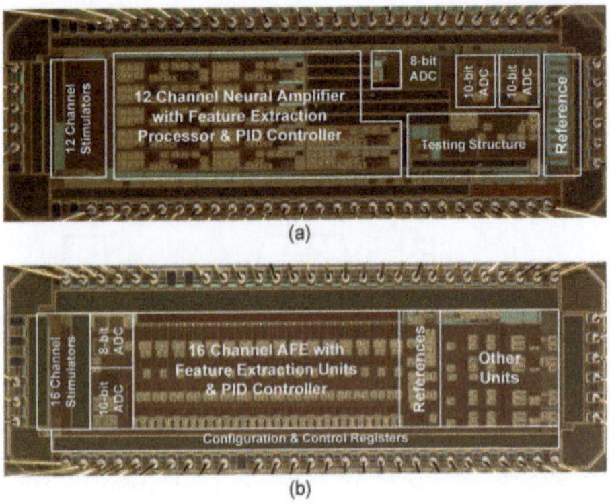

Fig. 6.49 The micrographs of the bidirectional neural interface with the proposed energy-efficient neural feature extraction and PID closed-loop controller. (**a**) shows the first version with 12 channels with debugging and testing structures. (**b**) shows the second version with 16 channels. Major building blocks are highlighted

The developed BMI devices have been used in a few experiments with freely behaving monkeys. The experimental results presented in the following analysis were conducted in monkey (*Macaca mulatta*) O, D, and F (8–12 kg). A study of hippocampal (HIPP) gamma-slow oscillation coupling during sedation and sleep is presented. The slow oscillation (SO) of non-rapid eye movement sleep plays a critical role in the consolidation of newly formed memories [322]. There is substantial behavioral evidence linking the amount of SO activity after learning to the strength of both procedural and declarative memories [323, 324]. These effects of SO on HIPP activity have been studied in rodents and cats, but have not been documented in primates. In this work, we recorded the HIPP field potentials during sedation and during natural sleep. In addition, electrical stimulation was delivered to HIPP afferents in the parahippocampal gyrus (PHG) during sedation and awake to study the effects of the sleep like SO on excitability.

Table 6.6 Key specifications of the bidirectional neural interface SoC

Analog front-end	LNA gain	40 dB
	LNA bandwidth	0.3 Hz–7 kHz
	LNA integral noise	4.57 μV
	LNA power	9 μW
	LNA NEF/PEF	4.77/41.1
	THD (10 mVpp input)	−61 dB
	CMRR/PSRR	81 dB/71 dB
	PGA + filters power/ch	8 μW
Energy extraction	Center frequency	1–200 Hz
	Tuning steps	64 natural log
	Quality factor	1–8
	Window length	10–500 ms
	Ex + PID power/ch	7 μW
Spike discriminator	Algorithm	Window discrimination
	Amplitude thresholds	6-bit
	Latency	10 μs
	Avg. power/ch	4 μW
ADC (volt mode/curr mode)	Sampling rate	1 MSps/250 kSps
	ENOB	9.1/7.9
	FoM (fJ/step)	34.2/10.7
	Power (at 200 kSps)	7 μW/0.5 μW
Stimulator	Stim. current	4 mA/200 μA
	Amplitude res.	6-bit
	Pulse width	1–255 μs
Total power	Chip power/ch	56 μW/ch
	MCU + wireless (avg)	8 mW

Figure 6.50 illustrates the implanted electrode array in the monkey's brain. The platinum electrode sites are shown as red boxes and the electrode trajectories are shown as black outlined rectangles. The dimensions of the electrode sites are indicated at the bottom. Typical MRI and CT image with the visible hippocampal array is shown in Fig. 6.50b. Colored regions indicate different neuroanatomical areas. Since the goal of this project is to identify potential closed-loop stimulation paradigms for modulating memory for human patients, clinical microelectrodes were used in this study, rather than the conventional microelectrodes. Single neurons cannot be collected from these electrodes, but the field potentials were sufficient to document the regionally specific SO and effective PHG-HIPP connectivity for the objective of this study.

Figure 6.51 shows the power spectrums recorded from the three electrode arrays in the hippocampus, entorhinal cortex, and medial septum, respectively. The signal was recorded using the developed BMI device on a Micro-SD card. The device was placed in the chamber during the sedation, and retrieved the next time the monkey was brought to the lab. Different brain states, from sedation, recovery, awake and

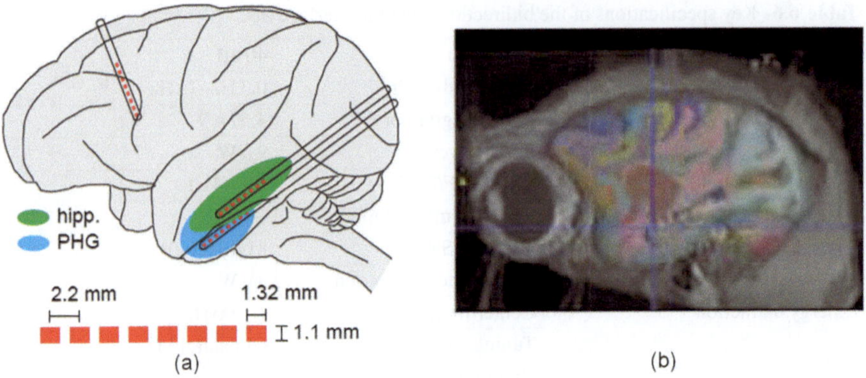

Fig. 6.50 (**a**) Illustration of the implanted depth electrode arrays. (**b**) MRI and CT image with the visible hippocampal array

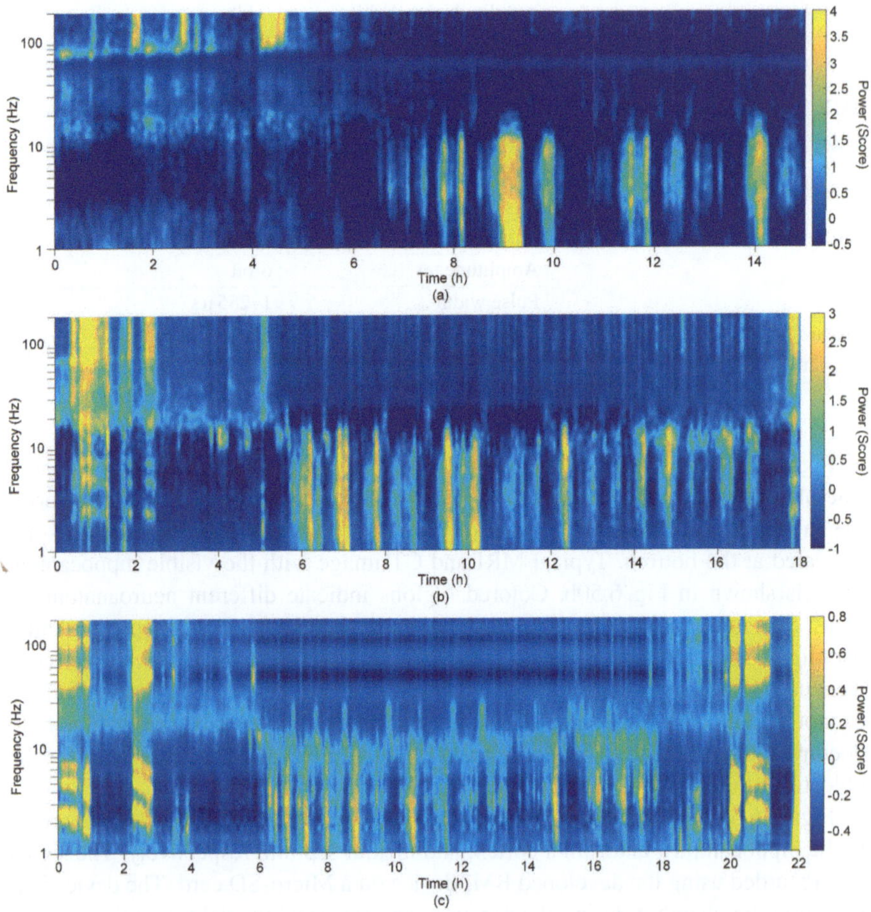

Fig. 6.51 The power spectrums of the long-term recordings of monkey D from the three electrode arrays in: (**a**) hippocampus, (**b**) entorhinal cortex, and (**c**) medial septum

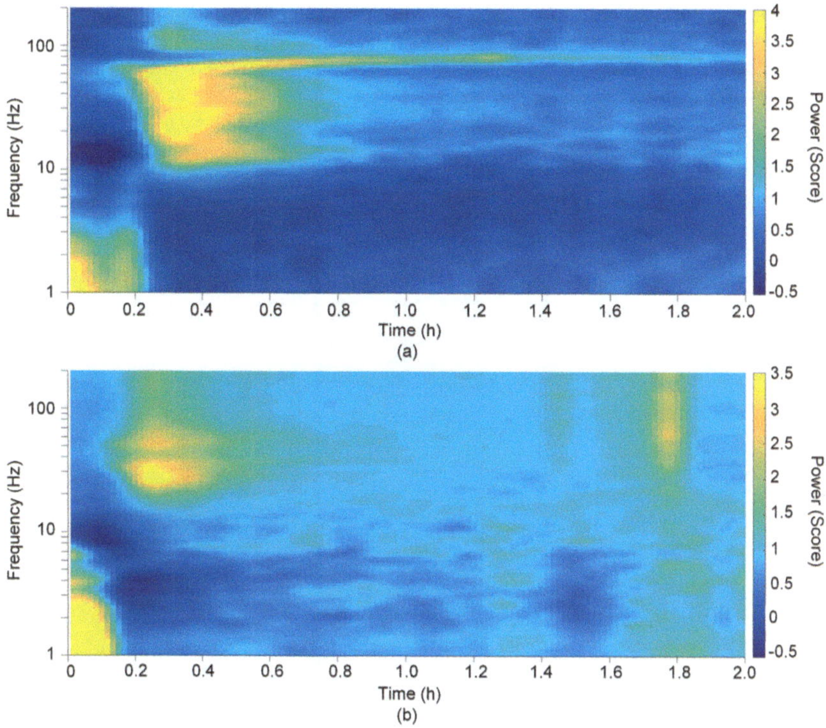

Fig. 6.52 The recorded spectrum of the recovery process from anesthesia in (**a**) monkey D and (**b**) monkey F

sleep can be told from the spectrums. Both time and frequency domain features and chewing artifacts verify the reliability of the recording.

Figure 6.52 shows the recorded spectrum of the recovery process from anesthesia in monkey D and monkey F, respectively. Both recordings show a period of an increase of the high-frequency oscillation after the sedation, which is the effect of the ketamine-dexmedetomidine.

Separate sessions were conducted with a focus on quantifying neural connectivity. In order to study during stimulation reversal and awake states, the developed bidirectional BMI device was configured to deliver a single bipolar charge-balanced pulse with an amplitude of 2 mA in every 30 s to the entorhinal cortex. The same device recorded the evoked response in the hippocampus. In this way, the PHG-HIPP connectivity in consistent states defined by oscillatory activity was studied. Figure 6.53 shows a stacked plot of 278 responses, aligned by the onset of the stimulation. The red arrow marks the stimulation time. Thanks to the fast artifact recovery design as presented in Chap. 5, the evoked potentials can be clearly seen from the recording without signal corruption. The experiment was repeated on Monkey D approximately 2 months. These results demonstrate the reproducibility and stability of the effects. Figure 6.54 shows the average waveform of the responses.

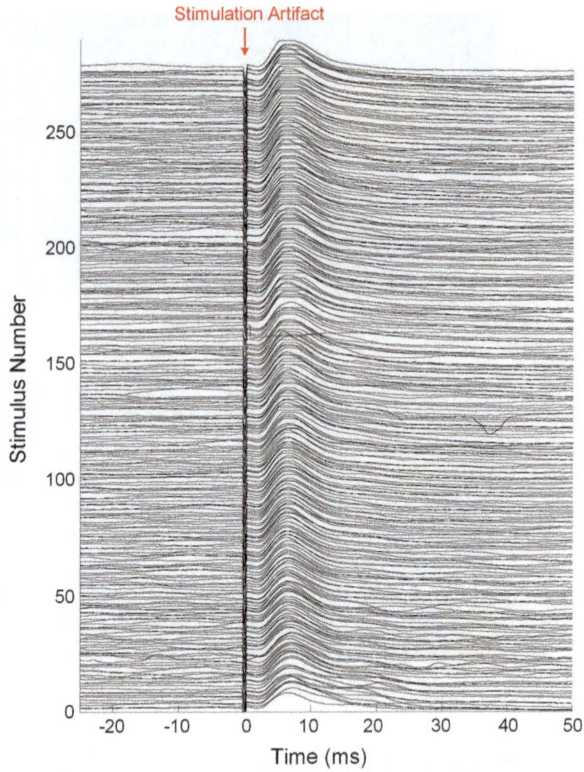

Fig. 6.53 The stacked plot of 278 stimulation triggered evoked potentials recorded using the developed BMI device

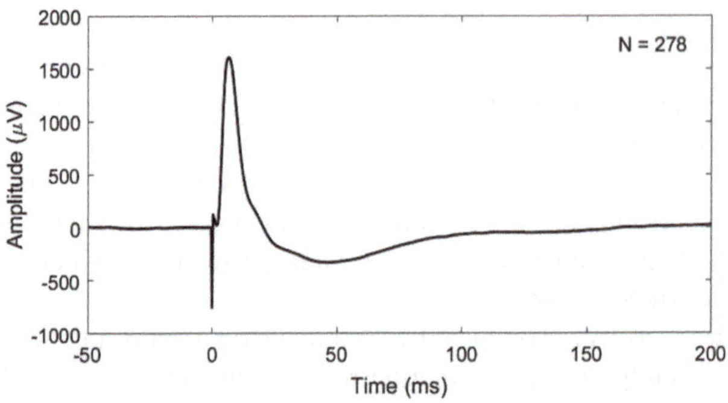

Fig. 6.54 The average response waveform of the stimulation triggered evoked potentials in the hippocampus

In the next experiment, a programmable stimulus pulse train was delivered to the medial septum. The stimulation frequency was switched between 20 to 80 Hz during one session. Figure 6.55 compares the evoked potentials (EP) from three states: sedation, recovery, and awake. The plots are generated by the average of an over 3-h recording with a stimulus pulse train delivered every 30 s. In all three states,

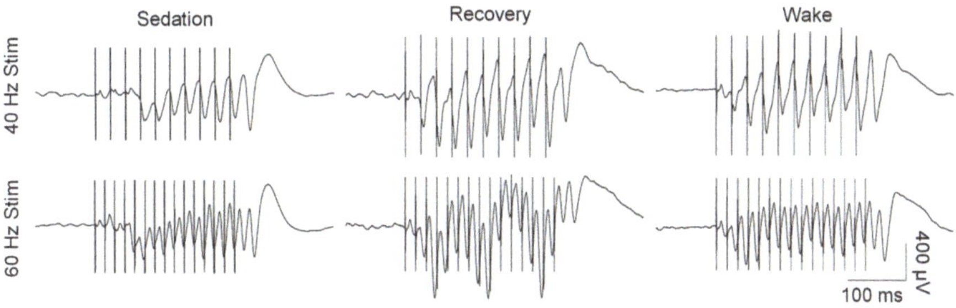

Fig. 6.55 The recordings of the stimulus-evoked potentials. Stimulus trains of 40 and 60 Hz in different brain states are shown. The plots show a triggered average over 3 h recording in total, with stimulation every 30 s

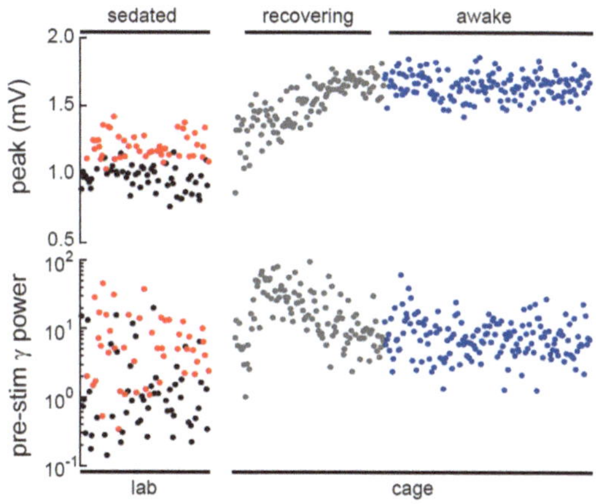

Fig. 6.56 The time course of EP peaks and pre-stimulus gamma power in across three behavioral states: sedated (*red, black circles*), recovery (*gray circles*), and awake (*blue circles*). Gamma (30–50 Hz) power was calculated in a 300-ms window preceding each stimulus

the stimulus pulse train evoked oscillations in the hippocampus at 40 and 60 Hz. The oscillation continues for at least one cycle after the stimulation pulse train. Due to the high stimulation current (2 mA) and the high compliance voltage (12 V), the battery typically lasts 3–4 h in the stimulation-recording sessions.

Figure 6.56 illustrates the Gamma-dependence of hippocampal EPs in sedated versus awake monkeys. The time course of EP peaks and pre-stimulus gamma power in monkey D across three behavioral states are shown: sedated (red, black circles), recovery (gray circles), and awake (blue circles). Gamma (30–50 Hz) power was calculated in a 300-ms window preceding each stimulus. The horizontal axis is scaled to stimulus number, not the absolute time. PHG stimuli during sedation in the lab were delivered every 5s by a commercial stimulator. PHG stimuli during recovery and awake periods in the home cage were delivered every 30 s by the custom designed BMI device. Stimulation amplitude was 0.5mA in both cases. The reversal agent, atipamezole, was given between the sedated and recovering states. In both monkeys, the recovery periods show a transient increase in HIPP gamma-band activity and steady increase in the peak amplitude of the EP.

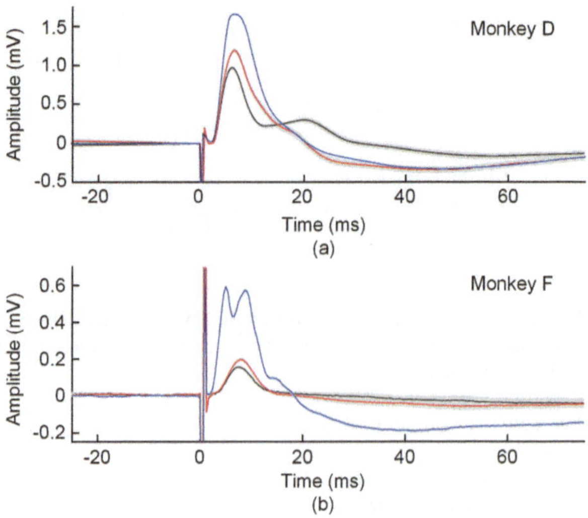

Fig. 6.57 Mean EPs during sedation (*red*, *black*) and awake (*blue*) for (**a**) monkey D and (**b**) monkey F. 95% confidence intervals on the mean are shown in *gray*

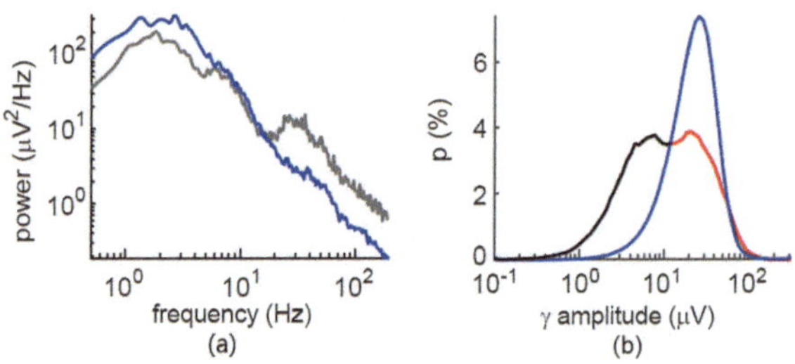

Fig. 6.58 (**a**) The power spectral density of the hippocampal recordings during recovery (*gray*) and awake (*blue*) states in monkey F. (**b**) Distribution of gamma amplitude across sedated (*red*, *black*) and awake (*blue*) recording sessions in monkey F

Thus, the gamma-dependence of the evoked HIPP responses was specific to the sedated state and the responses overall were weaker than during the awake state. The results were replicated in Monkey F using instrumentation recording. Figure 6.57 compares the EPs during sedation and awake for Monkey D and Monkey F. 95% confidence intervals on the mean are shown in gray.

Figure 6.58a shows the power spectral density of the hippocampal recordings during recovery (gray) and awake (blue) states in monkey F. Monkey F exhibited a similar HIPP oscillatory activity during recovery from sedation as Monkey D, with an increase in gamma power and a decrease in low-frequency power relative to the awake states. Figure 6.58b shows the distribution of gamma amplitude across sedated (red, black) and awake (blue) recording sessions in monkey F. The sedated distribution was colored black and red to highlight its bimodal nature and correspondence to the bimodal EP response amplitudes as in Figs. 6.56 and 6.57. The higher mode of the sedation distribution (red) aligned with the awake gamma amplitude distribution (blue), while the lower mode (black) was not presented in the awake distribution.

Table 6.7 Comparison with the state-of-the-art bidirectional neural interface designs

Reference	Rhew [48]	Cong [49]	Limnuson [45]	Chen [44]	Shulyzki [53]	Biederman [56]	This work
Publication	2014 JSSC	2014 ESSCIRC	2014 CICC	2014 JSSC	2015 TBioCAS	2015 JSSC	–
CMOS technology	180 nm	0.25 μm and 90 nm	0.35 μm	180 nm	0.35 μm	65 nm	180 nm
ch # of rec./stim.	4/8	32/16	1/1	8/1	256/64	64/8	16/16
AFE noise	6.3 μVrms	100 nV/rtHz	3.42 μVrms	5.23 μVrms	7.99 μVrms	7.5 μVrms	4.57 μVrms
AFE NEF	–	–	2.75	1.77	8.9	3.6	4.77
Bandwidth (Hz)	0.64–6k	0.5–1.7k	0.1–12.3k	0.1–10k	10–5k	10–8k	0.3–7k
ADC ENOB	5.6	12	10	9.57	–	8.2	9.1
Feature extraction	Custom DSP	Custom CPU	Custom DSP	Custom DSP	–	Custom DSP	Parallel analog
Neural features	Energy	FFT	–	Spectrum, Entropy	–	Spikes	Energy, Spikes
Feedback control	PI	–	–	–	–	–	PID
Stim. mode	Monopolar	Monopolar	Monopolar	Bipolar	Monopolar	Bipolar	Mono/bipolar
Stim. artifact rej.	LP filter	LP filter	Subtraction	–	Blanking	–	Pole-shifting
Output current	4410 μA	12 mA	100 μA	30 μA	20–250 μA	900 μA	4 mA

In summary, the study above shows that the ketamine-dexmedetomidine sedation in primates produces phase-amplitude coupling of gamma and slow oscillations in the PHG-HIPP network. This work presents the first study to directly compare the macaques HIPP field potentials in sleep and sedation. The custom developed BMI system played a key role in the presented experiment especially in freely behaving monkeys. The study suggests that future investigations of the SO in primates would best be conducted during natural slow-wave sleep rather than sedation, and these studies would be best studied directly through the BMI devices as developed in this work.

To conclude, a comparison with the recently reported designs of the bidirectional neural interfaces is listed in Table 6.7. Compared with the state-of-the-art designs, this work is the first reported wireless bidirectional, closed-loop BMI system used for long-term freely behaving animal experiments and investigation. This design shows a promising and practical solution for use in the future animal experiments based on primate models.

Chapter 7
Conclusion and Future Direction

7.1 Summary of the Work

This book has presented the analysis and design of BMI systems. To the best of our knowledge, this is the first work dedicated to studying bidirectional closed-loop BMI systems. The main motivation of this work is the fact that many significantly meaningful neuroscience experiments, especially in freely behaving animals, cannot be conducted without custom designed bidirectional closed-loop BMIs. With the close collaboration between neuroscientists and engineers, this work was able to identify and address several important and practical issues in BMI systems' design. The developed system has been successfully used in several animal experiments, resulting in significant new observations.

The main work and contributions of this book are summarized as follows. In the first chapter, a brief historical review of BMI development was given. A comprehensive survey and review with a focus on BMIs with bidirectional neural interface were presented. Consecutively, an overview of BMI system architecture was presented. Design considerations and key specifications were summarized. The configurations for various closed-loop operations were illustrated.

In the following three chapters, the analysis, design, and experimental results of the three main building blocks of a BMI system, namely the neural signal recording module, the neural feature extraction module, and the neural stimulator module were presented. In Chap. 2, neural recording front-end designs were introduced. The design of a general–purpose low-noise instrumentation amplifier and a low-power ADC were discussed. A novel pre-whitening neural amplifier was proposed to increase the equivalent dynamic range of the front-end. The pre-whitening processing takes advantage of the neural signal's characteristics, significantly relaxing the ADC design without sacrificing the signal quality. In addition, compressive sensing technique was used to reduce the wireless data rate of the recording front-end. A fully integrated wireless neural signal acquisition front-end was developed for chronic neural recording and BMI applications.

X. Liu, J. Van der Spiegel, *Brain-Machine Interface*,
https://doi.org/10.1007/978-3-319-67940-2_7

In Chap. 3, commonly used neural features for real–time closed-loop BMIs were summarized. The circuit implementation for energy-efficient feature extraction of local field potential and action potential was presented. An energy extraction circuit with a natural logarithmic domain tuning scheme was proposed to provide a sufficient frequency resolution for low-frequency brain oscillations with a minimum number of tuning steps. An ultra low-power action potential discriminator was designed in current-mode circuits. A matched filter was proposed to extract amplitude-phase coupled neural features. The performance of the matched filter was further improved by using in combination with the proposed pre-whitening amplifier. Experimental results showed that the proposed design improved the feature detection accuracy especially in low SNR cases.

Chapter 4 presented the analysis and design of neural stimulators. The background and mechanisms of neurostimulation were first reviewed, followed by a description of physical and electrical models of the electrode and electrolyte interface. An overview of electrical stimulator design was given, including a review of stimuli generation methods, stimulation waveform, and electrode configuration. Methods for achieving charge balancing were also discussed. A general–purpose neural stimulator was designed. In addition, a novel net-zero charge stimulation strategy was proposed. Instead of focusing on circuit matching and residue charge removal, this work attempts to achieve charge balancing by employing feedback. The developed chip has been validated in both in vitro and in vivo experiments.

Chapter 5 discussed the design of bidirectional closed-loop BMIs from two important perspectives: the stimulation artifacts and the closed-loop operation. Stimulation artifact is a known issue in simultaneous neural stimulation and recording. A long-lasting stimulation artifact blanks the recording front-end and corrupts the recorded signal. Prior work proposed different methods to address this problem, however, with constraints in their applications. In this work, stimulation artifacts in different electrode configurations and circuit topologies were studied. Conclusions and design recommendations were given. In addition, the mechanisms of closed-loop operation of BMIs were reviewed, followed by the design of a low-power general–purpose programmable PID controller. The PID controller was integrated in a bidirectional neural interface SoC.

Chapter 6 presented the BMI system integration and animal experiments. The design of a general–purpose experimental platform was first described. Custom communication protocols and user-friendly interfaces were developed and used in this work. Then, a novel watermaze experiment for the study of augmenting perception though modulated electrical stimulation was presented. A complete experimental system was designed, including a wearable waterproofed stimulator, an animal tracking system, and a computer based control interface. Different stimulation parameters and versatile neuromodulation algorithms can be configured in the system. Moreover, the design of a bidirectional neural interface device for the operation in a freely behaving animal was presented. Long-term experiments of neural stimulation and recording in monkeys during awake, sedated, and sleeping states were given. A study of hippocampal gamma-slow oscillation coupling using the developed system was also described. The design has shown a promising and practical solution for the future experiments in non-human primate models.

7.2 Future Direction

A decade ago, Mikhail A. Lebedev and Miguel A.L. Nicolelis predicted that future BMIs will have sensory feedback directly delivered to cortical or subcortical somatosensory areas, and closed-loop BMIs would be the ideal tool to restore motor functions [275]. This prediction has inspired this work. With the efforts of scientists and engineers around the world, we are stepping into the future of BMIs at an incredible pace. Nevertheless, several bottlenecks still need to be overcome.

1. **Interfacing**: The direct interface between the neuron and electronics remains a challenge, preventing a safe long-term neural stimulation and recording. Novel interfacing material and electronics still need to be developed;
2. **Wireless Communication**: Although a lot of efforts have been put in developing wireless neural recorders, a reliable solution for real–time streaming of neuron activities from high density microelectrode arrays is still highly desirable. The ideal solution would fully consider the trade-offs between the bandwidth and power consumption, with a minimum data corruption;
3. **On-chip Processing**: On-chip processing is important for reducing the wireless data rate, and more importantly, to support real–time closed-loop operation, which is the ultimate goal for the development of most BMI devices. The on-chip operation is usually much more reliable than streaming the data through a wireless link and relying on an external processing station. However, the limited power budget and on-chip resources place a significant challenge on on-chip neural signal processing design. With the help of artificial intelligence and advanced machine learning techniques, on-chip neural signal processing is one of the most promising research areas in the next few years;
4. **Power Consumption**: Low-power is always an important design consideration for implantable medical devices, for both extending battery life and minimizing the tissue damage caused by heat. Developing low-power circuit design techniques as well as exploring energy harvesting opportunities would be the path to overcome this challenge;
5. **Packaging**: Biocompatible packaging is critical in the developing of implantable BMI devices. An ideal implantable device would be fully sealed with only wireless interfaces for communication, programming, and battery recharging.

It should be noticed that aforementioned challenges and opportunities are mainly from the electrical engineering perspective. In addition, the development of fundamental neuroscience and neural engineering innovations has always been the main driving force in the BMI research. The biggest opportunity is the close collaboration between neuroscientists and electrical engineers, as well as scientists and engineers in all related fields. With further improvements in performance, reliability, and range of applications, the BMI technology would benefit a larger and larger population, revolutionize our way of interacting with the external world, and fundamentally help us better understand ourselves.

Bibliography

1. Neuron, Brain-computer interface. Retrieved Jan 2017. https://en.wikipedia.org/wiki/Neuron
2. Wikipedia and the Free Encyclopedia, Star. Retrieved Jan 2017. https://en.wikipedia.org/wiki/Star
3. E. Adrian, B. Matthews, The interpretation of potential waves in the cortex. J. Physiol. **81**, 440–471 (1934)
4. Wikipedia and the Free Encyclopedia, Brain-computer interface. Retrieved Jan 2017. https://en.wikipedia.org/wiki/Brain-computer_interface
5. J. Vidal, Toward direct brain-computer communication. Annu. Rev. Biophys. Bioeng. **2**(1), 157–180 (1973)
6. P.R. Kennedy, The cone electrode: a long-term electrode that records from neurites grown onto its recording surface. J. Neurosci. Methods **29**, 181–193 (1989)
7. S. Bozinovski, M. Sestakov, L. Bozinovska, Using EEG alpha rhythm to control a mobile robot, in *Proceedings of the Annual International Conference of the IEEE Engineering in Medicine and Biology Society*, vol. 3 (1988), pp. 1515–1516
8. G.B. Stanley, F. Li, Y. Dan, Reconstruction of natural scenes from ensemble responses in the lateral geniculate nucleus. J. Neurosci. **19** (18), 8036–8042 (1999)
9. J.K. Chapin et al., Real-time control of a robot arm using simultaneously recorded neurons in the motor cortex. Nature Neurosci. **2**(7), 664–670 (1999)
10. J. Wessberg et al., Real-time prediction of hand trajectory by ensembles of cortical neurons in primates. Nature **408**(6810), 361–365 (2000)
11. G. Schalk et al., BCI2000: a general-purpose brain-computer interface (BCI) system. IEEE Trans. Biomed. Eng. **51**(6), 1034–1043 (2004)
12. L.R. Hochberg et al., Reach and grasp by people with tetraplegia using a neurally controlled robotic arm. Nature **485**(7398), 372–375 (2012)
13. S.N. Flesher et al., Intracortical microstimulation of human somatosensory cortex. Sci. Transl. Med. **8**(361), 361ra141 (2016)
14. M.A. Lebedev, M.A.L. Nicolelis, Brain-machine interfaces: past, present and future. Trends Neurosci. **29**(9), 536–546 (2006)
15. J.C. Rothwell, M.M. Traub, B.L. Day, J.A. Obeso, P.K. Thomas, C.D. Marsden, Manual motor performance in a deafferented man. Brain **105**(3), 515–542 (1982)
16. S.J. Bensmaia , L.E. Miller, Restoring sensorimotor function through intracortical interfaces: progress and looming challenges. Nat. Rev. Neurosci. **15**, 313–325 (2014)
17. Wikipedia and the Free Encyclopedia, Control theory. Retrieved Jan 2017. https://en.wikipedia.org/wiki/Control_theory
18. C. Hammond, R. Ammari, B. Bioulac, L. Garcia, Latest view on the mechanism of action of deep brain stimulation. Mov. Disord. **23**(15), 2111–2121 (2008)

X. Liu, J. Van der Spiegel, *Brain-Machine Interface*,
https://doi.org/10.1007/978-3-319-67940-2

19. B. Rosin, M. Slovik, R. Mitelman, M. Rivlin-Etzion, S.N. Haber, Z. Israel, E. Vaadia, H. Bergman, Closed-loop deep brain stimulation is superior in ameliorating parkinsonism. Neuron **72**(2), 370–384 (2011)
20. A. Berenyi et al., Closed-loop control of epilepsy by transcranial electrical stimulation. Science **337**(6095), 735–737 (2012)
21. T. Jeanne et al., Closed-loop optogenetic control of thalamus as a tool for interrupting seizures after cortical injury. Nat. Neurosci. **16**(1), 64–70 (2013)
22. U. Rutishauser, A. Kotowicz, G. Laurent, A method for closed-loop presentation of sensory stimuli conditional on the internal brain-state of awake animals. J. Neurosci. Methods **215**(1), 139–155 (2013)
23. E.E. Fetz, Restoring motor function with bidirectional neural interfaces. Prog. Brain Res. **218**, 241–252 (2015)
24. C.T. Moritz, S.I. Perlmutter, E.E. Fetz, Direct control of paralysed muscles by cortical neurons. Nature **456**, 639–642 (2008)
25. T.H. Lucas, E.E. Fetz, Myo-cortical crossed feedback reorganizes primate motor cortex output. J. Neurosci. **33**(12), 5261–5274 (2013)
26. R.W. Eaton, T. Libey, E.E. Fetz, Operant conditioning of neural activity in freely behaving monkeys with intracranial reinforcement. J. Neurophysiol. **117**(3), 1112–1125 (2017)
27. Scopus, Retrieved Jan 2017. https://www.scopus.com
28. J. Mavoor et al., An autonomous implantable computer for neural recording and stimulation in unrestrained primates. J. Neurosci. Methods **148**(1), 71–77 (2005)
29. F. Heer, S. Hafizovic, W. Franks, CMOS microelectrode array for bidirectional interaction with neuronal networks. IEEE J. Solid State Circuits **41**(7), 1620–1629 (2006)
30. R. Blum, J. Ross, E.A. Brown, S.P. DeWeerth, An integrated system for simultaneous, multichannel neuronal stimulation and recording. IEEE Trans. Circuits Syst. I Regul. Pap. **54**(12), 2608–2618 (2007)
31. S. Venkatraman, K. Elkabany, J.D. Long, Y. Yao, J.M. Carmena, A system for neural recording and closed-loop intracortical microstimulation in awake rodents. IEEE Trans. Biomed. Eng. **56**(1), 15–22 (2009)
32. J. Rolston, R. Gross, S. Potter, Neurorighter: closed-loop multielectrode stimulation and recording for freely moving animals and cell cultures. IEEE Eng. Med. Bio. Soc. (EMBC) **2009**, 6489–6492 (2009)
33. J. Lee, H. Rhew, A 64 channel programmable closed-loop neurostimulator with 8 channel neural amplifier and logarithmic ADC. IEEE J. Solid-State Circuits **45**(9), 1935–1945 (2010)
34. F. Shahrokhi, K. Abdelhalim, The 128-channel fully differential digital integrated neural recording and stimulation interface. IEEE Trans. Biomed. Circuits Syst. **4**(3), 149–161 (2010)
35. S.F. Liang, F.Z. Shaw, C.P. Young, D.W. Chang, Y.C. Liao, A closed-loop brain computer interface for real-time seizure detection and control. IEEE Eng. Med. Bio. Soc. **2010**, 4950–4953 (2010)
36. S. Zanos, A.G. Richardson, L. Shupe, F.P. Miles, E.E. Fetz, The neurochip-2: an autonomous head-fixed computer for recording and stimulating in freely behaving monkeys. IEEE Trans. Neural Syst. I Rehabil. Eng. **19**(4), 427–435 (2011)
37. A.G. Rouse, S.R. Stanslaski, P. Cong, R.M. Jensen, P. Afshar, D. Ullestad, R. Gupta, G.F. Molnar, D.W. Moran, T. Denison, A chronic generalized bi-directional brain-machine interface. J. Neural Eng. **8**(3), 036018 (2011)
38. M. Azin, D.J. Guggenmos, S. Barbay, R.J. Nudo, P. Mohseni, A battery-powered activity-dependent intracortical microstimulation IC for brain-machine-brain interface. J. Solid State Circuits **46**(4), 731–745 (2011)
39. D. loi et al., Peripheral neural activity recording and stimulation system. IEEE Trans. Biomed. Circuits Syst. **5**(4), 368–379 (2011)
40. S. Stanslaski, P. Afshar, P. Cong, J. Giftakis, P. Stypulkowski, D. Carlson, D. Linde, D. Ullestad, A.-T. Avestruz, T. Denison, Design and validation of a fully implantable, chronic, closed-loop neuromodulation device with concurrent sensing and stimulation. IEEE Trans. Neural Syst. Rehabil. I Eng. **20**(4), 410–421 (2012)

41. U. Bihr, T. Ungru, H. Xu, J. Anders, J. Becker, M. Ortmanns, A bidirectional neural interface with a HV stimulator and a LV neural amplifier, in *IEEE International Symposium Circuits Systems (ISCAS)*, May 2013, pp. 401–404
42. K. Abdelhalim et al., 64-channel UWB wireless neural vector analyzer SOC with a closed-loop phase synchrony-triggered neurostimulator. IEEE J. Solid State Circuits **48**, 2494–2510 (2013)
43. X. Liu, B. Subei, M. Zhang, A.G. Richardson, T.H. Lucas, J. Van der Spiegel, The PennBMBI: a general purpose wireless brain-machine-brain interface system for unrestrained animals, in *2014 IEEE International Symposium on Circuits and Systems (ISCAS)*, May 2014, pp. 650–653. http://ieeexplore.ieee.org/document/6865219/
44. W.-M. Chen et al., A fully integrated 8-channel closed-loop neural-prosthetic CMOS SoC for real-time epileptic seizure control. J. Solid State Circuits **49**(1), 232–247 (2014)
45. K. Limnuson et al., A bidirectional neural interface SoC with an integrated spike recorder, microstimulator, and low-power processor for real-time stimulus artifact rejection, in *CICC* (2014)
46. G. Angotzi et al., A programmable closed-loop recording and stimulating wireless system for behaving small laboratory animals. Sci. Rep. **4**, Article No 5963 (2014)
47. M. Ballini, J. Muller, P. Livi, Y. Chen, U. Frey, A. Stettler, A. Shadmani, V. Viswam, I.L. Jones, D. Jackel, M. Radivojevic, M.K. Lewandowska, W. Gong, M. Fiscella, D.J. Bakkum, F. Heer, A. Hierlemann, J. Muller, P. Livi, Y. Chen, A. Stettler, A. Shadmani, V. Viswam, I.L. Jones, D. Jackel, M. Radivojevic, M.K. Lewandowska, W. Gong, M. Fiscella, D.J. Bakkum, A. Hierlemann, A 1024-channel CMOS microelectrode array with 26,400 electrodes for recording and stimulation of electrogenic cells in vitro. IEEE J. Solid State Circuits **49**(11), 2705–2719 (2014)
48. H. Rhew, J. Jeong, J. Fredenburg, S. Dodani, A fully self-contained logarithmic closed-loop deep brain stimulation SoC with wireless telemetry and wireless power management. IEEE J. Solid State Circuits **49**(10), 2213–2227 (2014)
49. P. Cong et al., A 32-channel modular bi-directional neural interface system with embedded DSP for closed-loop operation, in *ESSCIRC* (2014)
50. T.K. Nguyen et al., Closed-loop optical neural stimulation based on a 32-channel low-noise recording system with online spike sorting. J. Neural Eng. **11**(4), 046005 (2014)
51. X. Liu, B. Subei, M. Zhang, A.G. Richardson, T.H. Lucas, J. Van der Spiegel, The pennBMBi: design of a general purpose wireless brain-machine-brain interface system. IEEE Trans. Biomed. Circuits Syst. **9**(2), 248–258 (2015). http://ieeexplore.ieee.org/document/7055376/.
52. X. Liu, M. Zhang, A.G. Richardson, T.H. Lucas, J. Van der Spiegel, A 12-channel bidirectional neural interface chip with integrated channel-level feature extraction and PID controller, in *IEEE Biomedical Circuits and Systems Conference (BioCAS)* (2015)
53. R. Shulyzki, K. Abdelhalim, A. Bagheri, M.T. Salam, C.M. Florez, J.L.P. Velazquez, P.L. Carlen, R. Genov, 320-channel active probe for high-resolution neuromonitoring and responsive neurostimulation. IEEE Trans. Biomed. Circuits Syst. **9**(1), 34–49 (2015)
54. M.A. Bin Altaf, C. Zhang, J. Yoo, A 16-channel patient-specific seizure onset and termination detection SoC with impedance-adaptive transcranial electrical stimulator. IEEE J. Solid-State Circuits **50**(11), 2728–2740 (2015)
55. A.E. Mendrela et al., Enabling closed-loop neural interface: a bi-directional interface circuit with stimulation artifact cancellation and cross-channel cm noise suppression, in *2015 Symposium on VLSI Circuits (VLSI Circuits)* (2015)
56. W. Biederman, D.J. Yeager, N. Narevsky, J. Leverett, R. Neely, J.M. Carmena, E. Alon, J.M. Rabaey, A 4.78mm^2 fully-integrated neuromodulation soc combining 64 acquisition channels with digital compression and simultaneous dual stimulation. IEEE J. Solid State Circuits **50**(4), 1038–1047 (2015)
57. X. Liu, M. Zhang, A.G. Richardson, T.H. Lucas, J. Van der Spiegel, Design of a closed-loop, bi-directional brain machine interface system with energy efficient neural feature extraction and PID control. IEEE Trans. Biomed. Circuits Syst. **11**(4), 729–742 (2017). http://ieeexplore.ieee.org/abstract/document/7786863/

58. A. Abdi, H.-K. Cha, A bidirectional neural interface CMOS analog front-end ic with embedded isolation switch for implantable devices. Microelectronics J. **58**, 70–75 (2016)
59. A.E. Mendrela, J. Cho, J.A. Fredenburg, V. Nagaraj, T.I. Netoff, M.P. Flynn, E. Yoon, A bidirectional neural interface circuit with active stimulation artifact cancellation and cross-channel common-mode noise suppression. IEEE J. Solid State Circuits **51**(4), 955–965 (2016)
60. M. Shoaran, M. Shahshahani, M. Farivar, J. Almajano, A. Shahshahani, A. Schmid, A. Bragin, Y. Leblebici, A. Emami, A 16-channel 1.1mm^2 implantable seizure control SoC with sub-μw/channel consumption and closed-loop stimulation in 0.18μm CMOS, in *VLSI Symposium on Circuits* (2016), pp. 256–257
61. Y. Su, S. Routhu, K. Moon, S. Q. Lee, W. Youm, Y. Ozturk. A wireless 32-channel implantable bidirectional brain machine interface. Sensors **16**, 1–19 (2016)
62. M. Haas, A. Jens, O. Maurits, A bidirectional neural interface featuring a tunable recorder and electrode impedance estimation, in *Biomedical Circuits and Systems Conference (BioCAS)* (2016)
63. X. Liu, H. Zhu, M. Zhang, A.G. Richardson, S.Y. Sritharan, D. Ge, Y. Shu, T.H. Lucas, J. Van der Spiegel, A fully integrated wireless sensor-brain interface system to restore finger sensation, *IEEE International Symposium on (ISCAS) Circuits and Systems*, May (2017)
64. G. Santhanam, S.I. Ryu, B.M. Yu, A. Afshar, K.V. Shenoy, A high-performance brain-computer interface. Nature **442**, 195–198 (2006)
65. C.A. Chestek et al, Hermesc: low-power wireless neural recording system for freely moving primates. IEEE Trans. Neural Syst. I Rehabil. Eng. **17**(4), 330–338 (2009)
66. H. Miranda, V. Gilja, C.A. Chestek, K.V. Shenoy, T.H. Meng, Hermesd: a high-rate long-range wireless transmission system for simultaneous multichannel neural recording applications. IEEE Tans. Biomed. Circuits Syst. **4**(3), 181–191 (2010)
67. H. Gao et al., Hermese: a 96-channel full data rate direct neural interface in 0.13um CMOS. IEEE J. Solid-State Circuits **47**(4), 1043–1055 (2012)
68. M. Yin, D. Borton, J. Komar, N. Agha, Y. Lu, Wireless neurosensor for full-spectrum electrophysiology recordings during free behavior. Neuron **84**, 1170–1182 (2014)
69. D.A. Schwarz, M.A. Lebedev, T.L. Hanson, D.F. Dimitrov, G. Lehew, J. Meloy, S. Rajangam, V. Subramanian, P.J. Ifft, Z. Li, A. Ramakrishnan, A. Tate, K.Z. Zhuang, M.A.L. Nicolelis, Chronic, wireless recordings of large-scale brain activity in freely moving rhesus monkeys. Nat. Methods **11**(6), 670–676 (2014)
70. T. Jochum, T. Denison, P. Wolf, Integrated circuit amplifiers for multi-electrode intracortical recording. J. Neural Eng. **6**(1), 12001–12026 (2009)
71. I.H. Stevenson, K.P. Kording, How advances in neural recording affect data analysis. Nat. Neurosci. **14**(2), 139–142 (2011)
72. Human brain project framework partnership agreement proposal. June (2014)
73. R.R. Harrison, C. Charles, A low-power low-noise CMOS amplifier for neural recording applications. IEEE J. Solid State Circuits **38**(6), 958–965 (2003)
74. R. Yazicioglu, P. Merken, A 60 μW 60 nv/root hz readout front-end for portable biopotential acquisition systems. IEEE J. Solid State Circuits **42**(5), 1100–1110 (2007)
75. N. Van Helleputte, S. Kim, H. Kim, J.P. Kim, C. Van Hoof, R.F. Yazicioglu, A 160 ua biopotential acquisition IC with fully integrated ia and motion artifact suppression. IEEE Tans. Biomed. Circuits Syst. **6**(6), 552–561 (2012)
76. R. Harrison, P. Watkins, A low-power integrated circuit for a wireless 100-electrode neural recording system. IEEE J. Solid State Circuits **42**(1), 123–133 (2007)
77. W. Wattanapanitch, M. Fee, R. Sarpeshkar, An energy-efficient micropower neural recording amplifier. IEEE Tans. Biomed. Circuits Syst. **1**(2), 136–147 (2007)
78. T. Denison, K. Consoer, W. Santa, A. Avestruz, J. Cooley, A. Kelly, A 2 μW 100 nV/rtHz chopper-stabilized instrumentation amplifier for chronic measurement of neural field potentials. IEEE J. Solid State Circuits **42**(12), 2934–2945 (2007)
79. Q. Fan et al., A 1.8μW 60 nv/rtHz capacitively-coupled chopper instrumentation amplifier in 65 nm CMOS for wireless sensor nodes. IEEE J. Solid State Circuits **1**(2), 136–147 (2007)

80. X. Zou, L. Liu, J. Cheong, L. Yao, A 100-channel 1-mW implantable neural recording IC. IEEE Trans. Circuits Syst. I Reg. Pap. **60**(10), 2584–2596 (2013)
81. M.S. Chae, Z. Yang, M.R. Yuce, L. Hoang, W. Liu, A 128-channel 6 mW wireless neural recording ic with spike feature extraction and UWB transmitter. IEEE Trans. Neural Syst. I Rehabil. Eng. **17**(4), 312–321 (2009)
82. R. Muller, H.-P. Le, W. Li, P. Ledochowitsch, S. Gambini, T. Bjorninen, J.M. Rabaey. A minimally invasive 64-channel wireless uecog implant. IEEE J. Solid State Circuits **50**(1), 344–359 (2015)
83. R.R. Harrison, R.J. Kier, C.A. Chestek, V. Gilja, Wireless neural recording with single low-power integrated circuit. IEEE Trans. Neural Syst. I Rehabil. Eng. **17**(4), 322–329 (2009)
84. A. Borna, K. Najafi, A low power light weight wireless multichannel microsystem for reliable neural recording. IEEE J. Solid State Circuits **49**(2), 439–451 (2014)
85. P. Mohseni, K. Najafi, S. J. Eliades, X. Wang, Wireless multichannel biopotential recording using an integrated fm telemetry circuit. IEEE Trans. Neural Syst. I Rehabil. Eng. **13**(3), 263–271 (2005)
86. N.M. Neihart, R.R. Harrison, Micropower circuits for bidirectional wireless telemetry in neural recording applications. IEEE Trans. Biomed. Eng. **52**(11), 1950–1959 (2005)
87. W. Biederman, D. Yeager, A fully-integrated, miniaturized (0.125 mm^2) 10.5 μW wireless neural sensor. IEEE J. Solid State Circuits **48**(4), 960–970 (2013)
88. D.J. Yeager, J. Holleman, R. Prasad, J.R. Smith, B.P. Otis, Neuralwisp: a wirelessly powered neural interface with 1-m range. IEEE Tans. Biomed. Circuits Syst. **3**(6), 379–387 (2009)
89. R. Olsson, K. Wise, A three-dimensional neural recording microsystem with implantable data compression circuitry. IEEE J. Solid State Circuits **40**(12), 2796–2804 (2005)
90. E.C. Leuthardt, G. Schalk, D. Ph, J. Roland, D.W. Moran, Evolution of brain-computer interfaces: going beyond classic motor physiology. J. Neural Eng. **27**(1): 1–21 (2010)
91. C.M. Lopez, An implantable 455-active-electrode 52-channel CMOS neural probe, in *2013 IEEE International Solid-State Circuits Conference Digest of Technical Papers*, Feb 2013, pp. 288–289
92. European Committee for Electrotechnical Standardization, Medical electrical equipment - Part 2–47: particular requirements for the safety, including essential performance, of ambulatory electrocardiographic systems. ICE60601-2-47 (2015)
93. M. Steyaert, W. Sansen, C. Zhongyuan, A micropower low-noise monolithic instrumentation amplifier for medical purposes. J. Solid State Circuit **22**(6), 1163–1168 (1987)
94. J. Holleman, Design considerations for neural amplifiers, in *IEEE International Symposium on Circuits Systems (ISCAS)* (2016), pp. 6331–6334
95. R. Muller, S. Gambini, J.M. Rabaey, A 0.013 mm^2 5 μW, DC-coupled neural signal acquisition IC with 0.5V supply. IEEE J. Solid State Circuits **47**, 232–243 (2012)
96. D. Han, Y. Zheng, R. Rajkumar, G.S. Dawe, M. Je, A 0.45 V 100-channel neural-recording ic with sub-uw/channel consumption in 0.18μm CMOS. IEEE Trans. Biomed. Circuits Syst. **7**(6), 735–746 (2013)
97. K.A. Ng, E. Greenwald, Y.P. Xu, N.V. Thakor, Implantable neurotechnologies: a review of integrated circuit neural amplifiers. Med. Bio. Eng. Comput. **33**(4), 395–401 (2015)
98. B. Gosselin, Recent advances in neural recording microsystems. Sensors **11**(5), 4572–4597 (2011)
99. E. Bharucha, H. Sepehrian, B. Gosselin, A survey of neural front end amplifiers and their requirements toward practical neural interfaces. J. Low Power Electron. Appl. **4**, 268–291 (2014)
100. C. Toumazou, Novel current-mode instrumentation amplifier. IEEE Electron. Lett. **25**(3), 228–230 (1989)
101. F. Yazicioglu, Low-power interface circuits for bio-potential and physiological signal acquisition. ISSCC Short Course, Feb (2014)
102. B. Gosselin, M. Sawan, C.A. Chapman, A low-power integrated bioamplifier with active low-frequency suppression. IEEE Trans. Biomed. Circuits Syst. **1**, 184–192 (2007)

103. R. Rieger, M. Schuettler, D. Pal, C. Clarke, P. Langlois, J. Taylor, N. Donaldson, Very low-noise ENG amplifier system using CMOS technology. IEEE Trans. Neural Syst. I Rehabil. Eng. **14**(4), 427–437 (2006)
104. J. Holleman, B. Otis, A sub-microwatt low-noise amplifier for neural recording, in *IEEE Engineering in Medicine and Biology Society (EMBC)*, Aug (2007)
105. W.A. Smith, B.J. Mogen, E.E. Fetz, V.S. Sathe, B.P. Otis, Exploiting electrocorticographic spectral characteristics for optimized signal chain design: a 1.08μW analog front end with reduced ADC resolution requirements. IEEE Trans. Biomed. Circuits Syst. **10**, 1–10 (2016)
106. C.J. Deepu, X. Zhang, W. Liew, D. Liang, T. Wong, Y. Lian, An ECG-on-chip with 535 nw/channel integrated lossless data compressor for wireless sensors. IEEE J. Solid State Circuits **49**(11), 2435–2448 (2014)
107. N. Verma, A. Shoeb, J. Bohorquez, J. Dawson, J. Guttag, A.P. Chandrakasan, A micro-power eeg acquisition SoC with integrated feature extraction processor for a chronic seizure detection system. IEEE J. Solid State Circuits **45**(4), 804–816 (2010)
108. C.H. Chan, J. Wills, J. LaCoss, J.J. Granacki, J. Choma, A micro-power low-noise auto-zeroing CMOS amplifier for cortical neural prostheses, *IEEE 2006 Biomedical Circuits and Systems Conference on Healthcare Technology* (2006), pp. 214–217
109. A. Bagheri, M.T. Salam, J.L. Perez Velazquez, R. Genov, Low-frequency noise and offset rejection in dc-coupled neural amplifiers: a review and digitally-assisted design tutorial. IEEE Trans. Biomed. Circuits Syst. **11**(1), 161–176 (2016)
110. Y. Chen, D. Jeon, Y. Lee, Y. Kim, Z. Foo, I. Lee, N.B. Langhals, G. Kruger, H. Oral, O. Berenfeld, Z. Zhang, D. Blaauw, D. Sylvester. An injectable 64 nW ECG mixed-signal SoC in 65 nm for arrhythmia monitoring. IEEE J. Solid State Circuits **50**(1), 375–390 (2015)
111. J. Yoo, L. Yan, D. El-Damak, M.A. Bin Altaf, A.H. Shoeb, A.P. Chandrakasan, An 8-channel scalable EEG acquisition SoC with patient-specific seizure classification and recording processor. IEEE J. Solid State Circuits **48**(1), 214–228 (2013)
112. R. Sarpeshkar, *Ultra Low Power Bioelectronics* (Cambridge University Press, Cambridge, 2010)
113. K.J. Miller, S. Zanos. E.E. Fetz, M. den Nijs, J.G. Ojemann, Decoupling the cortical power spectrum reveals real-time representation of individual finger movements in humans. J. Neurosci. **29**, 3132–3137 (2009)
114. K.J. Miller, L.B. Sorensen, J.G. Ojemann, M. den Nijs. Power-law scaling in the brain surface electric potential. PLoS Comput Biol. **5**(12) (2009). doi:10.1371/journal.pcbi.1000609
115. B. Razavi, *Design of Analog CMOS Integrated Circuits* (McGraw-Hill Companies, Inc., New York, 2001)
116. T. Jang, M. Choi, S. Jeong, S. Bang, D. Sylvester, D. Blaauw. A 4.7nW 13.8ppm/°C self-biased wakeup timer using a switched-resistor scheme. ISSCC Digest of Technical Papers, Feb 2016, pp. 102–103
117. J. Venesty. J. Chen, Y. Huang, I. Cohen, Pearson correlation coefficient, in *Noise Reduction in Speech Processing*, March (2009)
118. J. Fredenburg, M.P. Flynn, ADC trends and impact on SAR ADC architecture and analysis, *Proceedings of the Custom Integrated Circuits Conference*, Nov 2015, pp. 1–8
119. T. Rabuske, J. Femandes, *Charge-Sharing SAR ADCs for Low-Voltage Low-Power Applications*. Analog Circuits and Signal Processing (Springer, Berlin, 2017)
120. S. Roshani, S. Roshani, *Low Power Mixed-Mode Circuit Design of SAR ADC* (Lambert Academic Publishing, Saarbrücken, 2013)
121. A. Agnes, E. Bonizzoni, P. Malcovati, F. Maloberti, A 9.4-ENOB 1v 3.8uw 100kS/s SAR ADC with time-domain comparator, in *Solid-State Circuit Conference (ISSCC), IEEE International*, Feb 2008, pp. 246–610
122. A. Agnes, E. Bonizzoni, P. Malcovati, F. Maloberti, An ultra-low power successive approximation A/D converter with time-domain comparator. Analog Integr. Circuits Signal Process. **64**(2), 183–190 (2010)
123. C. Liu, S. Chang, G. Huang, Y. Lin, A 10-bit 50-ms/s SAR ADC with a monotonic capacitor switching procedure. IEEE J. Solid State Circuits **45**(4), 731–740 (2010)

124. C.H. Kuo, C.E. Hsieh, A high energy-efficiency SARADC based on partial floating capacitor switching technique, in *Proceedings of IEEE ESS- CIRC*, (2011), pp. 475–478
125. M. Van Elzakker, E. Van Tuijl, A 10-bit charge-redistribution ADC consuming 1.9 μW at 1 ms/s. IEEE J. Solid State Circuits **45**(5), 1007–1015 (2010)
126. Y. Zhu, C.-H. Chan, U-F. Chio, S.-W. Sin, S.-P. U, R.P. Martins, F. Maloberti, A 10-bit 100-MS/s reference-free SAR ADC in 90 nm CMOS. IEEE J. Solid State Circuits **45**, 1111–1121 (2010)
127. S.W.M. Chen, R.W. Brodersen, A 6-bit 600-MS/s 5.3-mW asynchronous ADC in 0.13-μm CMOS. IEEE J. Solid State Circuits **41**(12), 2669–2680 (2006)
128. S.O. Driscoll, K.V. Shenoy, T.H. Meng, Adaptive resolution ADC array for an implantable neural sensor. IEEE Trans. Biomed. Circuits Syst. **5**(2), 120–130 (2011)
129. B. Malki, T. Yamamoto, B. Verbruggen, P. Wambacq, J. Craninckx, A 70 dB DR 10 b 0-to-80 MS/s current-integrating SAR ADC with adaptive dynamic range. IEEE J. Solid State Circuits **49**(5), 1173–1183 (2014)
130. V. Chaturvedi, T. Anand, B. Amrutur, An 8-to-1 bit 1-MS/s SAR ADC with VGA and integrated data compression for neural recording. IEEE Trans. Very Large Scale Integr. (VLSI) Syst. **21**(11), 2034–2044 (2013)
131. W. Tang, A. Osman, B. Kim, D. Goldstein, C. Huang, B. Martini, V. Pieribone, E. Culurciello, Continuous time level crossing sampling ADC for bio-potential recording systems. IEEE Trans. Circuits Syst. I Regul. Pap. **60**(6), 1407–1418 (2013)
132. Y. Lyu, C. Wu, A low power 10bit 500ks/s delta-modulated SAR ADC (dmSAR ADC) for implantable medical devices, in *IEEE International Symposium on Circuits and Systems (ISCAS)*, May 2013, pp. 2046–2049
133. G.Y. Huang, S.J. Chang, C.C. Liu, Y.Z. Lin, A 1-μw 10-bit 200-KS/s SAR ADC with a bypass window for biomedical applications, in *IEEE International Symposium on Circuits and Systems (ISCAS)*, vol. 47(11), May 2012, pp. 2783–2795
134. F.M. Yaul, A.P. Chandrakasan, A 10b SAR ADC with data-dependent energy savings using LSb-first successive approximation. IEEE J. Solid State Circuits **57**(12), 198–199 (2014)
135. P. Harpe, H. Gao, R. Van Dommele, E. Cantatore, A. Van Roermund, A 3nW signal-acquisition IC integrating an amplifier with 2.1 NEF and a 1.5fj/conv-step ADC, *IEEE International Solid-State Circuit Conference (ISSCC)*, vol. 58, Feb 2015, pp. 382–383
136. B. Razavi, The bootstrapped switch. IEEE Solid State Circuits Mag. **7**(3, Summer), 12–15 (2015)
137. F. Chen, A.P. Chandrakasan, V.M. Stojanovi, Design and analysis of a hardware-efficient compressed sensing architecture for data compression in wireless sensors. IEEE J. Solid State Circuits **47**(3), 744–756 (2012)
138. B. Gosselin, M. Sawan, An ultra low-power CMOS automatic action potential detector. IEEE Trans. Neural Syst. I Rehabil. Eng. **17**(4), 346–353 (2009)
139. S. Narasimhan, H.J. Chiel, S. Bhunia, Ultra-low-power and robust digital-signal-processing hardware for implantable neural interface microsystems. IEEE Trans. Biomed. Circuits Syst. **5**(2), 169–178 (2009)
140. K.G. Oweiss, A. Mason, Y. Suhail, A.M. Kamboh, K.E. Thomson, A scalable wavelet transform VLSI architecture for real-time signal processing in high-density intra-cortical implants. IEEE Trans. Circuits Syst. I Regul. Pap. **54**(6), 1266–1278 (2007)
141. Y. Yang, A.M. Kamboh, A.J. Mason, A configurable realtime DWT-based neural data compression and communication VLSI system for wireless implants. J. Neurosci. Methods **227**, 140–150 (2014)
142. X. Liu, M. Zhang, H. Zhu, A.G. Richardson, T.H. Lucas, J. Van der Spiegel, Design of a low-noise, high power efficiency neural recording front-end with an integrated real-time compressed sensing unit, in *IEEE International Symposium on (ISCAS) Circuits and Systems*, May (2015)

143. E.J. Candes, M.B. Wakin, An introduction to compressive sampling. IEEE Signal Process. Mag. **25**(2), 21–30 (2008)
144. D.L. Donoho, Compressed sensing. IEEE Trans. Inf. Theory **52**(4), 1289–1306 (2006)
145. Y. Oike, A. El Gamal, CMOS image sensor with per-column sigma-delta ADC and programmable compressed sensing. IEEE J. Solid State Circuits **48**(1), 319–328 (2013)
146. J. Zhang, Y. Suo, S. Mitra, S. Chin, S. Hsiao, R.F. Yazicioglu, T.D. Tran, R. Etienne-Cummings, An efficient and compact compressed sensing microsystem for implantable neural recordings. IEEE Tans. Biomed. Circuits Syst. **8**(4), 485–496 (2014)
147. X. Chen, Z. Yu, S. Hoyos, A sub-nyquist rate sampling receiver exploiting compressive sensing. IEEE Trans. Circuits Syst. Reg. Pap. **58**, 507–520 (2011)
148. S. Aviyente, Compressed sensing framework for EEG compression, in *IEEE Statistical Signal Processing* (2007), pp. 181–184
149. T. Xiong, J. Zhang, Y. Suo, D.N. Tran, R. Etienne-Cummings, S. Chin, T.D. Tran, A dictionary learning algorithm for multi-channel neural recordings, in *2014 IEEE Biomedical Circuits and Systems Conference (BioCAS)* (2014), pp. 9–12
150. T. Xiong, J. Zhang, Y. Suo, D.N. Tran, R. Etienne-Cummings, S. Chin, T.D. Tran, An unsupervised dictionary learning algorithm for neural recordings, in *IEEE International Symposium on Circuits and Systems (ISCAS)* (2015), pp. 1010–1013
151. R. Baraniuk, Compressive sensing, in *IEEE Signal Processing Magazine*, July 2007, pp. 118–121
152. M. Shoaib, N.K. Jha, N. Verma, Signal processing with direct computations on compressively sensed data. IEEE Trans. Very Large Scale Integr. VLSI Syst. **23**(1), 30–43 (2015)
153. X. Liu, M. Zhang, T. Xiong, A.G. Richardson, T.H. Lucas, P.S. Chin, R. Etienne-Cummings, T.D. Tran, J. Van der Spiegel, A fully integrated wireless compressed sensing neural signal acquisition system for chronic recording and brain machine interface. IEEE Trans. Biomed. Circuits Syst. **10**(4), 874–883 (2016). http://ieeexplore.ieee.org/document/7515159/
154. J.A. Tropp, A.C. Gilbert, Signal recovery from random measurements via orthogonal matching pursuit. IEEE Trans. Inf. Theory **53**(12), 4655–4666 (2007)
155. Y. Lu, W. Ki, A 13.56 MHz CMOS active rectifier with switched-offset and compensated biasing for biomedical wireless power transfer systems. IEEE Tans. Biomed. Circuits Syst. **8**(3), 334–344 (2014)
156. I.M. Filanovsky, H. Baltes. CMOS schmitt trigger design. IEEE Trans. Circuits Syst. I Fundam. Theory Appl. **41**(1), 46–49 (1994)
157. X. Li, C. Tsui, W. Ki, A 13.56MHz wireless power transfer system with reconfigurable resonant regulating rectifier and wireless power control for implantable medical devices. IEEE J. Solid State Circuits **50**(4), 978–989 (2014)
158. M. Kiani, M. Ghovanloo, A 13.56-Mbps pulse delay modulation based transceiver for simultaneous near-field data and power transmission. IEEE Trans. Biomed. Circuits Syst. **9**(1), 1–11 (2015)
159. D. Gangopadhyay, E.G. Allstot, A.M.R. Dixon, K. Natarajan, S. Gupta, D.J. Allstot, Compressed sensing analog front-end for bio-sensor applications. IEEE J. Solid State Circuits **49**(2), 426–438 (2014)
160. J. Zhang, S. Mitra, Y. Suo, A. Cheng, T. Xiong, F. Michon, M. Welkenhuysen, F. Kloosterman, P.S. Chin, S. Hsiao, T.D. Tran, F. Yazicioglu, R. Etienne-Cummings, A closed-loop compressive-sensing-based neural recording system. J. Neural Eng. **12**(3), 036005 (2015)
161. Wikipedia and the Free Encyclopedia, Feature extraction. Retrieved Jan 2017 https://en.wikipedia.org/wiki/Feature_extraction
162. J. Echauz, Feature extraction for brain computer interface. Lecture notes from BE521, the University of Pennsylvania, vol. 3, Jan 2013, pp. 1–2
163. W. Chen, X. Liu, B. Litt, Logistic-weighted regression improves decoding of finger flexion from electrocorticographic signals, in *International Conference of the IEEE Engineering in Medicine and Biology Society (EMBC)*, Aug (2014)

164. J. Donoghue, J. Sanes, N. Hatsopoulos, G. Gaal, Neural discharge and local field potentials oscillations in primate motor cortex during voluntary movements. J. Neurophysiol. **79**(1), 159–173 (1998)
165. A. Shoeb, S. Schachter, D. Schomer, B. Bourgeois, J. Guttag, Detecting seizure onset in the ambulatory setting: demonstrating feasibility, in *Proceedings of IEEE Engineering in Medicine and Biology Social Conference*, Sept 2005, pp. 3546–3550
166. A.-T. Avestruz, W. Santa, D. Carlson, R. Jensen, S. Stanslaski, A. Helfenstine, T. Denison, A 5 μW/channel spectral analysis ic for chronic bidirectional brain-machine interfaces. IEEE J. Solid State Circuits **43**(12), 3006–3024 (2008)
167. S. Debener, A. Beauducei, D. Nessler, B. Brocke, H. Heilemann, J. Kayser, Is resting anterior EEG alpha asymmetry a trait marker for depression. Neuropsychobiology **41**, 31–37 (2000)
168. X. Liu, J. Sacks, M. Zhang, A.G. Richardson, T.H. Lucas, J. Van der Spiegel, The virtual trackpad: an electromyography-based, wireless, real-time, low-power, embedded hand gesture recognition system using an event-driven artificial neural network. IEEE Trans. Circuits Syst. II (2016). http://ieeexplore.ieee.org/document/7769179
169. F. Zhang, A. Mishra, A.G. Richardson, B. Otis. A low-power ECoG/EEG processing IC with integrated multiband energy extractor. IEEE Trans. Circuits Syst. I Reg. Pap. **58**(9), 2069–2082 (2011)
170. R.R. Harrison, G. Santhanam, K.V. Shenoy, Local field potential measurement with low-power analog integrated circuit. IEEE Eng. Med. Bio. Soc. **6**, 4607–4670 (2004)
171. G. Gagnon-Turcotte, C.O.D. Camaro, B. Gosselin, Comparison of low-power biopotential processors for on-the-fly spike detection, in *Proceedings - IEEE International Symposium on Circuits and Systems*, May 2015, pp. 802–805
172. V. Shalchyan, W. Jensen, D. Farina, Spike detection and clustering with unsupervised wavelet optimization in extracellular neural recordings. IEEE Trans. Biomed. Eng. **59**(9). 2576–2585 (2012)
173. Wikipedia and the Free Encyclopedia, Principal component analysis. Retrieved Jan 2017. https://en.wikipedia.org/wiki/Principal_component_analysis
174. Wikipedia and the Free Encyclopedia, Independent component analysis. Retrieved Jan 2017. https://en.wikipedia.org/wiki/Independent_component_analysis
175. Wikipedia and the Free Encyclopedia, Genetic algorithm. Retrieved Jan 2017. https://en.wikipedia.org/wiki/Genetic_algorithm
176. E. Ciaccio, S. Dunn, M. Akay, Biosignal pattern recognition and interpretation systems. 2. methods for feature extraction and selection. IEEE Eng. Med. Biol. Mag. **12**, 106–113 (1993)
177. E. Stark, M. Abeles, Predicting movement from multiunit activity. J. Neurosci. **27**, 8387–8394 (2007)
178. B. Ahmed, A. Redissi, R. Tafreshi, An automatic sleep spindle detector based on wavelets and the teager energy operator. Conf. Proc. IEEE Eng. Med. Biol. Soc. **2009**, 2596–2599 (2009)
179. S.-C. Lin, D. Gervasoni . Defining global brain states using multielectrode field potential recordings, *Methods for Neural Ensemble Recordings* (CRC Press, Boca Raton, 2008)
180. S. Ohara, T. Mima, K. Baba, A. Ikeda, T. Kunieda, Increased synchronization of cortical oscillatory activities between human supplementary motor and primary sensorimotor areas during voluntary movements. J. Neurosci. **21**, 9377–9386 (2001)
181. A. Graef, M. Hartmann, C. Flamm, C. Baumgartner, M. Deistler, T. Kluge, A novel method for the identification of synchronization effects in multichannel ECoG with an application to epilepsy. Biol. Cybern. **107**, 321–335 (2013)
182. M.S. Lewicki, A review of methods for spike sorting: the detection and classification of neural action potentials. Netw. Comput. Neural Syst. **9**(4), 53–78 (1998)
183. S. Hiseni, C. Sawigun, W. Ngamkham, W.A. Serdijn, A compact, nano-power CMOS action potential detector, *Biomedical Circuits Systems and Conference* (2009), pp. 97–100
184. S. Gibson, J.W. Judy, D. Markovic, An FPGA-based platform for accelerated offline spike sorting. **215**, 1–11 (2013)

185. D.C. Tam, Real-time estimation of predictive firing rate. Neurocomputing **52**(4), 637–641 (2003)
186. A.C. Smith, J.D. Scalon, S. Wirth, M. Yanike, W.A. Suzuki, E.N. Brown, State-space algorithms for estimating spike rate functions. Comput. Intell. Neurosci. **2010**, 426539 (2010)
187. H. Scherberger, M.R. Jarvis, R.A. Andersen, Cortical local field potential encodes movement intentions in the posterior parietal cortex. Nature **46**(2), 347–354 (2005)
188. M. Spler, A. Walter, A. Ramos Murguialday, G. Naros, N. Birbaumer, A. Gharabaghi, W. Rosenstiel, M. Bogdan, Decoding of motor intentions from epidural ECoG recordings in severely paralyzed chronic stroke patients. J. Neural Eng. **11**(6), 066008 (2014)
189. G. Buzsaki, A. Draguhn, Neuronal oscillations in cortical networks. Science **304**, 1926–1930 (2004)
190. M. Penttonen, G. Buzsaki, Natural logarithmic relationship between brain oscillators. Thalamus Relat. Syst. **2**(2), 145–152 (2003)
191. P. Gerrard, R. Malcolm, Mechanisms of modafinil: a review of current research. Neuropsychiatr. Dis. Treat. **3**(3), 349 (2007)
192. R.R. Harrison, The design of integrated circuits to observe brain activity. Proc. IEEE **96**(7), 1203–1216 (2008)
193. P.M. Furth, H.A. Ommani, Low-voltage highly-linear transconductor design in subthreshold CMOS, *IEEE International Midwest Symposium Circuits Systems (MWSCAS)*, vol. 1 (1997), pp. 156–159
194. R. Sarpeshkar, *Ultra Low Power Bioelectronics: Fundamentals, Biomedical Applications, and Bio-Inspired Systems* (Cambridge University Press, Cambridge, 2010)
195. W.M.C. Sansen, *Analog Design Essentials* (Springer, Berlin, 2006)
196. R. Torrance, T. Viswanathan, J. Hanson. CMOS voltage to current transducers. IEEE Trans. Circuits Syst. **32**(11), 1097–1104 (1985)
197. J. Gak, M.R. Miguez, A. Arnaud, Nanopower OTAs with improved linearity and low input offset using bulk degeneration. IEEE Trans. Circuits Syst. I **61**(3), 689–698 (2014)
198. C.D. Salthouse, R. Sarpeshkar. A practical micropower programmable bandpass filter for use in bionic ears. IEEE J. Solid State Circuits **38**(1), 63–70 (2003)
199. A. Kumar, A wide dynamic range high-q high-frequency bandpass filter with an automatic quality factor tuning scheme. PhD Thesis, Georgia Institute of Technology (2008)
200. Wikipedia and The Free Encyclopedia. Digital biquad filter. Retrieved Jan 2017. https://en.wikipedia.org/wiki/Digital_biquad_filter
201. B. Gilbert, A precise four-quadrant multiplier with subnanosecond response. IEEE J. Solid State Circuits **3**(4), 365–373 (1968)
202. G. Han, E. Sanchez-Sinencio, CMOS transconductance multipliers: a tutorial. IEEE Trans. Circuits Syst. II, Analog Digit. Signal Process. **45**(12), 1550–1563 (1998)
203. S-C. Qin, R.L. Geiger, A 5V CMOS analog multiplier. IEEE J. Solid State Circuits **22**(6), 1143–1146 (1987)
204. H.G. Rey, C. Pedreira, R.Q. Quiroga, Past, present and future of spike sorting techniques. Brain Res. Bull. **119**, 106–117 (2015)
205. R.Q. Quiroga, Spike sorting. Curr. Biol. **22**(2), 45–46 (2012)
206. R.Q. Quiroga, L. Reddy, C. Koch, I. Fried, Decoding visual inputs from multiple neurons in the human temporal lobe. J. Neurophysiol. **98**(4), 1997–2007 (2007)
207. A.L. Hodgkin, A.F. Huxley, A quantitative description of membrane current and its application to conduction and excitation in nerve. J. Physiol. **117**(4), 500–544 (1952)
208. P. Dayan, L.F. Abbott, *Theoretical Neuroscience Computational and Mathematical Modeling of Neural Systems* (The MIT Press, Cambridge, 2001)
209. I. Obeid, P.D. Wolf, Evaluation of spike-detection algorithms for a brain-machine interface application. IEEE Trans. Biomed. Eng. **51**(6), 905–911 (2004)
210. M. Bahoura, J. Rouat, Wavelet speech enhancement based on the Teager energy operator. IEEE Signal Process. Lett. **8**(1), 10–12 (2001)
211. K.G. Oweiss, D.J Anderson, Noise reduction in multichannel neural recordings using a new array wavelet denoising algorithm. Neurocomputing **38**, 1687–1693 (2001)

212. A. Diedrich, W. Charoensuk, R.J. Brychta, A.C. Ertl, R. Shiavi, Analysis of raw microneurographic recordings based on wavelet de-noising technique and classification algorithm: wavelet analysis in microneurography. IEEE Trans. Biomed. Eng. **50**(1), 41–50 (2003)
213. J. Holleman, A. Mishra, A micro-power neural spike detector and feature extractor in 0.13μm CMOS (2008)
214. Z.S. Zumsteg, C. Kemere et al., Power feasibility of implantable digital spike sorting circuits for neural prosthetic systems. IEEE Trans. Neural Syst. I Rehabil. Eng. **13**(3), 4237–4240 (2005)
215. B. Schell, Y. Tsividis, A continuous-time ADC/DSP/DAC system with no clock and with activity-dependent power dissipation. IEEE J. Solid State Circuits **43**(11), 2472–2481 (2008)
216. L. Chen, B. Shi, C. Lu, A high speed/power ratio continuous-time CMOS current comparator, in *7th IEEE International Conference Electronics Circuits and Systems*, vol. 2 (2000), pp. 883–886
217. S. Al-Ahdab, R. Lotfi, W.A. Serdijn, A 1-v 225-nW 1KS/s current successive approximation ADC for pacemakers, in *6th Conference on Ph.D. Research in Microelectronics & Electronics* (2010), pp. 1–4
218. Wikipedia and The Free Encyclopedia, Mached filter. Retrieved Jan 2017. https://en.wikipedia.org/wiki/Matched_filter
219. G. Turin, An introduction to matched filters. IRE Trans. Inf. Theory **6**, 311–329 (1960)
220. F.C. Robey et al., A CFAR adaptive matched filter detector. IEEE Trans. Aerosp. Electron. Syst. **28**(1), 208–216 (1992)
221. I.G. Cumming, F.H. Wong, *Digital Processing of Synthetic Aperture Radar Data*. vol. 1(2) (Artech house, Boston, 2005)
222. N. Seto, S. Kawamura, T. Nakamura, Possibility of direct measurement of the acceleration of the universe using 0.1 Hz band laser interferometer gravitational wave antenna in space. Phys. Rev. Lett. **87**(22), 221103 (2001)
223. M. Al-Rawi, M. Qutaishat, M. Arrar, An improved matched filter for blood vessel detection of digital retinal images. Comput. Biol. Med. **37**(2), 262–267 (2007)
224. I. Rankman, R. Chandra, An algorithm for generating templates of neural waveforms, in *Proceedings of the 15th Annual International Conference of the IEEE* (1993)
225. S.C. Wu, A.L. Swindlehurst, Z. Nenadic, Matched subspace detector based feature extraction for sorting of multi-sensor action potentials, in *International Conference of the IEEE Engineering in Medicine and Biology Society* (2011)
226. A.M. Haas, M.H. Cohen, P.A. Abshires, Real-time variance based template matching spike sorting system, in *Life Science Systems and Applications Workshop* (2007), pp. 7–10
227. D.J. Krusienski, G. Schalk, D.J. McFarland, J.R. Wolpaw, A μ-rhythm matched filter for continuous control of a brain-computer interface. IEEE Trans. Biomed. Eng. **54**(2), 273–280 (2007)
228. G. Turin, An introduction to matched filters. IRE Trans. Inf. Theory **6**(3), 311–329 (1960)
229. Wikipedia and The Free Encyclopedia. Digital biquad filter. Retrieved Jan 2017. https://en.wikipedia.org/wiki/Cauchy-Schwarz_inequality
230. T.H. Chao, A.M. Tai, M.S. Dymek, F.T.S. Yu, Optimum correlation detection by prewhitening. Appl. Opt. **19**(14), 2461–2464 (1980)
231. A.B. Tort et al., Measuring phase-amplitude coupling between neuronal oscillations of different frequencies. J. Neurophysiol. **104**(2), 1195–1210 (2010)
232. M. Ghovanloo, K. Iniewski, Integrated circuits for neural interfacing: neural stimulation, *VLSI Circuits for Biomedical Applications* (2008), pp. 191–199
233. Wikipedia, Electrical brain stimulation. https://en.wikipedia.org/wiki/Electrical_brain_stimulation
234. S.A.P. Haddad, R.P.M. Houben, W.A. Serdijin, The evolution of pacemakers. IEEE Eng. Med. Bio Mag. **25**(3), 38–48 (2006)
235. A. Demosthenous, I.F. Triantis, X. Liu, Circuits for implantable neural recording and stimulation (2008), pp. 207–240
236. Wikipedia, Neurostimulation. https://en.wikipedia.org/wiki/Neurostimulation

237. Z.B. Kagan, A.K. RamRakhyani, G. Lazzi, R.A. Normann, D.J. Warren, In vivo magnetic stimulation of rat sciatic nerve with centimeter- and millimeter-scale solenoid coils. IEEE Trans. Neural Syst. I Rehabil. Eng. **24**(11), 1138–1147 (2016)
238. S.W. Lee, S. Fried, Enhanced control of cortical pyramidal neurons with micro-magnetic stimulation. IEEE Trans. Neural Syst. I Rehabil. Eng. **99** (2016). doi:10.1109/TNSRE.2016.2631446
239. R.D. Meyer, S.F. Cogan, T.H. Nguyen, R.D. Rauh, Electrodeposited iridium oxide for neural stimulation and recording electrodes. IEEE Trans. Neural Syst. I Rehabil. Eng. **9**(1), 2–11 (2001)
240. J. Chen, K.D. Wise, J.F. Hetke, S.C. Bledsoe, A multichannel neural probe for selective chemical delivery at the cellular level. IEEE Trans. Biomed. Eng. **44**(8), 760–769 (1997)
241. K. Paralikar, P. Cong, W. Santa, D. Dinsmoor, B. Hocken, G. Munns, J. Giftakis, T. Denison, An implantable 5mw/channel dual-wavelength optogenetic stimulator for therapeutic neuromodulation research. ISSCC Digest of Technical Papers, Feb 2010, pp. 238–240
242. K. Paralikar, P. Cong, O. Yizhar, L.E. Fenno, W. Santa, C. Nielsen, D. Dinsmoor, B. Hocken, G.O. Munns, J. Giftakis, K. Deisseroth, T. Denison, An implantable optical stimulation delivery system for actuating an excitable biosubstrate. J. Solid State Circuits **46**(1), 321–332 (2011)
243. K. Iniewski, *VLSI Circuits for Biomedical Applications* (Artech house Inc, Boston, 2008)
244. D.R. Merrill, M. Bikson, J.G.R. Jefferys, Electrical stimulation of excitable tissue: design of efficacious and safe protocols. J. Neurosci. Methods **141**(2), 171–198 (2005)
245. S.F. Cogan, Neural stimulation and recording electrodes. Annu. Rev. Biomed. Eng. **10**, 275–309 (2008)
246. D.A. Borkholder. Cell based sensors using microelectrodes. Ph.D. Dissertation, Stanford University (1998)
247. R.A. Blum, J.D. Ross, S.K. Das, E.A. Brown, S.P. DeWeerth, Models of stimulation artifacts applied to integrated circuit design, in *Proceedings of the 26th Annual International Conference of the IEEE EMBS*, Sept (2004)
248. J. Sit, R. Sarpeshkar, A low-power blocking-capacitor-free with less than 6 nA DC error for 1-mA full-scale stimulation. IEEE Trans. Biomed. Circuits Syst. **1**(3), 172–183 (2007)
249. S. Kelly, J. Wyatt, A power-efficient neural tissue stimulator with energy recovery. IEEE Trans. Biomed. Circuits Syst. **5**(1), 20–29 (2011)
250. L. Wong, S. Hossain, A. T. Jorgen Edivinsson, D. Rivas, and H. Haas. A very low-power CMOS mixed-signal IC for implantable pacemaker applications. J. Solid State Circuits **39**(12), 2446–2456 (2004)
251. B. Thurgood, D. Warren, N.M. Ledbetter, G.A. Clark, R.R. Harrison, A wireless integrated circuit for 100-channel charge-balanced neural stimulation. IEEE Trans. Biomed. Circuits Syst. **3**(6), 405–414 (2009)
252. M. Sivaprakasam, W. Liu, M. Humayun, J. Weiland, A variable range bi-phasic current stimulus driver circuitry for an implantable retinal prosthetic device. IEEE J. Solid State Circuits **40**(3), 763–771 (2005)
253. X. Liu, M. Zhang, H. Sun, A.G. Richardson, T.H. Lucas, J. Van der Spiegel, Design of a net-zero charge neural stimulator with feedback control, *2014 IEEE Biomedical Circuits and Systems Conference (BioCAS)*, Oct (2014)
254. M. Ghovanloo, Switched-capacitor based implantable low-power wireless microstimulating systems. *IEEE International Symposium on Circuits and Systems (ISCAS)*, pp. 2197–2200 (2006)
255. H. Lee, K.Y. Kwon, W. Li, A power-efficient switched-capacitor stimulating system for electrical/optical deep brain stimulation. **50**(1), 360–374 (2015)
256. X. Liu, A. Demosthenous, N. Donaldson, An integrated implantable stimulator that is fail-safe without off-chip blocking-capacitors. IEEE Trans. Biomed. Circuits Syst. **2**(3), 231–244 (2008)
257. M. Sahin, Y. Tie, Non-rectangular waveforms for neural stimulation with practical electrodes. J. Neural Eng. **4**, 227–233 (2007)

258. C.M. Zierhofer, I.J. Hochmair-Desoyer, E.S. Hochmair, Electronic de- sign of a cochlear implant for multichannel high-rate pulsatile stimulation strategies. IEEE Trans. Neural Syst. I Rehabil. Eng. **3**(1), 112–116 (1995)
259. M. Bugbee, N. de N Donalsdon, A. Lickel, N.J. Rijkhoff, J. Taylor, An implant for chronic selective stimulation of nerves. Med. Eng. Phys. **23**(1), 29–36 (2001)
260. J.D. Techer, S. Bernard, Y. Bertrand G. Cathebras, D. Guiraud. New implantable stimulator for the FES of paralyzed muscles, in *Proceedings of 30th European Solid-State Circuits Conference (ESSCIRC04)*, pp. 455–458 (2004)
261. K. Chen, Z. Yang, L. Hoang, An integrated 256-channel epiretinal prosthesis. IEEE J. Solid State Circuits **45**(9), 1946–1956 (2010)
262. K. Song, H. Lee, S. Hong. A sub-10 nA DC-balanced adaptive stimulator IC with multi-modal sensor for compact. IEEE Trans. Circuits Syst. **6**(6), 533–541 (2012)
263. M. Monge, M. Raj, M. Honarvar-nazari, H. Chang, Y. Zhao, J. Weiland M. Humayun, Y. Tai, A. Emami-neyestanak, A fully intraocular 0.0169 mm^2/pixel 512-channel self-calibrating epiretinal prosthesis in 65nm CMOS. ISSCC Digest of Technical Papers, Feb 2013, pp. 296–298
264. K. Sooksood, T. Stieglitz, M. Ortmanns, An active approach for charge balancing in functional electrical stimulation. IEEE Trans. Biomed. Circuits Syst. **4**(3), 162–170 (2010)
265. M. Ortmanns, A. Rocke, M. Gehrke, H.J. Tiedtke, A 232-channel epiretinal stimulator ASIC. IEEE J. Solid State Circuits **42**(12), 2946–2959 (2007)
266. E. Noorsal, K. Sooksood, H. Xu, A neural stimulator frontend with high-voltage compliance and programmable pulse shape for epiretinal implants. J. Solid State Circuits **47**, 244–256 (2012)
267. R.J. Baker, *CMOS Circuit Design, Layout, and Simulation*, 3rd edn. (IEEE Press, New York, 2010)
268. S. Kelly et al., A power-efficient voltage-based neural tissue stimulator with energy recovery, ISSCC Digest of Technical Papers vol. 26(2), (2004), pp. 579–588
269. J. Vidal, M. Ghovanloo, Towards a switched-capacitor based stimulator for efficient deep-brain stimulation, in *Conference on Proceedings of IEEE Engineering in Medicine and Biology Society* (2010)
270. L. Lapicque, Recherches, Quantitatives sur l'excitation electrique des nerfs traites comme une polarization. J. Physiol. (Paris) **9**, 622–635 (1907)
271. M. Ghovanloo et al., A compact large voltage-compliance high output-impedance programmable current source for implantable microstimulators. IEEE Trans. Biomed. Eng. **52**, 97–105 (2005)
272. F. Boi, T. Moraitis, V. De Feo, F. Diotalevi, C. Bartolozzi, G. Indiveri, A. Vato. A bidirectional brain-machine interface featuring a neuromorphic hardware decoder. Front. Neurosci. **10**(12), 1–15 (2016)
273. M. Lebedev, Augmentation of sensorimotor functions with neural prostheses. Opera. Med. Physiol. **2**(3), 211–227 (2016)
274. A. Wallach, D. Eytan, A. Gal, C. Zrenner, S. Marom, Neuronal response clamp. Front. Neuroeng. **4**(4), 3 (2011)
275. M. Lebedev, M.L. Nicolelis, Brain-machine interfaces: past, present and future. Trends Neurosci. **29**(9), 536–546 (2006)
276. L. Rossi, G Foffani, S Marceglia, F Bracchi, S Barbieri, A Priori, An electronic device for artefact suppression in human local field potential recordings during deep brain stimulation. J. Neural Eng. **4**(2), 96–106 (2007)
277. G.a. DeMichele and P.R. Troyk. Stimulus-resistant neural recording amplifier, in *IEEE Engineering in Medicine and Biology Society (EMBC)*, Sept (2003)
278. E.A. Brown, J.D. Ross, R.A. Blum, Y. Nam, B.C. Wheeler, S.P. DeWeerth, Stimulus-artifact elimination in a multi-electrode system. IEEE Trans. Biomed. Circuits Syst. **2**(1), 10–21 (2008)
279. C. Ethier, E.R. Oby, M.J. Bauman, L.E. Miller, Restoration of grasp following paralysis through brain-controlled stimulation of muscles. Nature **485**, 368–371 (2012)

280. C.E. Bouton, A. Shaikhouni, N.V. Annetta, M.A. Bockbrader, D.A. Friedenberg, D.M. Nielson et al., Restoring cortical control of functional movement in a human with quadriplegia. Nature **533**, 247–250 (2016)
281. S. Musallam, B.D. Corneil, B. Greger, H. Scherberger, R.A. Andersen, Cognitive control signals for neural prosthetics. Science **305**, 258–262 (2004)
282. H.-V.V. Ngo, T. Martinetz, J. Born, M. Mlle, Auditory closed-loop stimulation of the sleep slow oscillation enhances memory. Neuron **78**(3), 545–553 (2013)
283. D.A. Wagenaar, S.M. potter, Real-time multi-channel stimulus artifact suppression by local curve fitting. J. Neurosci. Methods **120**(2), 113–120 (2002)
284. Toshiba, Toshiba photocoupler InGaAs infrared & photo-transistor. Datasheet, 10 (2015) https://toshiba.semicon-storage.com/info/docget.jsp?did=15284&prodName=TLP292-4
285. Burr Brown. High speed FET-input instrumentation amplifier. Datasheet, March 1998. http://www.ti.com/lit/ds/symlink/ina111.pdf
286. Wikipedia and the Free Encyclopedia. Control theory. Retrieved Jan 2017. https://en.wikipedia.org/wiki/Control_theory
287. Wikipedia and the Free Encyclopedia, Feedback. Retrieved Jan 2017. https://en.wikipedia.org/wiki/Feedback
288. Wikipedia and the Free Encyclopedia, Frequency compensation. Retrieved Jan 2017. https://en.wikipedia.org/wiki/Frequency_compensation
289. K.J. Astrom, T.H. Hagglund, New tuning methods for PID controllers, in *Proceedings of the 3rd European Control Conference* (1995)
290. Wikipedia and the Free Encyclopedia, PID controller. Retrieved Jan 2017. https://en.wikipedia.org/wiki/PID_controller
291. J. Liu, H.K. Khalil, K.G. Oweiss, Neural feedback for instantaneous spatiotemporal modulation of afferent pathways in bi-directional brain-machine interfaces. IEEE Trans. Neural Syst. I Rehabil. Eng. **19**, 521–533 (2011)
292. J. Daly, J. Liu, M. Aghagolzadeh, K. Oweiss, Optimal space-time precoding of artificial sensory feedback through mutichannel microstimulation in bi-directional brain-machine interfaces. J. Neural Eng. **9**(6), 065004 (2012)
293. K. Kurosawa, R. Futami, T. Watanabe, N. Hoshimiya, Joint angle control by FES using a feedback error learning controller. IEEE Trans. Neural Syst. I Rehabil. Eng. **13**, 359–371 (2005)
294. M. Arsiero, H.-R. Lscher, M. Giugliano, Real-time closed-loop electrophysiology: towards new frontiers in in vitro investigations in the neurosciences. Arch. Ital. Biol. **145**(3–4), 193–209 (2007)
295. I. Lita, D.A. Visan, I.B. Cioc, FPAA based PID controller with applications in the nuclear domain, in *32nd International Spring Seminar on Electronics Technology* (2009)
296. V. Aggarwal, M. Meng , U.M. O'Reilly, A self-tuning analog proportional-integral-derivative (PID) controller, in *First NASA/ESA Conference on Adaptive Hardware and Systems (AHS'06)*, Dec 2006, pp. 12–19
297. L. Samet, N. Masmoudi, M.W. Kharrat, L. Kamoun, A digital PID controller for real time and multi loop control: a comparative study, in *IEEE International Conference on Electronics, Circuits and Systems. Surfing the Waves of Science and Technology* (1998), pp. 291–296
298. L.H. Keel, J.I. Rego, S.P. Bhattacharyya, A new approach to digital PID controller design. IEEE Trans. Autom. Control **48**(4), 687–692 (2003)
299. W.K. Ho, C.C. Hang, J.H. Zhou, Performance and gain and phase margins of well-known pi tuning formulas. IEEE Trans. Control Syst. Technol. **3**(2), 245–248 (1995)
300. K.H. Ang, G. Chong, Y. Li, PID control system analysis, design, and technology. IEEE Trans. Control Syst. Technol. **13**(4), 559–576 (2005)
301. V. Gilja et al., A high-performance neural prosthesis enabled by control algorithm design. Nat. Neurosci. **15**(12), 1752–1757 (2012)
302. Atmel, 8/16-bit Atmel XMEGA microcontroller. Sept 2014. http://www.atmel.com/devices/ATXMEGA128A4U.aspx

303. X. Liu, M. Zhang, J. Van der Spiegel, A low power multi-mode CMOS image sensor with integrated on-chip motion detection, in *2014 IEEE International Symposium on Circuits and Systems (ISCAS)*, May (2014)
304. X. Liu, M. Zhang, J. Van der Spiegel, A low-power multifunctional CMOS sensor node for an electronic façade. IEEE Trans. Circuits Syst. I Regul. Pap. **61**(9), 2550–2559 (2014)
305. Taxes Instruments, Micropower, single-supply, CMOS instrumentation amplifier. INA2321 datasheet (2000)
306. Atmel, At32uc3c series complete. Atmel Datasheet, Oct (2000)
307. Analog Devices, 3-axis, ±2g/±4g/±8g/±16g digital accelerometer. June (2009)
308. Nordic Semiconductor, nRF24L01 + single chip 2.4GHz transceiver product specification v1.0. Sept (2008)
309. S.K. Talwar, S. Xu, E.S. Hawley, S.A. Weiss, K.A. Moxon, J.K. Chapin, Rat navigation guided by remote control. Nature **417**, 37–38 (2002)
310. E.E. Thomson, R. Carra, M.A. Nicolelis, Perceiving invisible light through a somatosensory cortical prosthesis. Nat. Commun. **4**, 1482 (2013)
311. R. Morris, Developments of a water-maze procedure for studying spatial learning in the rat. J. Neurosci. Methods **11**, 47–60 (1984)
312. S. Torres Maldonado, A.G. Richardson, X. Liu, S. DeLuccia, Y. Ghenbot, J. Van der Spiegel, T.H. Lucas, Augmenting perception through direct electrical stimulation of adult somatosensory cortex, in *American Association of Neurological Surgeons*, April (2017)
313. Texas Instruments, Ultralow noise, high PSRR, fast RF 100-mA low-dropout linear regulators. June 2008. http://www.ti.com/product/TPS791
314. Linear Technology, Dual micropower DC/DC converter with positive and negative outputs. Datasheet. http://www.linear.com/product/LT1945
315. Linear Technology, 100mA regulated charge-pump inverters in thinsot. Datasheet. http://www.linear.com/product/LTC1983
316. Texas Instruments, High-precision, low-noise, rail-to-rail output 11-MHz JFET Op Amp, Aug 2016. http://www.ti.com/product/OPA2140
317. A.G. Richardson, P.K. Weigand, S.Y. Sritharan, T.H. Lucas, A chronic neural interface to the macaque dorsal column nuclei. J. Neurophysiol. **115**(5), 2255–2264 (2016)
318. S. Sritharan, A. Richardson, P.W.I. Planell-Mendez, X. Liu, H. Zhu, M. Zhang, J. Van der Spiegel, T. Lucas, Somatosensory encoding with cuneate nucleus microstimulation: detection of artificial stimuli, in *International Conference of the IEEE Engineering in Medicine and Biology Society* (2016)
319. Texas Instruments, Tiva TM4c123GH6Pm microcontroller. Datasheet, 10, 2007–2013. http://www.ti.com/product/TM4C123GH6PM
320. Open Source, Fatfs - generic fat file system module. Retrieved (2017). http://elm-chan.org/fsw/ff/00index_e.html
321. Wikipedia and the Free Encyclopedia, Iic (inter-integrated circuit). Retrieved Jan 2017. https://en.wikipedia.org/wiki/I%C2%B2C
322. S. Diekelmann, J. Born, The memory function of sleep. Nat. Rev. Neurosci. **11**(2), 114–126 (2010)
323. R. Huber, M.F. Ghilardi, M. Massimini, G. Tononi, Local sleep and learning. Nature **430**(6995), 78–81 (2004)
324. L. Marshall, H. Helgadottir, M. Molle, J. Born, Boosting slow oscillations during sleep potentiates memory. Nature **444**(7119), 610–613 (2006)

Index

X. Liu, J. Van der Spiegel, *Brain-Machine Interface*,
https://doi.org/10.1007/978-3-319-67940-2

MIX
Papier aus verantwortungsvollen Quellen
Paper from responsible sources
FSC® C105338

If you have any concerns about our products,
you can contact us on
ProductSafety@springernature.com

In case Publisher is established outside the EU,
the EU authorized representative is:
Springer Nature Customer Service Center GmbH
Europaplatz 3, 69115 Heidelberg, Germany

Printed by Libri Plureos GmbH
in Hamburg, Germany